D1827760

COMPREHENSIVE ANALYTICAL CHEMISTRY

ELSEVIER SCIENTIFIC PUBLISHING COMPANY
335 JAN VAN GALENSTRAAT
P.O. BOX 211, AMSTERDAM, THE NETHERLANDS

AMERICAN ELSEVIER PUBLISHING COMPANY, INC.
52 VANDERBILT AVENUE
NEW YORK, NEW YORK 10017

LIBRARY OF CONGRESS CARD NUMBER: 58-10158

ISBN 0-444-41163-1

WITH 134 ILLUSTRATIONS AND 16 TABLES

PRINTED IN THE NETHERLANDS

COMPREHENSIVE ANALYTICAL CHEMISTRY

ADVISORY BOARD

Contributors to Volume IV

R.W. Frei, Sandoz Ltd., CH-4002 Basle

M.M. Frodyma, National Science Foundation, Washington, D.C.

G.F. Kirkbright, Chemistry Department, Imperial College of Science & Technology, London

V.T. Lieu, Department of Chemistry, California State College, Long Beach, California

I.L. Marr, Department of Chemistry, The University, Aberdeen

M. Sargent, Chemistry Department, Imperial College of Science & Technology, London

Wilson and Wilson's

COMPREHENSIVE ANALYTICAL CHEMISTRY

Edited by

G. SVEHLA, PH.D., D.SC., F.R.I.C.

Reader in Analytical Chemistry
The Queen's University of Belfast

VOLUME IV

Instrumentation for Spectroscopy
Analytical Atomic Absorption and Fluorescence Spectroscopy
Diffuse Reflectance Spectroscopy

ELSEVIER SCIENTIFIC PUBLISHING COMPANY
AMSTERDAM   OXFORD   NEW YORK
1975

WILSON & WILSON'S

COMPREHENSIVE ANALYTICAL CHEMISTRY

VOLUMES IN THE SERIES

Preface

In *Comprehensive Analytical Chemistry* the aim is to provide a work which, in many instances, should be a self-sufficient reference work; but where this is not possible, it should at least be a starting point for any analytical investigation.

It is hoped to include the widest selection of analytical topics that is possible within the compass of the work, and to give material in sufficient detail to allow it to be utilised directly, not only by professional analytical chemists, but also by those workers whose use of analytical methods is incidental to their other work rather than continual. Where it is not possible to give details of methods, full reference to the pertinent original literature is made.

All the contributions to Volume IV are connected with spectroscopy. The aim of the chapter on instrumentation for spectroscopy (Chap. 1) is to assist the spectroscopist in selecting the proper instrument and/or the proper experimental conditions for his measurement. The contributions on atomic absorption and fluorescence spectroscopy (Chap. 2) and on diffuse reflectance spectroscopy (Chap. 3) cover modern techniques widely used nowadays in analytical laboratories. As usual, these contributions are written by outstanding internationally known experts in their fields. Contributions on other spectroscopic and optical methods will be published in further volumes.

Dr. C.L. Graham of the University of Birmingham assisted in the production of the present volume; his contribution is acknowledged with many thanks.

July, 1974.   G. Svehla

Contents

Chapter 1

Instrumentation for spectroscopy

I.L. MARR

1. Introduction

(A) SCOPE

In these days of increasing specialisation it becomes increasingly difficult for a chemist to be fully conversant with the theory of optics, of electronics, of instrument design, and so on, yet more chemists, and analysts in particular, are using complex pieces of equipment routinely in the course of their work. The description by Julius [1] of the infrared spectrometer which he built and then used to investigate the absorption spectra of organic compounds (in 1888) makes fascinating reading, but very few chemists today have either the time or the "know-how" for such occupations and the majority rely on the wide range of excellent, commercially available equipment. This chapter will not attempt to reverse this situation; rather, it will try to help the chemist find his way through a sometimes difficult and confusing field. It is hoped that he can then appreciate the advantages and disadvantages of different pieces of equipment, the kind of things which can go wrong in, and the limitations of, the devices and instruments which he uses.

So many different topics are dealt with in the following pages that discussion has had to be restricted to the most important facts. An attempt has been made to give references to the more detailed and most readily available sources wherever possible. Interesting though original papers may be, they are not always the best starting point for the newcomer to a field who must rely on textbooks for a clearer exposition of the problem, the answer, and the reasons. The author has consulted many such books and the reader will be referred to some of them in the course of this chapter. At this stage, attention is drawn to two texts [2,3] which will not be quoted in connection with any particular topic, but which are very relevant to the subject as a whole.

(B) THE NATURE OF LIGHT — WAVELENGTH STANDARDS

The electromagnetic theory of light successfully combines the two older wave and corpuscular theories. For many practical purposes, however, it is more convenient to consider light as a wave phenomenon with a corresponding wavelength related to the energy of the individual photons by

$$E = h\nu = hc/\lambda \tag{1}$$

where E is the energy in Joules, h is Planck's constant, c is the velocity of light in vacuo (2.99793×10^8 m.sec^{-1}) [4] and λ is the wavelength in metres. So precisely can the wavelengths of different spectral lines be compared by interferometry that the unit of length, the metre, is now defined [5] as being equal to 1.65076373×10^6 wavelengths of the ^{86}Kr red line at 605.6 nm. The wavelength of this line was determined relative to that of the previous standard, the cadmium red line at 643.8 nm, by Terrien [6] who found its stability to be at least ten times better than that of the cadmium line.

Another convenient wavelength standard which has been accurately checked against these references is the ^{198}Hg vapour lamp [7]. McNally [8] showed that this isotope could be produced at better than 99.6% purity by irradiation of ^{197}Au in a reactor and was therefore suitable for the construction of a narrow-line wavelength standard. The wavelength of the green line is given as 546.07532 nm in air [9] and 546.27024 nm in vacuo.

The difference between these two values reminds one that wavelength as such is not constant since the velocity of light varies according to the medium through which it is passing. If v is the velocity of light in a transparent medium (such as air or glass) then the ratio c/v is called the refractive index of the medium, with the symbol n. Since the frequency of the radiation is constant we can write

$$\lambda_{vac} = \lambda_{air} n_{air} \tag{2}$$

The value of n for air has been determined with high precision at

2

various wavelengths by several workers whose results have been used by Edlén [10] to compute the relationship

$$(n-1)10^8 = 6432.8 + 2.949810 \times 10^6 (146 - \sigma^2)^{-1}$$
$$+ 2.5540 \times 10^4 (41 - \sigma^2)^{-1}$$

where σ is the wavenumber in μm^{-1} of the radiation concerned. For theoretical purposes in spectroscopy it would be best to work in terms of frequency in units of sec^{-1}; but since it is wavelength which is measured and since it can be measured with a higher accuracy than c, it is more convenient to use the wavenumber, σ, with the unit μm^{-1}, calculated from the wavelength in vacuo. The Kayser has been proposed [11] as a name for this unit but was finally dropped after some years of debate [12].

In infrared spectroscopy measurements of wavelength are made by first calibrating the instrument on accurately known absorption bands of some common gases such as H_2O, NH_3 or CO_2 and then using a set of tables to convert directly to wavenumbers [13]. A number of emission lines of the noble gases have been measured by interferometric techniques with high precision and are suitable for calibration purposes in the visible and near infrared [14]. For ultraviolet and visible spectrophotometry [15] two very convenient solid standards are available in the holmium (Corning 3130) and didymium (Chance ON 12) glasses which give a number of sharp absorption bands throughout the spectrum (see Table 1).

TABLE 1

Holmium and didymium glasses*

	Wavelength (nm)	Wavenumber (cm^{-1})		Wavelength (nm)	Wavenumber (cm^{-1})
Holmium	241.5	41,410	Holmium	536.2	18,650
Holmium	279.4	35,790	Didymium	573	17,450
Holmium	287.5	34,780	Didymium	586	17,060
Holmium	333.7	29,970	Holmium	637.5	15,690
Holmium	360.9	27,710	Didymium	685	14,600
Holmium	418.4	23,900	Didymium	741	13,490
Holmium	453.2	22,070	Didymium	803	12,540

* Reproduced by courtesy of Pye—Unicam Ltd.

(C) THE SPECTRUM

The units in common use and their inter-relationships are best seen from a diagram such as Fig. 1. Also shown are the approximate boundaries of the spectral regions into which, for purely practical purposes, the complete spectrum may be divided.

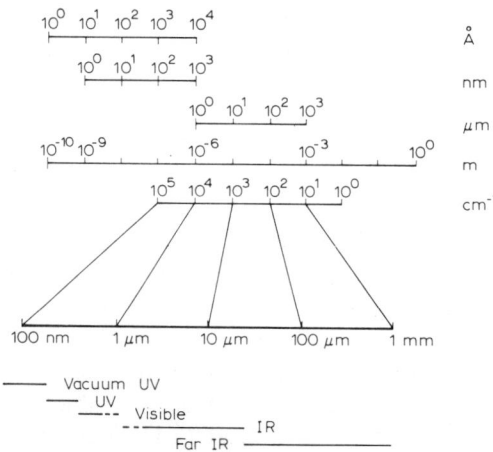

Fig. 1. Wavelength units and spectral regions.

Vacuum ultraviolet extends from about 200 nm to the region of soft X-rays. At wavelengths lower than 200 nm, oxygen begins to absorb, hence the spectrometer must be evacuated. For most practical purposes, a lower limit of about 60 nm is set by the sources available (helium discharge). A useful extension of the easily accessible ultraviolet to around 185 nm (limited at this point by the transparency of silica) can be obtained by flushing the instrument with high-purity nitrogen. This enables the emission spectra of several light elements to be recorded.

Ultraviolet (UV) extends from 200 nm to the short wavelength limit of visibility at about 400—450 nm. Quartz optics must be used since glass is opaque over the part of this region below about 350 nm.

Visible light extends from the limit mentioned to about 750 nm but instruments for this region commonly take in wavelengths up to about 1 μm.

4

Infrared (IR) is taken to mean light from 2 μm to 15 μm for general purpose work, particularly for organic chemistry. Only a few instruments cover the gap from 0.75 μm to 2 μm but the extension to longer wavelengths is dependent mainly on the choice of materials for prisms and windows.

Far infrared extends to about 1 mm beyond which microwave techniques take over from optics. Though the coverage by prism and grating instruments is being extended, the farthest IR is more usually studied by interferometric techniques.

(D) MOLECULAR AND ATOMIC SPECTRA

A second way of subdividing the spectrum for the sake of convenience is according to the nature of the transitions giving rise to the spectra — whether they be of rotational, vibrational or electronic origin. The energy changes involved correspond to quanta of radiation in the far IR, IR and UV/visible, respectively. A glance at the spectrum of benzene vapour (Fig. 38, p. 53) will remind one that high-energy transitions do not take place without accompanying lower-energy transitions: in this case the interaction of vibrational modes of 520 and 923 cm^{-1} with an electronic transition gives rise to a complex spectrum. For detailed discussions on this subject the reader is referred to the books by Bauman [16] and by Walker and Straw [17].

(E) OBSERVATION OF SPECTRA

The transitions which give rise to spectra must be made to occur before they can be observed. The manner in which the necessary energy is supplied to the molecule, atom or ion is determined by the nature of the phenomenon one wishes to observe.

Absorption of monochromatic radiation occurs when the quanta possess the energy corresponding to a transition, so that the energy is supplied by the source, selected by a monochromator if need be, partially absorbed by the sample and partially received by the detector (Fig. 2(a)). The technique of absorption spectroscopy is applicable to any material in any state provided that it is not completely opaque, in which case it must be diluted in one way or another, or a thinner sample used.

Emission occurs from a molecule or atom when an excess of ener-

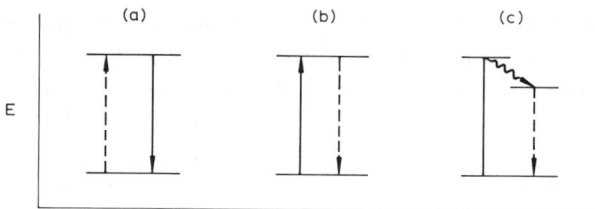

Fig. 2. (a) Absorption; (b) emission; (c) fluorescence.

gy is supplied to the species from its environment such as a flame or an electric discharge. Very often there is also sufficient energy to break chemical bonds, so that only rather simple molecules such as CaO, C_2, CH, CN etc. can be studied in this way (Fig. 2(b)).

Fluorescence and phosphorescence are phenomena associated with transitions between more than one excited state for a species. After excitation to a higher level, an electron drops by a non-radiative process to an intermediate level and then to the ground state giving rise to emission at a longer wavelength than that of the exciting radiation. In fluorescence the processes occur very rapidly, but in phosphorescence the life-time of the intermediate state is relatively long (seconds or even minutes) with the result that emission continues after excitation has ceased. These phenomena are most conveniently studied with equipment using two monochromators (Fig. 2(c)).

(F) TERMINOLOGY

The wide range of instruments and techniques used in modern spectroscopy has brought with it an equally wide range of terms, some of which are rather similar and hence tend to be used loosely in everyday speech. Attempts have been made to clarify this situation [18—20].

Photometer. An instrument for measuring the intensity of a beam of light or for comparing the intensities of two beams.

Spectrophotometer. An instrument with which the absorption or emission of a sample may be studied at any specified wavelength (in practice with a narrow band of wavelengths).

Spectroscope. An instrument through which one looks to see a spectrum due to the light of some external source. By calibration, this may become a spectrometer.

Spectrometer. An instrument for measuring the wavelength of radiation. This may also be accomplished with a spectrograph.

Spectrograph. An instrument in which a spectrum of light emitted, and possibly also absorbed, by a sample is recorded on a photographic plate which is then known as a *spectrogram.*

Monochromator. A device for selecting a narrow band of wavelengths from, for example, a "white" source, so that measurements may be made with monochromatic light. Strictly speaking, only certain atomic line sources used in conjunction with a good monochromator can deliver radiation which can reasonably be considered as monochromatic, though in many cases lasers approach the ideal even more closely.

The term spectrometer has come to be used very loosely to mean any instrument built round a monochromator. As many designs for optical systems may find application in more than one technique, it is perhaps useful to refer to all members of this group as spectrometers and this term will be used in this wider sense in this chapter.

Transmittance. This is the fractional intensity, T, of a beam of monochromatic light of a given wavelength transmitted by a medium

$$T = I/I_0 \quad \text{or} \quad \%T = 100 I/I_0$$

for percent transmittance, where I_0 is the incident intensity and I the transmitted intensity.

Absorbance. Formerly known as *optical density* or *extinction*, this is defined as

$$A = \log I_0/I = \log 1/T$$

According to the Beer—Lambert law, absorbance may also be expressed in terms of the molar concentration, C, of a solute in a transparent solvent with a path length, l, by

$$A = \epsilon C l$$

where ϵ is the *molar absorptivity* (formerly molar extinction coefficient).

2. Sources

(A) EMISSION OF ELECTROMAGNETIC RADIATION AND ITS SPECTRAL ENERGY DISTRIBUTION

Herschel's discovery of infrared radiation in 1800 was a consequence of his attempt to compare the heating and illuminating powers of the different colours of light obtained by passing sunlight through a prism [21]. By the end of the 19th century, the spectral distribution of energy emitted from hot bodies had been widely investigated and put on a firm quantitative basis. The total amount of energy emitted by a body was found experimentally to obey the relationship

$$W = \sigma T^4 \tag{3}$$

where W is in watts, T the temperature in $^\circ$K, and σ is the Stefan—Boltzmann constant [22], 5.672×10^{-16} $Jm^{-2} deg^{-4} sec^{-1}$.

Further experimental work by Paschen and considerations based on the kinetic theory of gases by Wien showed that for a hot "black body" the distribution of energy in the visible and near infrared was given by

$$E = c'\lambda^{-5} \exp(-c''/\lambda T) \tag{4}$$

where c' and c'' are constants. This is now known as the Wien Distribution Law [23].

The function (4) has a maximum value at some wavelength, λ_{max} depending on the temperature, which is given by the Wien Displacement Law

$$\lambda_{max} T = 2.880 \times 10^3 \ \mu m.deg \tag{5}$$

where T is in $^\circ$K and λ is in μm. This product was determined first by Lummer and Pringsheim [24] as 2.94×10^3 in 1900.

For long-wave radiation, the energy is related to the temperature by the Rayleigh—Jeans Distribution Law [25]

$$E_\lambda = 8\pi k T \lambda^{-4} \tag{6}$$

It remained for Planck to derive these relationships from theoretical considerations using the theory of electromagnetic radiation according to which the energy of a photon emitted by a resonator is proportional to its frequency. The distribution of energy for a black body is then given by [26]

$$E_\lambda = 8\pi hc\lambda^{-5}(\exp(hc/\lambda kT)-1)^{-1} \tag{7}$$

It will be noticed that eqns. (4) and (6) are simplified forms of the cases where $hc/\lambda kT$ is greater than about 5 and less than unity, respectively. Planck's constant, h, was calculated by him to be 6.55×10^{-27} erg.sec which is quite close to the currently accepted value of 6.624×10^{-34} Joules.sec. The calculated distribution curves for four common sources are shown in Fig. 3. For comparison, the curve for a hot body at $1000°$K is also shown.

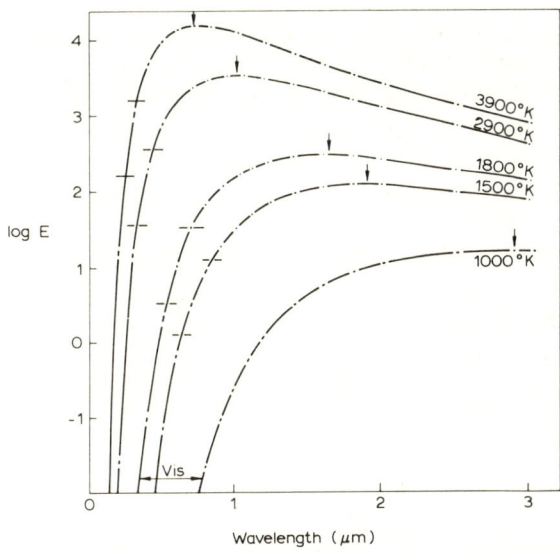

Fig. 3. Calculated energy distribution curves for four common sources. $1500°$K, Globar; $1800°$K, Nernst filament; $2900°$K, tungsten ribbon; $3900°$K, carbon arc.

Atomic line spectra

At about the same time, attention was being given to the series of lines which were observed in the emission spectra of many atoms, particularly of hydrogen, helium and the alkali metals. Ritz [27] produced a generalised form of Rydberg's earlier equation, which showed that frequencies of the lines could be expressed by the equation

$$\nu = N[(s+a)^{-2} + (r+b)^{-2}] \tag{8}$$

where N was a constant, a and b were integral constants for the particular element, and s and r integers, being different for each line. Bohr [28] in his new theory of atomic structure provided a theoretical explanation of many of the spectroscopic observations and showed that the frequency of a photon emitted by an electron transferring from one orbital to another could be written as

$$\nu = R_H \left(\frac{1}{n_1^2} - \frac{1}{n_2^2} \right) \tag{9}$$

for the hydrogen atom, where R_H is now known as the Rydberg constant for hydrogen (1.0967758×10^5 cm^{-1}) and n_1 and n_2 are integers with $n_2 \geq (n_1 + 1)$. His model fitted the experimental data for a few light elements very well. In Table 2 are listed the series of lines for hydrogen with the names of their discoverers and the wavelengths of the first two lines of each series and of the converging limit.

Moreover Bohr's theory predicted that if a photon possessing the exact amount of energy to cause a transition should interact with an atom, then the energy should be absorbed and the excitation occur. This is the principle of atomic absorption spectroscopy.

TABLE 2

Spectral series for hydrogen

Series	n_1	$\lambda(n_1 + 1)$	$\lambda(n_1 + 2)$	$\lambda(n_2 \rightarrow \infty)$
Lyman	1	121.6 nm	102.6 nm	91.2 nm
Balmer*	2	656.3 nm (C)	486.1 nm (F)	364 nm
Paschen	3	1.875 μm	1.282 μm	820 nm
Bracket	4	4.05 μm	2.63 μm	1.46 μm
Pfund	5	7.46 μm	4.67 μm	2.28 μm

* Letters C and F denote Fraunhofer lines in the sun's spectrum.

(1) Band spectra

A group of unresolved, closely spaced lines in a spectrum is known as a band (either emission or absorption). The lines may be unresolved because of limitations in the spectrometer, but in many cases the environment of the active species is responsible. Whilst electronic spectra of molecules may be relatively simple, the interaction of vibrational transitions can produce many lines. As lines become broader with increase in pressure (pressure broadening) and temperature (Doppler broadening), the closely spaced lines merge into one another resulting in a band. A similar effect is seen when a gaseous species is dissolved in a solvent. The spectrum of benzene in cyclohexane (Fig. 4) has lost the fine structure of the gas-phase spectrum (Fig. 38, p. 53) though both were recorded on the same instrument under the same conditions.

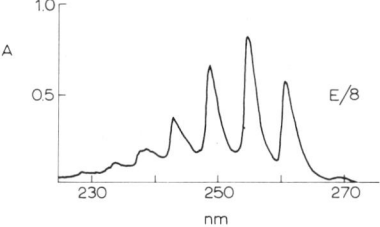

Fig. 4. Absorption spectrum of 3.4×10^{-3}M benzene in cyclohexane (10 mm).

(2) Coherent radiation — lasers

Light emitted from any source of excited atoms or molecules is due to energy transitions taking place within these particles which then act as electric dipoles. As they act independently, and as excitation and relaxation are random events, the timing of any one event is not related to that of any other and the photons emitted by the bulk material, while having the same energy (in the case of an atomic emission line) are not in phase. Such radiation is said to be incoherent. When emission from a number of excited atoms is controlled so that it occurs at a definite time, the total emission may be in phase and the radiation is said to be coherent. This is achieved in a device called a laser which was first demonstrated successfully by Maiman [29] in 1960. The word is an acronym — the letters being derived from: light amplification by stimulated emission of radiation (by analogy with the maser, developed some years earlier for microwave frequencies).

The laser is, in practice, of little or no value as an amplifier because the noise level is too high, but with optical feedback from a pair of parallel plane mirrors at either end of a resonant cavity, it can become an optical oscillator maintaining oscillation by stimulated emission of radiation through a mechanism similar to that of fluorescence.

In the case of ruby which was used by Maiman (aluminium oxide containing about 0.1% chromium(III)) the chromium ions may be excited by irradiation with blue-green light to the $^4F_2$ state (see Fig. 5) which has a life-time of the order of a few milliseconds. After the internal loss of energy to a ^2E state (a doublet with a spacing of 29 cm^{-1}) fluorescent radiation at 694.3 nm is emitted and the ion returns to the ground state. The normal population distribution of the $^4A_2$ and $^4F_2$ states is such that most ions are in the ground state and the intensity of fluorescence is proportional to the intensity of the exciting radiation. When population inversion occurs, i.e. when there are more ions in the excited state than there are in the ground state, emission can be stimulated by a quantum of energy corresponding to a wavelength of 694.3 nm. The excited ions would then all relax to the ground state by stimulated emission and the total light output would be in phase and therefore coherent. A high degree of stimulation is achieved by using mirrors with a very high reflectivity at the wavelength of operation and by adjusting the path length between the mirrors to a whole number of wavelengths. However, there still exists the problem of how to cause population inversion,

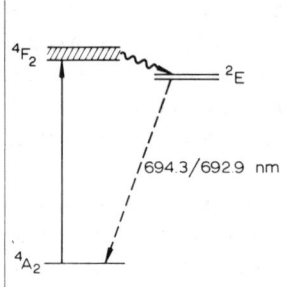

Fig. 5. Diagram for chromium(III) in alumina (ruby).

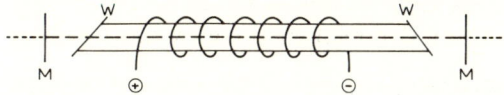

Fig. 6. Schematic representation of laser apparatus. M, adjustable mirror; W, windows fitted at Brewster angle.

since very large amounts of energy must be supplied in a short space of time, in fact, in less than the life time of the excited state. A xenon flash discharge tube is a convenient source with a high output in the desired spectral region. It can be operated at high currents for short intervals of time. The output of such a laser consists of a series of short pulses and the power depends largely on the power-handling capabilities of the flash tube and its power supply (Fig. 6).

(a) Helium—neon gas laser

This belongs to the class of four-level lasers and is now one of the most widely used, mainly because it can be operated continuously rather than pulsed. The gas mixture of helium and neon in the ratio of 10:1 at about 1 mm Hg supports a continuous discharge producing metastable helium atoms in the $2\,^3S$ state (Fig. 7). These can in turn transfer their energy to neon atoms, exciting them to the 2S state which has a higher energy and longer life time than the 2P state. Stimulated emission between the two occurs when population inversion is achieved, at relatively low pumping (i.e. exciting) powers since the 2P state is not normally populated to any great extent. The radiation is strongest at 1.153 μm though other lines in this region

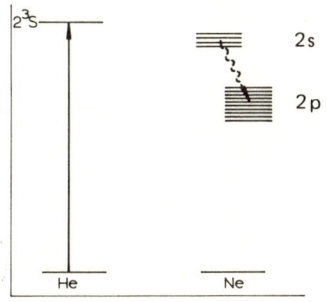

Fig. 7. Diagram for helium—neon mixture.

may be observed. This laser was first described by Javan et al. [30] in 1961. Its optical properties and construction were outlined by Herriott [31].

A year later, White and Rigden [32] reported operation of the He-Ne laser at 632.8 nm, corresponding to the $3 \, s_2 \to 3 \, p_4$ transition of neon pumped by the metastable $2 \, {}^1S$ helium atoms. The power output of 0.1 mW for about 60 W input was low and oscillation at this wavelength was achieved by using narrow bandwidth reflective mirrors tuned to around 635 nm.

(b) Giant-pulse lasers

Lasing normally takes place as soon as population inversion exists, so that output power is dependent on the input power. But if the reflecting mirrors are absent, or rotated out of alignment, stimulated emission cannot occur and a high concentration of excited atoms is built up. When one of the mirrors is rotated continuously, the geometrical conditions for lasing are satisfied for a very short space of time, once in each revolution. A pulse of very intense coherent radiation is produced. Switching may also be done with a Kerr cell.

(c) Properties of laser radiation

The coherency of the light has been discussed. But for the chemist there are perhaps two other features of greater interest: the perfectly parallel beam, with a typical diameter of 1 mm (which means it can be focussed to a very fine point and hence a large amount of energy released in a very small area); and the narrow spectral bandwidth (about 0.05 cm^{-1} for the He—Ne laser) which is a good approximation to monochromatic radiation, at a useful power. The radiation is also highly polarised by the windows fitted at the Brewster angle (θ, where $\tan \theta = n$).

(d) Laser Raman spectroscopy

The small bandwidth of a laser emission line — much less than that of the 435.8 nm mercury line — suggested that it might be a useful source for high-resolution Raman spectroscopy. Although the power available from the early lasers was not very great, the concentration of the energy into a small and very intense beam was more than

14

sufficient compensation because spectra could be recorded photo-electrically with a scanning spectrometer. Higher signal levels were obtained by including the sample cell between the mirrors of the laser but the experimental difficulties associated with this approach precluded its general acceptance. Small volumes of gases may be contained in multiple-reflectance cells. Power levels of around 50 mW from the larger commercially available He—Ne lasers now enable good spectra to be recorded in a matter of a few minutes with a resolution of a few wavenumbers. Because the line (at 632.8 nm) lies in the red, it is not absorbed by yellow and brown solutions as is the mercury line mentioned above. An added attraction is the possi-bility of recording Raman spectra with polarised light. By simply rotating the laser about its optical axis, the angle of polarisation with respect to the sample may be varied continuously. A good introduc-tion to the scope of this field is given by Evans [33] and by Hendra and Vear [179].

(e) Laser microprobe

In this instrument the high power in the laser beam is concen-trated on a small area of solid sample on a microscope stage position-ed a short distance from the entrance slit to a spectrograph. A small amount of the sample, selected by the operator viewing through the microscope, is vaporised and the emission spectrum of the hot va-pour is recorded with the spectrograph. That such a technique could be usefully employed for quantitative analysis was shown first by Runge et al. [34] in 1964. A single 20 nsec pulse from a rotating-prism giant-pulse ruby laser vaporised enough sample to produce a spectro-gram. Coefficients of variation of the order of 5% were found for the determination of nickel and chromium in steel. The size of the pit left after excitation was about 0.4 mm diameter so this early model could hardly qualify for the term "microprobe"; but better control of the energy available has improved this aspect. Commercial instru-ments are available.

(3) Requirements of sources for spectroscopy

It has already been mentioned that most spectroscopic techniques fall into one of two categories: emission or absorption. It has also been pointed out that the essential difference between them is that in the former the source acts as a supplier of energy to the sample to stimulate emission, whereas in the latter the source and the sample are physically separate. To the first class of sources belong flames, arcs and sparks, and sometimes discharge tubes. To the second, all continuous emitters, black-body radiators and, for atomic absorption spectrophotometry, atomic line emitters such as hollow-cathode lamps. Important factors affecting the choice of a particular source will include the following.

Spectral range. The need to change the source during the recording of a spectrum complicates the design of the equipment and of course calls for more power supplies. The wider its spectral range the more generally useful a source is.

Intensity. A higher intensity of emission from a source will result in a better signal-to-noise ratio from the detector and will facilitate the use of narrower slits which in turn may make possible a higher resolution.

Stability. The constancy of the emission intensity of a source is one of the major factors determining the accuracy that may be achieved in single-beam spectrophotometry. It is, of course, heavily dependent on the stability of the power supply, particularly for a black-body radiator (such as a tungsten filament lamp) operating on the short wavelength side of its λ_{max}. Long-term stability is not so important when the source is used with a double-beam instrument.

Mechanical factors. A long life is desirable to minimise lost time and the possible need for realignment when a source is replaced. A short running life also means increased running costs. The size of the actual emitter is also important and must be considered in terms of the dimensions of the entrance slit of the spectrometer (which may be quite large in the case of a far-infrared instrument).

The remainder of this section will describe briefly some of the more commonly used sources and a few others with more limited applications. The list cannot be comprehensive and the reader is referred to more detailed discussions for sources for the vacuum UV [35] and IR [36].

(B) ATOMIC LINE SOURCES

(1) Flames

Flames are used in analytical chemistry primarily as a means of

16

decomposing substances sprayed into them in the form of an aerosol to produce an atomic vapour which may be studied either by emission (flame spectrophotometry) or by absorption (atomic absorption spectrophotometry). In the latter case the source will usually be a hollow-cathode lamp.

The flame temperature is an important parameter, since the extent of dissociation of refractory materials (e.g. metal oxides), of ionisation of free atoms, and the proportion of atoms in an excited state all depend on this factor. As the equilibria are governed by the Boltzmann Distribution Law

$$\frac{n^*}{n} = \frac{g^*}{g} \exp(-E/kT) \tag{10}$$

the ratio of the numbers of atoms in the excited and ground states, n^*/n is very sensitive to changes in temperature★. The population of the excited state is usually very small and only for a few elements such as the alkali and alkaline earth elements are the emission intensities large enough to be measured easily. From a practical point of view, very careful control of experimental variables is essential if this type of source is to meet the requirement of stability.

The choice of fuel for the flame is affected not only by the temperature required, but also by the emission spectrum of the products of combustion. The common radicals found in hydrocarbon flames and their emission bands are shown in Fig. 8. The conditions chosen to record this spectrum were such as to emphasise these bands, and in a flame as normally used they would not be so predominant, but clearly this background spectrum could make the detection and measurement of weaker lines in some parts of the spectrum very difficult. A hydrogen-oxygen flame is better from this stand-point, but the high combustion velocity necessitates the use of special burners. Fuel-rich flames, while tending to be more luminous, do provide a reducing atmosphere which helps to break up more refractory molecules. Mavrodineanu and Boiteux [37] have written an excellent treatise on the subject of flames in which applications to analytical chemistry are emphasised. Many aspects of the chemistry of flames are covered in a book by Dean and Rains [38].

★ g^* and g are statistical weighting factors for the two states.

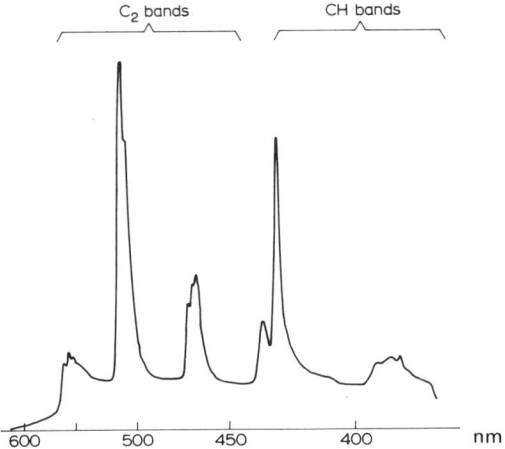

Fig. 8. Emission spectra of some common radicals in flames. Methyl ethyl ketone sprayed into a fuel-rich air—acetylene flame. Recorded with a Unicam SP 900 flame spectrophotometer.

(2) Arcs and sparks

These sources are used in spectrographic analysis because they provide sufficient energy to dissociate any sample into its constituents and also to excite the atoms so that emission results. In addition ionisation very often occurs and the emission spectra of the ions are added to those of the atoms. Spectrographs with high resolution are therefore essential to separate, and allow identification of, the many lines in the spectra. The emission spectrum of an iron arc is often recorded on the same plate to facilitate identification of the elements present in the standard. It thus provides a set of standard wavelengths for comparison.

Direct current arc. This requires the simplest power supply, giving up to about 10 A at about 50 V when running. The heavy current heats the anode, by bombardment with electrons, to a high temperature thus vaporising it and any sample it contains. Maximum emission, however, occurs in the region near the cathode where the electrons combine with the ions to give neutral atoms again. Because the reproducibility is not good enough for quantitative work, the d.c. arc is best used for qualitative analysis.

Alternating current arc. The high voltage (some thousands of volts) used for this arc makes it self-starting. It also results in much better reproducibility for quantitative work because the arc always

18

strikes a different part of the sample. Somewhat simpler spectra are obtained than with the d.c. arc and sensitivities are very high.

Alternating current spark. High excitation energies are available in the a.c. spark which, though supplied with current at 50 Hz, actually produces several sparks in each cycle. This source is said to give the best reproducibility and for metal analysis can be used non-destructively.

Further details on these and other related sources will be found in Vol. V, Chap. 1.

(3) Gas discharge tubes

The origin of atomic spectra and the effect of pressure broadening on many-line spectra to give an apparently continuous emission have already been discussed. Lyman [39], for example, found a continuum for helium under electrical discharge of several thousand volts, ranging from 25.6 to around 90 nm. In the case of hydrogen, the flash discharge showed reversal at the wavelengths now known by his name [40]. The "white" radiation from this source was apparently from the glass of the discharge tube, not the hydrogen.

(a) Xenon arc lamp

Schulz [41] investigated the emission spectra of neon, argon, krypton and xenon at pressures up to 37 atmospheres and currents up to 30 A, in quartz tubes. The continuum for xenon extended from about 240 nm well into the visible. Commercial lamps [42] running at high power (150 or 500 W) are available. The arc, struck between a conical tungsten anode and a compact spiral tungsten wire cathode sealed into a quartz envelope, is only about 4 mm high for the 500 W version. Anderson [43] has shown that the output from the xenon arc lamp is a good approximation to sunlight, having a useful output from just above 200 nm to 1.8 μm, with a maximum at 1 μm. There is considerable line emission superimposed on the continuum from about 450 nm upwards. This source finds wide application for fluorescence spectrometry.

Wilkinson and Tanake [44] have described the emission spectra of the noble gases when excited by an electrodeless microwave discharge at 2,450 MHz. Xenon gave a useful output from around 150 nm to about 225 nm, and krypton from about 125 nm to 160 nm, but with a much lower intensity.

(b) Hydrogen and deuterium lamps

For use as a wavelength standard, a high voltage discharge tube containing hydrogen at low pressure may be used. The glow is concentrated in a long capillary joining the two electrode compartments and an increased intensity is thus achieved. For vacuum UV work, where a continuum is required, a high voltage flash discharge may be used. Garton [45] has described a demountable Lyman source for the region 105—200 nm, which will pass large pulses of current. Modern recording instruments, however, require a steady light output rather than a pulsed one. This requirement has been met by a high-intensity a.c. arc between a heated nickel cathode and an anode which is also the entrance slit of the monochromator [46]. The pressure used was about 1 mm of mercury and the current about 2.5 A. Molecular line emission covered the vacuum-UV region from 90 to 180 nm.

Lower powers are adequate for a useful output in the near UV.

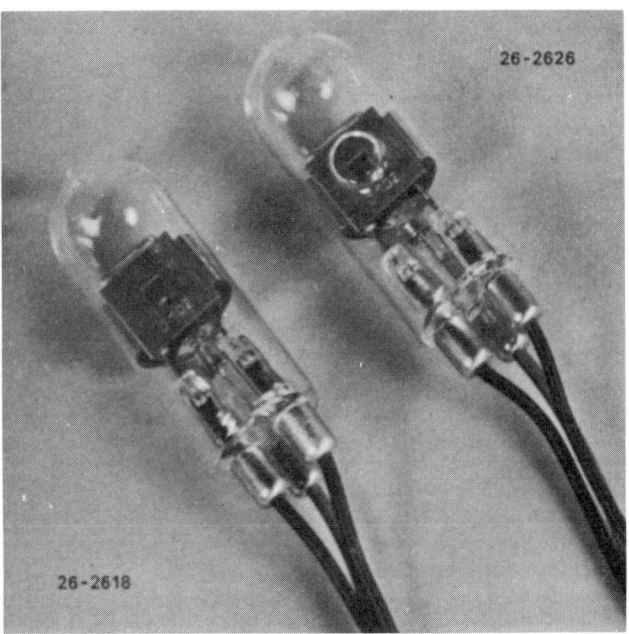

Fig. 9. Deuterium lamps. Model 2626 has a Suprasil window for transmission down to 165 nm. Reproduced by courtesy of Ealing Optics Ltd.

Currents of 300 mA at about 100 V maintain the cathode at a sufficiently high temperature to sustain emission so that heating is needed only for starting. A continuum from about 180 to 350 nm has, in addition, some of the lines in the visible such as those at 486.1 and 656.3 nm, which are useful for checking the calibration of the spectrophotometer. A quartz envelope must be used and the lower limit is determined mainly by the transparency of the quartz. In recent years deuterium has been used in place of hydrogen resulting in a three- to five-fold increase in intensity for the same power consumption and giving a slight extension to 370 nm. With a Suprasil window the range is extended to about 160 nm (Fig. 9).

(c) Mercury vapour discharge lamps

Reference has already been made to the use of the [198] Hg vapour discharge lamp as a wavelength standard [7—9]. Because of the high efficiency obtainable, the high-pressure lamp is widely used for general illumination and, with suitable filters, as a convenient line-source at several wavelengths, the most important being 253.7, 365.0, 404.7, 435.8, 546.1 nm. A high efficiency with most of the energy radiated at 253.7 nm is obtained with a low-pressure lamp. Quartz envelopes must be used for both high- and low-pressure lamps.

The relatively simple spectrum, the ease with which one line may be isolated, and the high powers available constitute major advantages which make the mercury lamp an obvious choice for use as a source for Raman spectroscopy. Wood described [47] two versions in 1929 and the so-called Toronto lamp, named after the University where it was developed, was reported in 1952. The discharge tube takes the form of a long spiral completely surrounding the sample vessel [48]. Stoicheff later designed a water-cooled model with liquid mercury electrodes running at a lower pressure. A narrower emission line [49] was produced.

Growing interest in far-IR spectroscopy has created a further use for mercury lamps. McCubbin and Sinton [50] used a high-pressure lamp for the region 100—700 μm where conventional black-body radiators have a very low output. Although at 350 μm about three-quarters of the energy came from the mercury (found by comparison of the intensities before and immediately after switching off), at 145 μm, three-quarters came from the quartz tube containing the discharge. Black polyethylene (a suspension of carbon black in polyethylene) is a very convenient filter transmitting beyond 100 μm (100 cm^{-1}). Comparison with a black-body radiator running at the same power [51] showed an improvement by a factor of three at 100 μm, and six at 200 μm, for the high-pressure mercury lamp.

Vapour discharge lamps are commercially available for a number of elements with reasonably high vapour pressures (alkali metals, cadmium, thallium, indium, gallium, and zinc) and have been used as sources for atomic fluorescence spectroscopy.

(4) Hollow-cathode lamps

Since the advent of commercial equipment for atomic absorption spectrophotometry, hollow-cathode lamps have become available for many elements and few people would think of making their own (even if the cost of those available is still rather high). But over 50 years ago, Paschen described a lamp with an aluminium cathode [52] in a helium atmosphere with which he observed the emission lines due to aluminium [53]. Schüler developed two forms of lamp, one in which the metal whose line spectrum was sought was heated to 800—900°C in a small furnace which served as the cathode, in an evacuated envelope [54] and the other in which a low pressure of argon carried the discharge, with the emission resulting from the sputtered atoms of the cathode being excited by collisions with the activated argon atoms [55]. This is the form now commonly used (Fig. 10). Full instructions for making these lamps are to be found in the book by Elwell and Gidley [56]. Schüler also showed that low-temperature and low-pressure operation was essential if narrow lines were required, a point later stressed by Walsh and co-workers [57] who showed that even vapour discharge lamps ("Wotan" by Osram) could be used for atomic absorption work if they were under-run.

A later development from Sullivan and Walsh was the high-intensity hollow-cathode lamp in which an electrical discharge also occurs between a second pair of partly shielded electrodes [58].

Fig. 10. Hollow-cathode lamp.

22

An iron hollow-cathode lamp has been proposed as a convenient many-line standard wavelength source, as the lines are very narrow [59]. Neon at about 3 mm pressure was found preferable to argon for filling, as the latter produced too many of its own lines.

(5) Electrodeless discharge tubes

High-frequency excitation of spectra (at 50 Mc/s) was utilised by Fenner [60] for the determination of selenium, cadmium, and mercury in trace amounts. The relatively recent availability of microwave power units has made it possible to concentrate sufficient energy in a small sealed tube (of quartz) in a resonant cavity, to excite a number of non-metals and semi-metals. Much work has been done by West and coworkers in the last few years to establish the working conditions and reliability of these sources which have the advantages of simplicity and low cost. The preparation of sources for selenium and tellurium is described [61]. Elements which are not sufficiently volatile to be used in these sources may be added as volatile compounds such as halides. At the time of writing, the stability and useful life of some of these tubes is not as good as might be hoped for, but no doubt improvements will be made. In a recent paper, Headridge and Richardson compare results obtained using hollow-cathode and electrodeless discharge tubes [180].

(C) THERMAL SOURCES — BLACK-BODY RADIATORS

(1) The tungsten filament lamp

This most widely used source has a convenient spectral range covering the whole of the visible and extending from about 320 nm, where the glass cuts off, to about 3 μm, though this upper limit has been extended to around 6 μm by fusing into the glass envelope a slice of periclase obtained by cleavage from a large crystal [62]. Sapphire has also been used, but this cannot be fused into the glass and has to be cemented on which generally gives rise to trouble at the joint. The power requirements of low-voltage lamps are easily met by solid-state power supplies so that high stability may be readily achieved. For most purposes in routine spectrophotometry a constant-voltage transformer or a large capacity 12 V lead accumulator gives satisfactory results. At 450 nm the output varies with the 8th power of the current, so the requirements are quite critical.

A tungsten ribbon running at 2,900°K in an inert atmosphere has been used in conjunction with an alkali halide window as a source for IR measurements up to 14 μm, free from all absorption bands due to water and carbon dioxide in the atmosphere [63]. The energy distribution for this and the following black-body sources are shown in Fig. 3, p. 9.

(2) Quartz—iodine lamp

When iodine vapour is present in the envelope, the sputtered tungsten atoms form compounds with the iodine, which subsequently are decomposed on contact with the hot filament. In this way the tungsten is recycled and darkening of the envelope and weakening of the filament are avoided. Higher running temperatures for the filament are also possible because the life is not shortened so seriously and because the quartz envelope used can withstand greater thermal stress than a glass one. A useful light output down to about 250 nm is available [64].

(3) Nernst filament

This is a very commonly employed source in IR spectrometers and, since it runs at about 1,800°K, it has a wide spectral output. It is a mixture of refractory metal oxides — typically 90% ZrO_2, 7% Y_2O_3, and 3% Er_2O_3 — and possesses a negative temperature coefficient of resistance which necessitates some form of stabilising circuitry such as an iron wire barreter in series with it. Platinum wires cemented into the ends of the rod make electrical contact for passing the heating current. Preheating is needed because it does not conduct when cold. A power consumption of 100 W is typical for a filament 20—30 mm long and 1.5—2 mm diameter. A modified filament for use up to 200 μm has been made by wrapping one with pieces of Welsbach mantle and firing them into position [65]. This filament should be left running because repeated heating and cooling leads to fracture of either the filament or the electrical connections.

24

(4) Globar

This is an electrically heated rod of silicon carbide run at about $1,500°K$ which, because of its positive thermal coefficient of resistance, is self-stabilising. A typical power consumption for an element 50 mm long and 5 mm in diameter would be 200 W. It approximates to a black-body at wavelengths below 15 μm. It is not as efficient as the Nernst filament at longer wavelengths but is often preferred because of its greater physical size which makes focussing less critical and which matches the wide entrance slits of spectrometers for the far IR.

(5) Carbon rod

As an alternative to the Globar, a carbon rod heated to $2,100°K$ in vacuo offers advantages of higher output beyond 10 μm and larger physical size [66]. Unfortunately, the running life is not very long and the high power consumption makes some external cooling necessary.

(6) Carbon arc

Rupert and Strong pointed out that because the temperature of the carbon arc is so much higher than that of other sources (say $3900°K$), then according to the Stefan—Boltzmann Law, a much greater total emission should result with particular gain in the longer-wavelength region. The large increase in energy means that narrower slits can be used and a much better resolving power attained [67]. Jaffe has described a modified microscope-illumination carbon arc lamp with automatic feed mechanism and suitable stabilising circuitry which he used as a source for IR spectroscopy [68].

3. Spectrometers, Spectrographs and Monochromators

These terms have already been defined and a very brief idea given of the purpose to which each instrument is best suited. We shall see that changing requirements of instrument designers and users have largely dictated the type of optical systems. Another result is the present use of instruments which are optically simpler than those

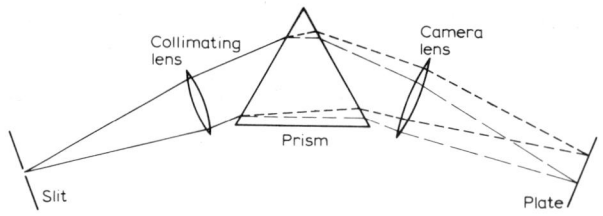

Fig. 11. The prism spectrograph.

which were common half a century ago. The widespread use of monochromators in modern scientific equipment oriented towards spectrum scanning has meant a return to optical systems where the dispersing element may be easily rotated so as to vary the wavelength passing through the exit slit without the necessity of altering any other part of the optical system.

The splitting of "white" light into its spectral components is known as dispersion, and may be achieved by the use of a *prism* or a *grating*, or sometimes a combination of the two, and, less efficiently, by using filters. It will be convenient to discuss first the simple prism spectroscope (Fig. 11) as used by Kirchhoff and Bunsen [69] in their early spectrochemical work leading to the discovery of the heavier alkali metals.

(A) THE PRISM SPECTROSCOPE

A collimating lens is used to produce a parallel beam of light from the diverging beam at the narrow entrance slit. This parallel beam then suffers deviation (change of direction) and dispersion (separation according to wavelength) on passage through the prism. The parallel beams are then focussed by the camera lens (in a spectroscope) to give an image of the entrance slit if monochromatic radiation is analysed or a series of images if the radiation consists of light of differing wavelengths. We have seen that radiation from a hot body will give a continuous spectrum as first described by Newton [70] in 1704. The camera plate may be replaced by an eyepiece (the pair of lenses forming a telescope) so that the observer can see the spectrum, or by a metal plate with a second narrow exit slit which can be moved across the spectrum and behind which is placed a photoelectric detector.

26

It should be remembered that the concepts of "line" and "band" spectra arise from the choice of a narrow slit as entrance aperture for a spectrometer rather than, for example, a pin-hole, so that a line image is recorded on the photographic plate. The importance of the slit will be discussed in a later section.

An autocollimating spectroscope is one which uses the same focussing system for both collimator and telescope. The commonest example is the Littrow spectrometer.

(1) Refraction

When a ray of light enters a more dense medium at an angle other than the perpendicular it will change direction. Snell's Law states that the refractive index of the medium is given by

$$n = \frac{\sin i}{\sin r} \tag{11}$$

when the ray passes from a vacuum into that medium (Fig. 12). The refractive index of all materials varies with wavelength, being general-

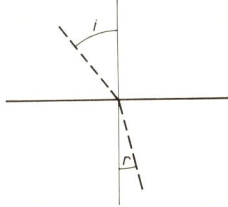

Fig. 12. Refraction of monochromatic radiation.

ly greater for shorter wavelengths. This is responsible for dispersion, since the angle r will therefore become smaller for shorter wavelengths.

(2) The prism

In the case of a parallel beam of light passing through a prism, the deviation will be such as to bend the beam even further, the total *angle of deviation* being shown as Δ in Fig. 13. It can be shown that as the angle i_1 varies, there is a position of minimum value for Δ

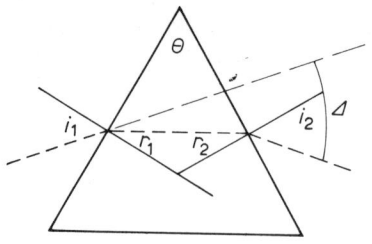

Fig. 13. Deviation with a prism.

which is the case of *minimum deviation*. It can also be shown that, for this case, where $i_1 = i_2$

$$\Delta = 2i_1 - \theta \tag{12}$$

where θ is the prism angle. Since $r_1 = r_2 = \theta/2$, it follows that

$$n = \frac{\sin i}{\sin(\theta/2)} = \frac{\sin \dfrac{\Delta + \theta}{2}}{\sin(\theta/2)} \tag{13}$$

This equation may be used to find n for any material which is transparent and which can be fashioned into a prism as long as the wavelength of the light is known.

If we differentiate eqn. (13) with respect to Δ, we can see how Δ will change with variation in n and hence with variation in wavelength, λ.

$$\frac{dn}{d\Delta} = \frac{\cos\left(\dfrac{\Delta + \theta}{2}\right)}{2\sin(\theta/2)} = \frac{\sqrt{\left(1 - \sin^2\left(\dfrac{\Delta + \theta}{2}\right)\right)}}{2\sin(\theta/2)} = \frac{\sqrt{(1 - n^2\sin^2(\theta/2))}}{2\sin(\theta/2)} \tag{14}$$

i.e. for a 60° prism

$$\frac{d\Delta}{dn} = \frac{1}{\sqrt{(1 - n^2/4)}} \tag{15}$$

One empirical relationship which has been found to hold over small changes in wavelength is the Hartmann dispersion formula

$$n - n_0 = a/(\lambda - \lambda_0) \tag{16}$$

28

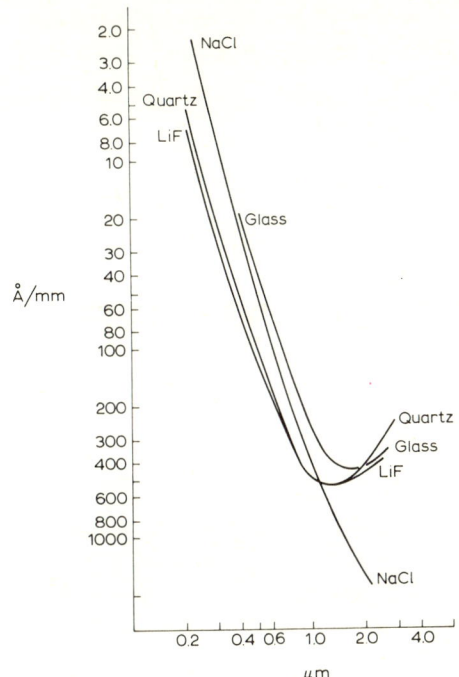

Fig. 14. Reciprocal linear dispersion as a function of wavelength for some common prism materials in a double-pass Littrow mount with a focal length of 270 mm.

where a is a constant (with the dimension of length) over a short spectral region around wavelength λ_0. More complicated relationships are discussed by Partington [71]. Differentiation of eqn. (16) gives

$$\frac{dn}{d\lambda} = \frac{a}{(\lambda - \lambda_0)^2} \qquad (17)$$

which may be combined with eqn. (15) to give a formula for the *angular dispersion* for a prism

$$\frac{d\Delta}{d\lambda} = \frac{d\Delta}{dn}\frac{dn}{d\lambda} = \frac{a}{\sqrt{((1 - n_0^2/4)(\lambda - \lambda_0)^2)}} \qquad (18)$$

The dispersion is quoted in radians/Å but it is more useful to refer to the *reciprocal linear dispersion* of a spectrograph containing a particular prism and with a focal length, f, of the camera lens. This may be written as

$$\frac{d\lambda}{dx} = \frac{1}{f} \frac{d\lambda}{d\Delta} \tag{19}$$

in units such as Å/mm. Figure 14 shows the variation of reciprocal linear dispersion with wavelength for a number of common prism materials used in a double-pass Littrow mount with a focal length of 270 mm. It can be seen that glass is a better prism material than quartz over the region of its transparency and that rock-salt is particularly bad around 2 μm.

(3) Resolving power of a prism spectrograph

One may make use of Fermat's principle, which states that the optical paths for all rays associated with a given wave-front are equal, to calculate the theoretical resolving power of a prism and therefore of a spectrograph. With reference to Fig. 15, for the complete path

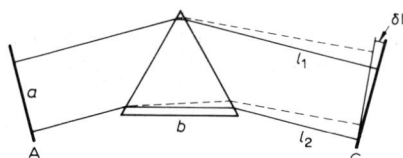

Fig. 15. Resolving power of a prism.

from plane A (collimator lens) to plane C (camera lens) the difference in path lengths is given by

$$l_1 - l_2 = nb \tag{20}$$

where b is the length of the base of the prism. If we now consider the difference in path length δl for two wavelengths λ and $(\lambda + \delta\lambda)$, we can write

$$\delta l = \delta n b \tag{21}$$

30

The change in deviation is then given by

$$\delta\Delta = \delta n b / a \qquad (22)$$

At the limit of resolving power, this minimum change in the angle of deviation is the diffraction limit of the aperture which is λ/a. We can now write

$$\delta n b = \lambda \qquad (23)$$

The resolving power of the prism is defined as $R = \lambda/\delta\lambda$, so that

$$R = \lambda/\delta\lambda = b \frac{\delta n}{\delta\lambda} = b \frac{dn}{d\lambda} \qquad (24)$$

The important point here is that the resolving power increases with size of prism. The resolving powers of several different prism materials for IR spectrometers have been calculated [72].

(4) Transparency of optical materials

The dispersion given by a prism is obviously going to determine, to a large extent, the performance of a spectrograph, but the transparency of the material may be a much more serious problem. The approximate transparent spectral ranges of some common prism materials are given in Table 3.

The lower wavelengths of transparency of the alkali halides are limited by the charge-transfer transitions (effectively photochemical redox reactions involving oxidation of the halide ion). The fluoride ion, with the highest oxidation potential of the series, has the highest frequency of charge-transfer band and hence transmits farthest into the UV. A fuller description of these spectra, measured first by Hilsch and Pohl [73], is given by Walker and Straw [77].

Absorption of IR radiation by ionic solids is due to lattice vibrations of the ions, which give rise to the so-called *Reststrahlen* bands. At these wavelengths, reflectivity becomes very high and the refractive index changes markedly, actually falling below unity (*anomalous dispersion*). Czerny [78] reported experimental data for sodium and potassium chlorides and Hohls [79] for lithium and sodium fluorides. Because of the relatively large masses of the constituent ions,

TABLE 3

Transparencies of materials for prisms and windows

Material	Lower limit[a] (nm)	Useful upper limit[b] (μm)
LiF	120	5
NaCl	175	15
KCl	180	
KBr	210	25
KI	250	
CsI	250	54
AgCl		28
CaF_2	125	8
BaF_2	200	11
SiO_2 (fused)	175	2.5
Glass (borosilicate)	310	2.5
Glass (crown)	350	2.5

[a] For alkali halides, taken from the absorption spectra published by Hilsch and Pohl [73].
[b] From Dodd [75] and from data supplied by Ealing Beck Ltd.
An excellent source of collected data are the Tables of Landolt—Börnstein [76].

caesium iodide shows a Reststrahlen band at very long wavelengths and the material is usefully transparent up to 54 μm [80].

The increasing interest in far-IR spectrophotometry has occasioned the need for filters for this region for use with grating spectrometers. A number of suitable filters have been proposed — so-called Reststrahlen filters — consisting of finely powdered ionic crystalline solids suspended in polyethylene [81] which is itself transparent above about 14 μm. (This material, containing suspended carbon black, may even be formed into lenses for far-IR work.) Short-wave cut-off (50% transmission for a 10% suspension) occurs at the following approximate wavelengths.

Material	LiF	SrF_2	CaF_2	NaF	BaF_2
Cut-off (μm)	45	60	75	75	90

Figure 14 showed the effect of variation of refractive index with wavelength on the dispersion of a prism. A convenient measure of the relative dispersions of two materials is the difference in the refractive index for light of different wavelengths, say the hydrogen C

TABLE 4

Refractive indices of glasses

Glass	n_D	$(n_F - n_C)$
Crown	1.52	0.009
Flint	1.58	0.014
Heavy flint	1.65	0.019
Silica	1.54	0.013

and F lines (see Table 2 p. 10). Some very approximate values for common optical materials are given in Table 4. A very extensive list is included in the Landolt—Börnstein Tables [82].

(5) Lenses in spectrographs

The most desirable property of a prism material, the variation of refractive index with wavelength, is a considerable inconvenience when the same material is to be used for a lens, and the difficulties involved in producing an achromatic lens increase with the spectral bandwidth it must transmit. Compensation in a spectrograph can be made by tilting the plate (and bending it slightly) to allow for the shorter focal length at shorter wavelengths. In the case of a mono-chromator it is not convenient to move the exit slit, but is preferable to rotate the prism in order to scan a spectrum. The difficulties of designing automatic focussing instruments have led to the widespread use of mirrors in place of lenses since they are not only achromatic when used as focussing elements, but have high reflectivity over a very wide spectral region.

(6) The Littrow spectrometer (1863)

Great economy of space and materials and an improved perfor-mance resulted from Littrow's modification to the prism spectro-scope [83,84] in which the light is passed twice through the prism (in his original instrument through a train of four prisms) and so is doubly dispersed. A further advantage is that, when natural quartz is used for the prism, the effect of birefringence (differing refractive index according to direction of the beam relative to the crystallo-graphic axes) which caused Cornu to design his cemented double

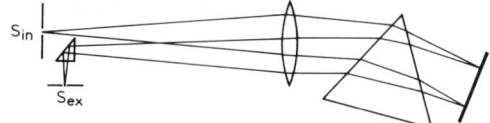

Fig. 16. Littrow spectrometer.

prism of suitably cut matching prisms, is cancelled. Rotation of the prism and mirror as one unit (see Fig. 16) enables one to select any wavelength at the exit slit S_{ex}. There have been many modifications to this basic design and it remains one of the most widely used in modern spectrophotometers.

Littrow himself mentioned [84] that Duboscq had, at about the same time, also described a spectroscope in which the light was doubly dispersed by passage through one prism in two directions; but the single prism was a $30°/60°$ one and a face was silvered so that the auxiliary mirror was dispensed with. It is interesting to note that many commercial instruments supposedly employing the Littrow design are, in fact, using that due to Duboscq with the modification that the lens is replaced by a spherical (as in the Beckman DU spectrophotometer [85]) or parabolic mirror (as in some Pye-Unicam spectrophotometers [86]). The paper by Cary and Beckman gives an interesting account of the considerations which led to their design [85]. While an off-axis parabolic mirror should be used to eliminate spherical aberration, they show that, for the resolution required of their instrument, the apparently wider slit-width resulting from the use of a spherical mirror is acceptable and makes the instrument

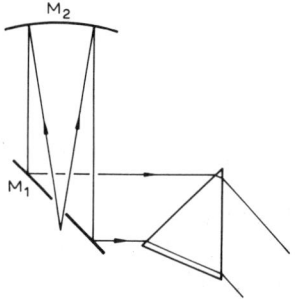

Fig. 17. Pfund spectrometer. M_1, Plane mirror with aperture; M_2, paraboloidal mirror.

cheaper to produce. Pfund [87] has shown that on-axis parabolic mirrors can be used satisfactorily in a single-pass spectrometer and offer an attractive compromise between cost and performance (Fig. 17).

(7) The Wadsworth mounting (1894)

The biggest improvement made by Wadsworth to the Littrow instrument was not the "multiple transmission" arrangement passing the beam six times through one prism and maintaining the optics at minimum deviation for all wavelengths [88], but the replacement of the lens, with its associated chromatic aberration, by a spherical concave mirror [89]. He claimed that, as the angle of reflection at the mirror was never greater than 1° (the focal length of the mirror was 1.75 m), spherical aberration was not serious. He also devised a system of coupled cranks to maintain the prism at the angle of minimum deviation as it was rotated (Fig. 18). In a subsequent paper

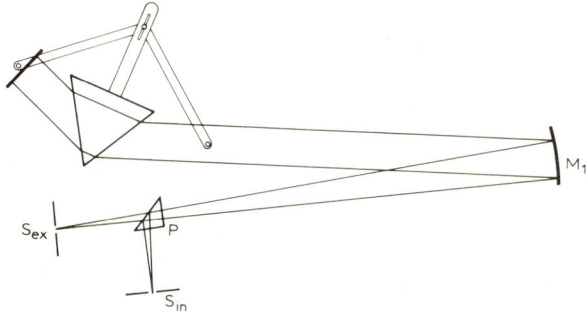

Fig. 18. Wadsworth spectrometer. M_1, Spherical concave mirror tilted so as to pass returning beam to slit S_{ex} lying above the reflecting prism P.

he described all the possible combinations of one plane mirror and a prism, of which the Littrow form is but one. Another of the combinations was that used by Pfund [87] which has already been mentioned.

(8) The Féry prism (1910)

This most unusual optical element is a prism with two curved

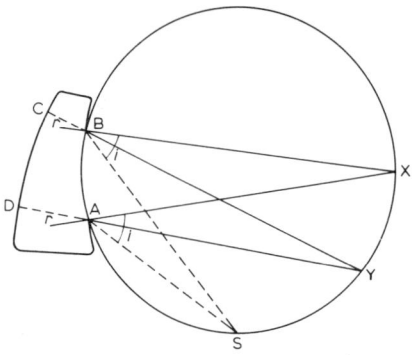

Fig. 19. Féry prism. AB and CD are arcs of circles with centres X and Y respectively. S = slit with coincident image for only one wavelength; *i* and *r* are angles of incidence and refraction, respectively.

faces, both cylindrical, one convex and the other concave (Fig. 19) [90]. Used in a Littrow-type spectrometer, it achieves autocollimation and dispersion all in one piece of glass. Losses through reflection and absorption and chromatic aberration are minimised, though astigmatism remains. A small commercial instrument using a Féry prism has been on the market for some years and covers the region 320—1000 nm [91].

(9) The Pellin—Broca prism (1899)

Of the many unusually shaped prisms which have been described, one which is used — in direct vision spectroscopes — is the Pellin—Broca prism [92]. This produces a constant deviation of 90° as it is

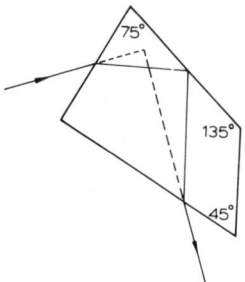

Fig. 20. Pellin—Broca prism.

rotated. It is used, for example, in the old Hilger wavelength spectrometer where the collimator and telescope can remain fixed while the spectrum is scanned (Fig. 20).

(10) The Rumsey "economical" prism (1962)

Faced with the problem of providing a prism with as large an aperture as possible (for spectral studies on Aurora displays) and as little weight as possible (for reasons of cost and space), Rumsey developed this prism [93]. It has one quarter of the weight of a conventionally mounted 60° prism with the same aperture and base. A further benefit is the improved transparency because of the shorter light path (Fig. 21).

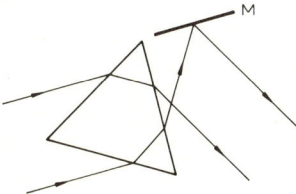

Fig. 21. Rumsey "economical" prism. M is a plane mirror.

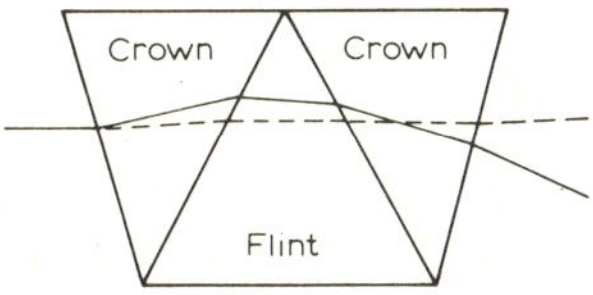

Fig. 22. Amici prism.

(11) The Amici prism

An arrangement for obtaining reasonably high dispersion with zero deviation has found application in small pocket "direct-vision" spectroscopes and consists of a number of prisms alternately of crown and flint glass (the former with lower refractive index and much lower dispersion) as shown in Fig. 22.

(12) Van Cittert spectroscope

A combination of two simple spectroscopes as in Fig. 23 results in a device with zero deviation, since the second prism recombines all rays from the first. If, however, a movable slit is placed between the two spectroscopes in such a way that it serves both as the exit slit for the first and the entrance slit for the second, then a constant-deviation monochromator results. Scanning is accomplished simply by

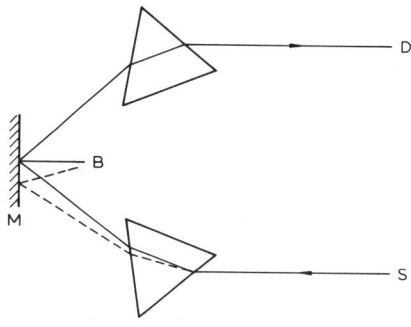

Fig. 23. The van Cittert spectroscope. S, Source; D, detector; M, movable mirror; B, slit blade.

lateral movement of the slit. Hardy designed a more compact version in which the slit was replaced by a mirror and a slit [94] and this form was used in a recording spectrophotometer developed commercially in 1938 [95].

(13) Two interesting prism spectrometers

From the many spectrometers described in the literature, two models are mentioned here because the author felt there was something a little different which made the papers worth further consultation. So many materials now available in a very high state of purity may be used for making prisms that one may easily forget that this was not always the case. At the same time it may come as a surprise to many to learn that the IR spectra of many organic compounds had been recorded under low resolution before the end of last century. The description by Julius [1] of his rock-salt prism IR spectrometer (1888) includes a wealth of practical details. At the other end of the spectrum, a design for a vacuum UV spectrograph of 1931, barely 30 cm long, using miniature fluorite lenses and prism, reminds one of the difficulties then common in obtaining a large enough piece of a crystal of sufficient optical purity [96].

(B) THE DIFFRACTION GRATING

We shall first consider diffraction of a parallel beam of monochromatic light passing through a transmission grating, i.e. a series of

38

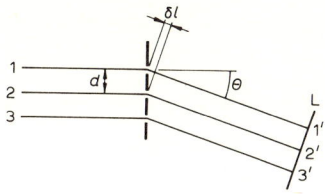

Fig. 24. Diffraction grating.

narrow, closely spaced, parallel slits, as in Fig. 24. If the wave-front, L, travelling at an angle θ is reinforced after passing through the grating, so that the rays following the paths 1—1′, 2—2′ and 3—3′ are all in phase, then the distance δl must equal a whole number of wavelengths. Hence

$$n\lambda = \delta l = d \sin \theta \qquad (25)$$

This is known as the grating equation. It can obviously be extended to cope with rays incident at an angle ϕ to the normal of the grating, so that

$$n\lambda = d(\sin \theta + \sin \phi) \qquad (26)$$

Light passing very close to any edge will suffer diffraction and the effect is most easily observed if the light passes between two edges very close together, in other words through a narrow slit. This relatively simple case, applied to parallel rays, is known as Fraunhofer Diffraction. Considering the wave nature of light, one can show (see, for example, Brown [97]) that the intensity at an angle of diffraction θ is given by

$$I_\theta = A_\theta^2 = \frac{\sin^2 \alpha}{\alpha^2} \frac{\sin^2 N\beta}{\sin^2 \beta} \qquad (27)$$

where

$$\alpha = \frac{\pi}{\lambda} w \sin \theta \qquad (w = \text{width of slit}) \qquad (28)$$

and

$$\beta = \frac{\pi}{\lambda} d \sin \theta \qquad (d = \text{spacing between slits}) \qquad (29)$$

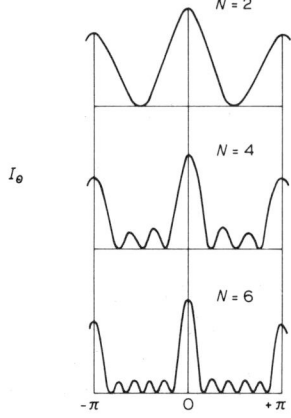

Fig. 25. Plots of I_θ vs. θ for different values of N.

The effects of the two parameters w and d may be separated if we write

$$I_\theta = f(\alpha) \, f'(N\beta) \tag{30}$$

in which $f(\alpha)$ is the shape factor, $f'(N\beta)$ the interference factor, and N the number of slits. The nature of the diffraction pattern changes markedly as N increases, particularly while N is relatively small. Figure 25 shows the effect on I_θ as a function of θ with varying values of N, for a given wavelength. Principal maxima of I_θ will occur when $\beta = n\pi$, n being an integer. Thus we can derive the grating equation (25) by substituting for β in (29). A point of practical importance is that θ is smaller for light of shorter wavelengths, which is the reverse of the case for the prism.

(1) Dispersion of a grating

As defined in the case of a prism, the dispersion is given by $d\theta/d\lambda$ which in this case can be obtained by differentiating eqn. (25)

$$\frac{d\theta}{d\lambda} = \frac{n}{d} \cos\theta \tag{31}$$

The dispersion is thus proportional to the order of the diffraction pattern and inversely proportional to the spacing.

40

(2) Resolving power of a grating

Figure 25 demonstrates how the "line-width" decreases as N increases. A change from maximum to minimum for the function $f'(N\beta)$ occurs for a change in β of π/N. Taking this as the half-width of the line and differentiating eqn. (29)

$$\delta\beta_{\frac{1}{2}} = \frac{\pi}{\lambda} d \cos \theta \quad \delta\theta_{\frac{1}{2}} = \frac{\pi}{N} \tag{32}$$

i.e.

$$\delta\theta_{\frac{1}{2}} = \frac{\lambda}{Nd \cos \theta} \tag{33}$$

We consider two lines of wavelengths λ and $(\lambda + \delta\lambda)$ to be resolved when the maximum of one coincides with the minimum of the other, so that

$$d \sin \theta = n(\lambda + \delta\lambda) \quad \text{for a maximum} \tag{34}$$

and

$$d \sin(\theta + \delta\theta) = n\lambda \quad \text{for a minimum} \tag{35}$$

If we expand eqn. (35) and make the approximations valid for very small values of $\delta\theta$, so that $\sin \delta\theta = \delta\theta$ and $\cos \delta\theta = 1$, then

$$d \sin \theta + d\delta\theta \cos \theta = n\lambda \tag{36}$$

Substituting for $d \sin \theta$ from eqn. (34) and for $\delta\theta$ from eqn. (35)

$$n(\lambda + \delta\lambda) + \lambda/(N \cos \theta) = n\lambda \tag{37}$$

which simplifies to

$$\lambda/\delta\lambda = nN/\cos \theta \approx \text{length of grating} \tag{38}$$

As an example, a 100 mm grating with 600 lines/mm would have a resolving power of 60,000 (considerably better than a prism) and would give a first-order diffraction for wavelength 830 nm and second-order for 415 nm etc. at $\theta = 30°$. A point worth noting here is that the resolving power is, to a first approximation, independent of the order in which the grating is used.

(3) Reflection gratings

The discussion just presented has considered the case of a transmission grating, but the same results hold for the reflection grating. In practice, the latter type is generally used because it is much easier to obtain a material which will reflect uniformly over a wide spectral range than one which will transmit uniformly. The early gratings were ruled, with fine diamond points, on speculum metal (an alloy of copper and tin in the ratio seven parts to three), but sometimes on glass. Later rulings were made on copper plates prepared for half-tone printing [98] and finally on aluminium deposited by sputtering on to optically flat glass [99]. A brief history of the ruling of gratings was given by Harrison some years ago [100].

(4) Blazing

Ames, writing in 1889, complained of his difficulties in getting a good diamond. He also remarked how variable the gratings he produced were with respect to the relative brilliance of the spectra in different orders [101]. However, it was not until 1910 that Wood and Trowbridge [102] published their work demonstrating the importance of the shape of the groove. They drew attention to the fact that no metal was actually removed in the process of ruling a grating on metal. In the early gratings, ruling simply resulted in a series of parallel scratches alternating with reflecting strips. Most of the light was then reflected at the angle for specular reflection and the intensities of the diffracted beams were very low. But when a grating is ruled as a continuous series of troughs with a saw-tooth wave-form, the reflection angle can be chosen (depending on the angle of the groove) so that a high proportion of the incident light is diffracted into a given order for a given wavelength. Such a grating is said to be blazed. Wood and Trowbridge gave the name *echellette* to the gratings they ruled, which had relatively large spacings, and were used in the IR region.

Michelson [103] described in 1898 a device related to the grating called a *transmission echelon*, which could give very high resolution

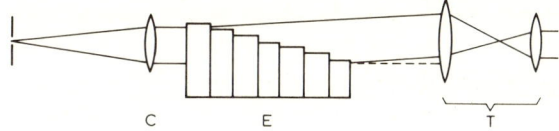

Fig. 26. Transmission echelon spectroscope. C, Collimator; E, echelon; T, telescope.

when analysing light of a very narrow spectral bandwidth by operating in a very high order (see eqn. (38)). The first echelon consisted of only seven plates each 5 mm thick (Fig. 26). When placed between a collimator and a telescope it was capable of separating the sodium D lines very easily and of allowing the Zeeman effect (splitting of the lines in a magnetic field) to be observed. A later, more practical instrument consisted of 20 plates, staggered in 1 mm steps, each being 18 mm thick. This type of spectroscope is ideal for the analysis of very simple line spectra. In more complex cases, a prism monochromator must be used as a filter to isolate the line or small group of lines.

The *reflecting echelon* [104] developed some years later by Williams, had the advantage of being useful over a much wider spectral range, but the difficulties of preparing and of stacking plates of uniform thickness (to within 1/20 of a wavelength) prevented the device from gaining wide acceptance.

In 1949 Harrison reported the successful ruling of what he termed an *echelle* [105] which would serve the purpose of the echelon but be easier to make. It is essentially a reflection grating with widely spaced grooves, operating in a high order for short wavelengths, so that the shape of the groove must be very carefully controlled. He pointed out that as far fewer grooves must be ruled, the time required to make an echelle is a matter of hours instead of the weeks for a big grating with high resolution. It is therefore easier to maintain constancy in the ruling engine and the diamond. Echelles are normally used in conjunction with a prism in a large spectrograph, replacing the Littrow mirror with the ruling oriented so as to give high resolution of closely spaced lines at right angles to the full spectrum. The prism thus selects the narrow spectral bandwidth which allows full use to be made of the high resolving power of the echelle.

(5) Rotation of a grating in a scanning monochromator

Martin has provided solutions to the problems of how to drive a grating when a scan linear in wavelength or linear in wavenumber is required [106]. The grating equation (25) may be expanded in terms of θ and ϕ, and $(\theta-\phi)$ replaced by $\delta\theta$, giving

$$2\,d\,\sin(\theta - \delta\theta/2)\,\cos(\delta\theta/2) = n\lambda \tag{39}$$

Referring to Fig. 27, we can follow the argument for the case of linear-wavelength scanning. G is the grating, connected to an arm AB which rests on the moveable plate P driven by the micrometer M.

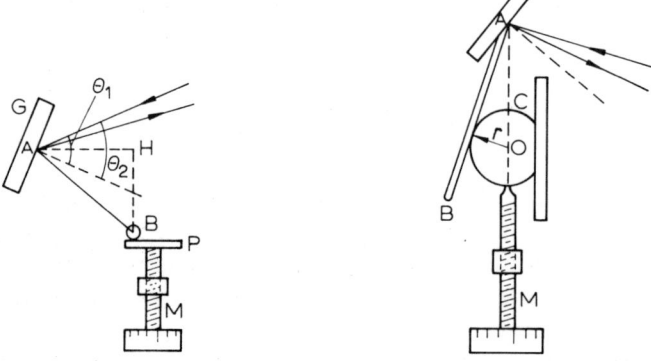

Fig. 27. Linear wavelength drive for a grating (Martin [106]). Reproduced with permission.

Fig. 28. Linear wavenumber drive for a grating (Martin [106]). Reproduced with permission.

The line AH is drawn perpendicular to the direction of travel of the micrometer. The arm AB is adjusted so that the angle BAH = $(\theta - (\delta\theta/2))$ for a given wavelength. Since sin BAH = BH/AB, and AB remains constant

$\lambda \propto \sin(\theta - (\delta\theta/2))$

i.e.

$\lambda \propto$ BH = travel of micrometer

44

For a linear-wavenumber scan the arrangement of Fig. 28 is adopted. The arm AB is held in contact with a sliding sphere C of radius r driven by the micrometer in the direction OA. The angle BAO is this time made equal to $(\theta - (\delta\theta/2))$, so that

$$\lambda \propto \sin BAO = r/AO$$

and

$$1/\lambda \propto AO = \text{travel of micrometer}$$

(6) Development of the grating spectroscope

The father of the diffraction grating could well be said to be Professor H.A. Rowland who invented a machine for making a near-perfect screw thread and then used this in his machine for ruling gratings. Reporting the developments in 1882, he made reference to several good gratings of about 15,000 lines per inch and to his best with some 43,000 [107]. His second major contribution to this field was the discovery that a concave grating (i.e. one ruled on the inner surface of a spherically ground and polished mirror) could be used without additional optics in a fully focussing achromatic spectrograph. The geometry is shown in Fig. 29. The circle, the diameter of which equals the radius of curvature of the grating and on whose circumference lie the grating, the slit and the curved photographic plate, is now known as the *Rowland circle*.

His instruments were large [108], a radius of 3.3 m, with plates up

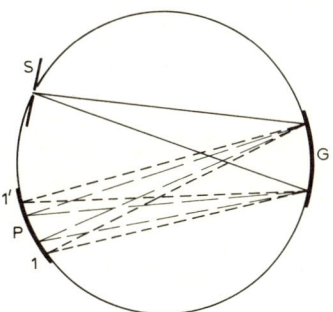

Fig. 29. Rowland's concave grating mounting. S, Slit; G, concave grating; P, plate; 1, and 1′, first-order spectra.

to 50 cm long. It was perhaps good fortune that the grooves should have been spaced evenly along the chord of the concave surface (which was what resulted from his ruling engine) rather than evenly along the arc, as the former arrangement eliminates spherical aberration and coma.

Ruling techniques have been further improved in recent years [109] and the ease with which replicas can be taken has made good gratings readily and cheaply available. Wood [98] commented in 1944 that customers specifying an original grating were given a replica because they were better — the plastic moulding was cemented to an optically flat plate and then aluminised, whereas the originals were ruled on copper and were usually not so flat.

Many different mountings for gratings, both plane and concave, have been described in the literature, each aiming to overcome some drawback of an earlier design. It is an interesting reflection on scientific progress that most commercial instruments nowadays follow designs varying but little from two of the earliest mounts.

(7) The Abney mount (1886)

This is a variation of the Rowland circle principle in which the plate and the concave grating remain fixed and the slit moves [110]. Abney described it as being portable, but it is usually dismissed as being very inconvenient because of its cumbersome nature. He also described a mounting for a plane grating using a pair of 2.1 m radius spherical mirrors (Fig. 30) which is not unlike the later development by Czerny and Turner.

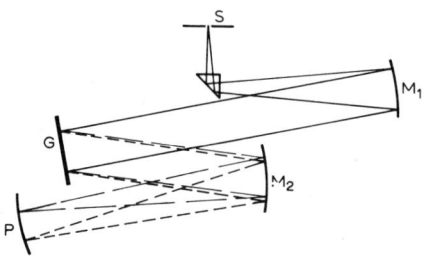

Fig. 30. The Abney mount. G, Plane grating; M_1 and M_2, spherical concave mirrors.

(8) The Eagle mount (1910)

Eagle [111] overcame the difficulties involved in using a Rowland mounting, where the whole room had to be light-proof and maintained at constant temperature, by putting everything in a long, narrow box with the grating at one end and the slit and plate at the other (Fig. 31) along with the controls for changing the angle of the

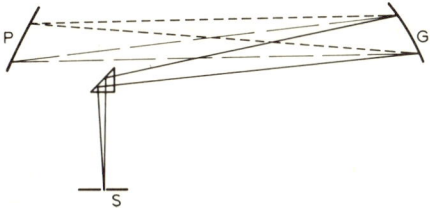

Fig. 31. The Eagle mount. G, Concave grating.

grating and its position. As it is, in effect, a Rowland mounting, the movements of the grating involved in scanning a spectrum are rather complex, which is mainly why this design is not widely used today.

(9) The Runge and Paschen mounting (1897)

The Rowland mounting suffers from astigmatism which can be troublesome if the distribution of light along the slit (e.g. of an image formed on the slit) is to be investigated. If parallel light falls on the concave grating, the spectra are focussed on the circumference of a circle with a diameter equal to the focal length of the grating (Fig. 32) and the astigmatism is overcome [112].

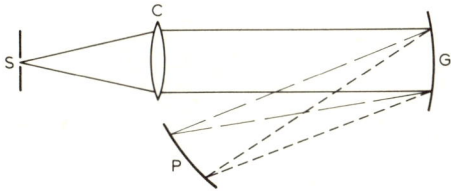

Fig. 32. The Runge and Paschen mounting. C, Collimator lens; G, concave grating.

(10) The Wadsworth mounting (1896)

It would be difficult to think of a design for a spectroscope using a grating or a prism which Professor Wadsworth of Chicago had not already described. Of his many ideas, the arrangement of Fig. 33, which was intended for mounting in an astronomical telescope, is

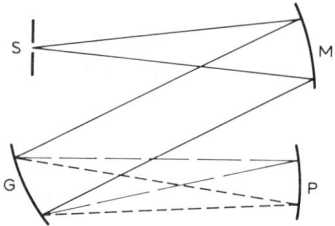

Fig. 33. The Wadsworth mounting. M, Spherical concave mirror; G, concave grating.

commonly associated with his name [113]. The mirror throws a parallel beam of light on to the grating which then focuses all spectra on a curved plate of radius $r/4$ so that a mounting similar to Rowland's is used, but with the dimensions reduced by half. This mounting is astigmatic. Meggers and Burns described an instrument using a very similar arrangement in which aberration was minimised by keeping the slit and the plate very close so that the angle of reflection remained very small [114].

(11) The Ebert spectrograph (1899)

Ebert pointed out that with spectrographs of high resolving power

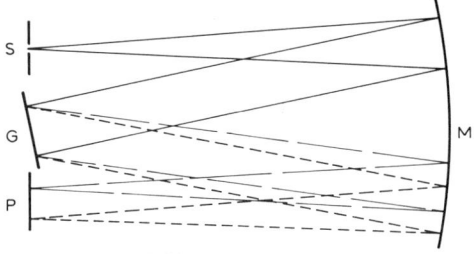

Fig. 34. The Ebert spectrograph. M, Single large spherical concave mirror; G, plane grating.

such as were then becoming available using gratings, it was desirable to minimise the loss of intensity by reducing the number of reflecting surfaces [115]. He proposed two designs, one with a lens and the second using a 1.2 m focal length concave, spherical mirror (Fig. 34) and both using a plane grating. The spherical aberration in the latter case was not serious for large instruments with long focal length mirrors. The simplicity of this design has attracted attention again in recent years and it is one of the most widely used in modern instruments.

(12) The Czerny—Turner spectrograph (1930)

These two workers examined the various existing mirror—grating instruments [116] and showed that when two spherical mirrors are used with a plane grating, the spherical aberration from the two mirrors will cancel in one arrangement but be additive in the alternative scheme (Fig. 35). Ebert's spectrograph happened to employ the

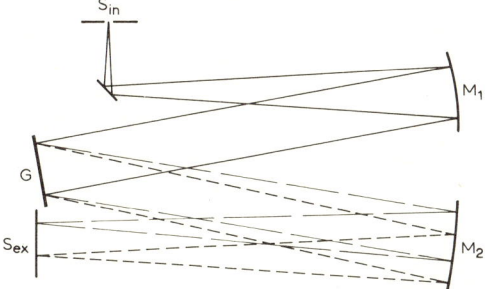

Fig. 35. The Czerny—Turner spectrograph. G, plane grating.

better arrangement. The small size of the mirrors results in a very compact monochromator which has found wide application in spectrochemical instruments. Rosendahl has reported a further study of this system using ray-tracing techniques to reduce coma practically to zero [117]. By making M_1 of slightly shorter focal length than M_2 coma can be corrected for, and resolution much improved [181,182].

(13) The Fastie—Ebert spectrograph (1952)

This is an improved version of Ebert's instrument with astigmatism minimised by the use of curved slits [118]. With an instrument only 75 cm long containing a 75 mm grating with 1200 lines/mm, Fastie obtained a resolution of 0.05 Å in the second order. As with the Czerny—Turner mounting, scanning involves only rotation of the grating so that all the other parts of the optics remain fixed.

(14) The Pfund—Hardy mount

This mounting avoids the reflection of rays at an angle deviating from the normal to the concave mirrors (the main source of aberration with spherical mirrors) by using Pfund's idea of a plane mirror

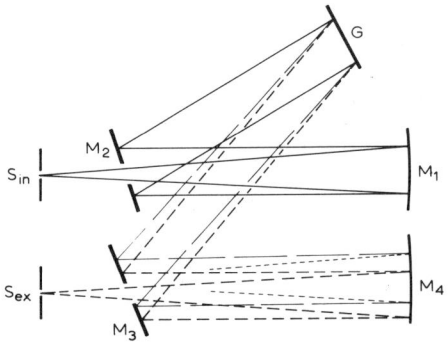

Fig. 36. The Pfund—Hardy mount. M_1 and M_4, concave spherical mirrors; M_2 and M_3, plane mirrors; G, plane grating.

on the axis but containing a small aperture for the diverging or converging ray [87]. Otherwise this mounting (Fig. 36) is similar to the one just described.

(15) Grating or prism?

With the high quality of modern optical components and the availability of many synthetic materials for prisms, the choice of prism or grating will not be decided by manufacturing limitations. The salient points connected with each are summarised briefly below. A paper by Churchill [119] gives some useful criteria for judging spectrographs and a recent review has covered the field of IR spectrophotometers [120].

50

	Gratings	Prisms
Dispersion	Uniform, depends on the order	Non-uniform, depends on the material
Resolution	May be very high	Moderate
Cost	Relatively cheap for replicas	May be expensive, but depends on material
Spectral range	Very wide, but several gratings will be needed to cover a wide range. To avoid overlap, order selection, by means of a filter or second prism, is usually necessary.	More restricted, but with a number of prisms a wide range can be covered.

(16) Overlapping spectra

The wider the spectral range covered the more serious becomes the problem of overlap of spectra of different orders. Though blazing can help to reduce this, the effect is still troublesome, particularly when accurate quantitative intensity measurements are to be made, as in spectrophotometry. Ames [101] overcame the difficulty by using a series of coloured aqueous solutions as filters. The first high-resolution IR spectrophotometer was described in 1918 by Sleator [121] who used a rock-salt prism as primary analyser followed by a 600 lines/mm grating. A similar arrangement is used in the Cary Model 14 spectrophotometer (described along with many other instruments by Bauman [16]) but in this case for a second important reason — the reduction of stray light to a very low value enabling the accurate measurement of high absorbances to be carried out. This latter problem is of some considerable importance for measuring the absorption spectra of single crystals (on a microscope stage) and in such cases double monochromators are usually used [122—124].

(17) Vacuum UV

The high reflectivity of aluminium which is used to coat gratings would suggest their usefulness for vacuum-UV spectrographs. A spectrometer based on the Rowland mounting but with two slits (one for the photomultiplier) has been described [125]. In order to get higher resolution, gratings are being used in higher orders rather than with closer spacings [126]. Ericson and Edlén [127] were able to measure wavelengths of emission lines down to about 10 nm working at a 10° take-off angle and in a very high order. It should be mentioned that vacuum spectroscopy techniques are also useful in the IR, particularly for studies on the spectra of water vapour and carbon dioxide, or in regions where these species absorb [128].

(C) THE SLITS

A spectrograph is essentially an image-forming optical instrument. When the object was the sun, early observers saw the familiar band of colours making up the spectrum and realised that this could be considered as a very large number of lines, all of different colours and lying very close together. The need for a very narrow entrance slit to a spectrometer to enable the colours to be separated more sharply was then obvious. The slit has two very important roles in all spectrometers — to control the amount of energy entering the instrument and to vary the spectral resolution of the instrument. In the preceding section it was shown that the theoretical resolving power of a prism or grating could be calculated quite simply. It is convenient to distinguish from this theoretical value, a practical value for the complete spectrometer — the *resolution*.

(1) Energy distribution about a given selected wavelength

For rather narrow slits (but not so narrow that the formation of interference patterns is serious) the distribution of energy being passed by a monochromator set at a wavelength λ_0 is given by the diagram in Fig. 37. The width at half-height is known as the *spectral band-width* and comprises 75% of the transmitted light. It is a useful

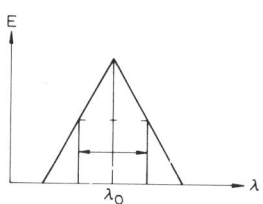

Fig. 37. Spectral band width passed by a slit.

indication of the performance of a spectrometer. As the sensitivity of detectors and the output of sources is not constant over wide ranges of wavelength, it is useful to be able to open and close the slit so as to obtain suitable working conditions for the detector. In a thorough discussion on the effect of slit width on the width of spectral lines,

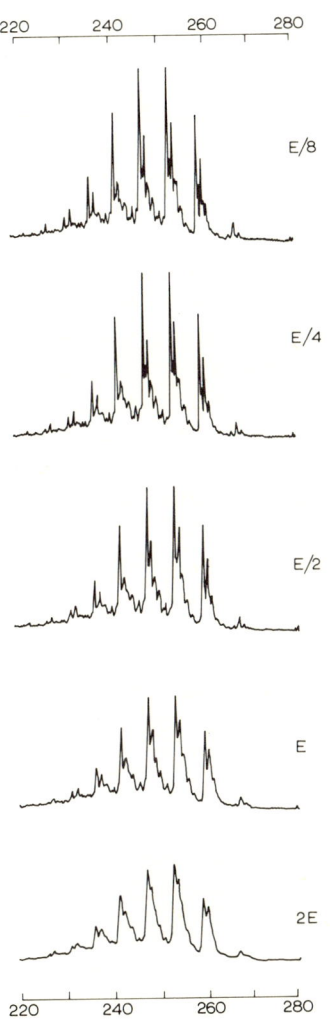

Fig. 38. UV absorption spectra of benzene vapour (10 mm path). For significance of E values see Fig. 40.

References pp. 88—93

van Cittert [129] showed that for maximum resolution the product of the slit width and the angular aperture of the collimator should be less than unity and that use of a very narrow slit simply reduces the available energy to the detector. The series of spectra of benzene vapour in Fig. 38 illustrates the effect of slit width on the resolution attainable with a modern recording spectrophotometer. At the smallest slit width (corresponding to E/8) very little improvement over the next (E/4) is realisable, but some noise is evident in the background.

(2) Line curvature

As long ago as 1874, Grubb [130] described the slightly curved lines seen in spectra with his spectroscope. This curvature arises because, under the experimental conditions, only rays from the centre of the slit (i.e. on the axis of the collimator) pass through the prism at minimum deviation, the others suffering somewhat greater deviation with the result that the lines are all curved with the convex side towards longer wavelengths. For spectrographic work this effect is not serious, but in a monochromator the loss of energy and lower resolution due to the poor overlap of the curved image on a straight slit can be quite significant. Roemer and Oetjen [131] investigated this phenomenon and produced two designs for curved slits with bilateral movement which largely overcame the difficulty.

For a prism spectrograph, the radius of curvature, R, of a line at a given wavelength is given by

$$R = \tfrac{1}{2}n^2 f(n^2 - 1)^{-1} \cot i \tag{40}$$

where n is the refractive index of the prism material for that wavelength, f is the focal length of the lens or mirror and i is the angle of incidence on the prism of the radiation from the slit. It is clear that R will vary with wavelength, but as it is fairly large anyway (of the same order as f) the difference is small for most spectrometers over quite a wide range and a compromise will be satisfactory. It is interesting to note that the zero-dispersion double monochromator used by Hardy [94,95] does not suffer from this trouble as the error of one spectrometer is cancelled by the second.

Curvature also arises in grating instruments, in which case the radius is given by

$$R = \tfrac{1}{2}f \cot i \tag{41}$$

(3) Bilateral slits

It is important that a change of slit width should not effect a change in wavelength of a spectrometer, or, more strictly, a change in the wavelength of maximum intensity. For this reason bilateral slits are commonly used so that both jaws move towards or away from the centre line symmetrically. An arrangement which maintains the jaws strictly parallel irrespective of the separation is shown in Fig. 39.

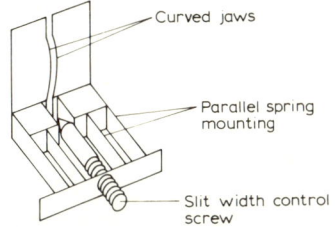

Curved jaws

Parallel spring mounting

Slit width control screw

Fig. 39. Slit mechanism.

In most instruments entrance and exit slits are coupled to ensure the best resolution for any given amount of energy. In some cases (e.g. the Pye—Unicam SP 600) only one pair of jaws is used, but the "entrance" and "exit" slits use different parts of the jaws.

(4) Constant energy and constant band-width

Some manual spectrophotometers work on a "constant-energy" principle so that an adjustment of the slits must be made for every change in wavelength. The spectral band width will then not be constant, but will be determined by the output of the source, the sensitivity of the detector, and the dispersion of the prism. In the case of a grating instrument it is easier to work at constant band width, since the dispersion is constant, as long as the sensitivity of the detector is adequate. Recording instruments normally work at the smallest band-width compatible with the demands of the detecting system. Figure 40 shows the variation of band-width with wavelength for the Pye—Unicam SP 800 recording spectrophotometer which has automatically programmed slits (controlled by a cam on the prism drive).

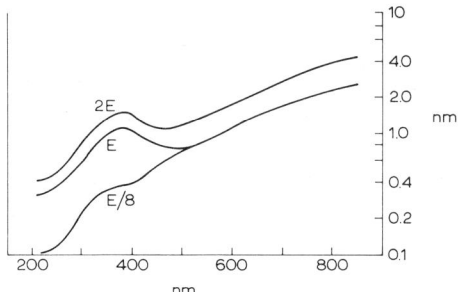

Fig. 40. Variation of band-width with wavelength (both in nm) of the Unicam SP 800 spectrophotometer with automatically programmed slits. Reproduced by courtesy of Pye—Unicam Ltd.

(5) Practical points affecting choice of slit width

The possibility of over-riding a slit control programme is a valuable feature of a recording spectrophotometer, but even when this does not exist an appreciation of the limitations imposed under certain circumstances by the band width of an instrument is very important.

Resolution. Figure 38 (p. 53) shows quite clearly that fine structure in a spectrum can only be studied when slits are sufficiently narrow to realise the full resolving power of the dispersing element.

Wavelength measurement. Under low resolution, an unsymmetrical band, such as those recorded for benzene in solution in Fig. 4 (p. 11) will appear to exhibit maximum absorbance at a slightly different wavelength from the true one since the detector is effectively integrating the absorption spectrum over a small range.

Absorbance measurements. For the same reason, absorbance measurements on narrow bands can be misleading. As a practical guide, the effective band width should not be greater than the peak width at the apex if Beer's law is to be obeyed. This is also demonstrated in Fig. 38 (p. 53). When the absorption line width is less than the bandwidth of the spectrometer (which may be the case for gases in the IR) pressure broadening due to the presence of a second gaseous species can have a very marked effect on the recorded absorbance. Discrepancies of several hundred percent can occur when air is admitted to a cell containing a sample gas at low pressure, the absorbance then becoming much greater.

56

(6) Care of slits

As these are very delicate mechanisms they must be handled with great care. Dirt or dust on the jaws will give rise to striations on plates in a spectrograph and will cause irreproducible absorbance measurements with a spectrophotometer. Forcing slits to close completely can cause damage if specks of hard dust are caught. For this reason slits are normally made self-closing (under pressure of a light spring) and are opened, as in Fig. 39 (p. 55), by the action of a screw. Edisbury has much of value and interest to say on this and related topics in spectrophotometry [132].

(7) Accessories for spectrograph slits

The discussion on slits has been largely from the point of view of spectrophotometry. The spectrographer will often prefer to work with fixed slits for the sake of simplicity, but there are two important accessories which make photographic work more convenient, extend the range of concentrations that may be determined, and improve the precision.

Diaphragm. There is little point in using a whole plate to record one spectrum; indeed it is useful to record a reference spectrum from, for example, an iron arc, as well as that of the sample. The simple *fishtail diaphragm* (Fig. 41(a)) reduces the height of the slit to any desired value when placed immediately in front of the slit. The photographic plate must in this case be raised or lowered between exposures in order to separate the spectra. The possibility of a small lateral shift occurring during the movement of the plate is eliminated by using the *Hartmann diaphragm* (Fig. 41(b)) which enables exposures to be taken using successive small sections of the slit. Sometimes the sections may have a slight overlap to facilitate identification of spectral lines. Line curvature may be more serious with this type of diaphragm than with the fishtail diaphragm.

Sectors. In place of a diaphragm one may substitute a rotating

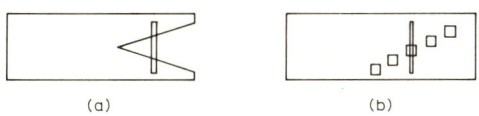

(a)   (b)

Fig. 41. (a) Fishtail diaphragm; (b) Hartmann diaphragm.

sector, so that different parts of the slit are exposed for different time intervals. This extends the useful range of the photographic plate considerably and makes possible semi-quantitative analysis by inspection of the plate. It is important that a sector be rotated rather quickly since the intensity of an arc often varies with time. When an a.c. spark or arc is used, the speed of rotation of the sector should not be a multiple or sub-multiple of the a.c. frequency, so that stroboscopic effects on the intensity of the arc are not observed.

Stepped rotating sector. With this device the length of exposure increases by factors of two while the slit length exposed decreases in uniform steps. It is desirable to cut two complete cycles on one sector so that it is well balanced and will run smoothly. (Fig. 42(a)).

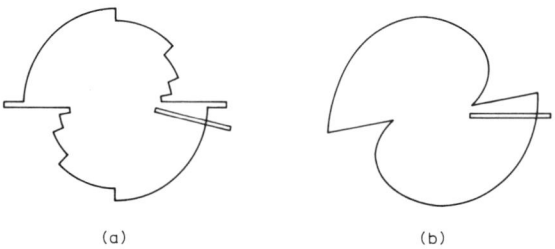

(a)   (b)

Fig. 42. (a) Stepped rotating sector; (b) logarithmic rotating sector.

Logarithmic rotating sector. With this type of sector the length of the exposure is still increased in multiples, but smoothly. Thus, the sector is cut with a logarithmic cam edge (Fig. 42(b)). The form of the spiral is given by

$$-\log \alpha = C + kl$$

where α is the angle of rotation, C the minimum radius, and l the length of the slit exposed.

(D) FOURIER TRANSFORM SPECTROSCOPY

This represents a completely different approach to the problem of spectrum analysis, using mathematical computation of intensities measured on a non-dispersive interferometer to arrive at an absorption spectrum. The idea is not new; indeed, Michelson [133] showed in 1891 that it was possible, if not practical, at that time. It is, however, only in recent years, since the high-speed digital computer has become a fairly readily available tool for the chemist, that Fourier transform spectroscopy has become a practical and useful technique. In this section a brief treatment of the principles and the theory will be followed by a necessarily tentative comparison with conventional grating spectroscopy in the far-IR, the region in which the interferometric spectrometer offers its greatest potentialities. Martin [134] has devoted a chapter of his recent book to this technique and Richards [135] has written an informative introduction to the field; both make some mention of commercially available instruments. A more detailed account of the theory is given by Strong and Vanasse [136].

The instrument used for the measurements is basically a Twyman—Green [137] interferometer in which the addition of a collimating concave mirror (to a simple Michelson instrument) produces a parallel beam of radiation and presents a plane wave-front to the beam-splitter. Figure 43 shows diagramatically the optics of an instrument developed by Gebbie and Stone [138] to cover the region $10-110$ cm^{-1}. Light from the source, S, (a high-pressure mercury vapour lamp) is collimated by the concave mirror to fall, after one

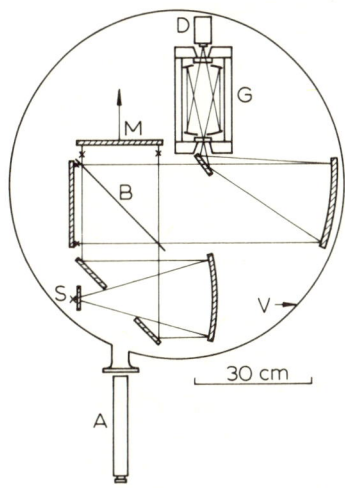

30 cm

Fig. 43. Optical layout of an interferometric spectrometer. From Gebbie and Stone [138]. S, Mercury lamp source; M, moving mirror; D, Golay detector; G, triple-pass gas cell; B, film beam divider; V, vacuum tank; A, auto collimator. Reproduced with permission.

reflection, on the beam-splitter, B. This is a very thin film of polyester ("Mylar"), polyethylene or similar material transparent in the far-IR, which ideally has a reflectance equal to its transmittance over the spectral range to be covered. The light is either reflected to the fixed plane mirror then back through the beam-splitter to the condenser mirror, or transmitted to the movable mirror, M, then reflected to the condenser. The light finally passes through the sample compartment, G, to the Golay detector, D.

There will generally be a difference in path lengths for the two beams, dependent on the position of the mirror, M. If this difference is denoted by Δ, and if the source provides a monochromatic beam of intensity I_0 at frequency ν, then the intensity incident on the detector will be a function of Δ given by

$$I_\Delta = I_0 \left(1 + \cos \frac{2\pi\Delta}{\lambda}\right)$$
$$= I_0(1 + \cos 2\pi\nu\Delta) \tag{42}$$

where ν is in cm^{-1} and Δ in cm. For light from a continuous source ("white light") the signal must be integrated over all the frequencies for each value of Δ, and I_Δ can be expressed in terms of $I(\nu)$, the source intensity after absorption by the sample and optics, as a function of frequency, by

$$I_\Delta = \int_0^\infty I(\nu)[1 + \cos 2\pi\nu\Delta]\, d\nu \tag{43}$$

$$= \tfrac{1}{2}I_{\Delta=0} + \int_0^\infty I(\nu) \cos 2\pi\nu\Delta\, d\nu \tag{44}$$

where $I_{\Delta=0}$ is the intensity measured for zero path difference. A direct plot of I_Δ against Δ is called an interferogram. An example is shown in Fig. 44.

The relationship (44) may be rearranged, making use of the Fourier integral theorem, to give

$$I(\nu) = 4 \int_0^\infty (I_\Delta - \tfrac{1}{2}I_{\Delta=0}) \cos 2\pi\nu\Delta\, d\Delta \tag{45}$$

so that the intensity of the source at any frequency ν, after absorption by the sample, is expressed as a function of the intensity measured by the detector and the distance moved by the mirror. To compute an absorption spectrum, a blank run is first carried out to

60

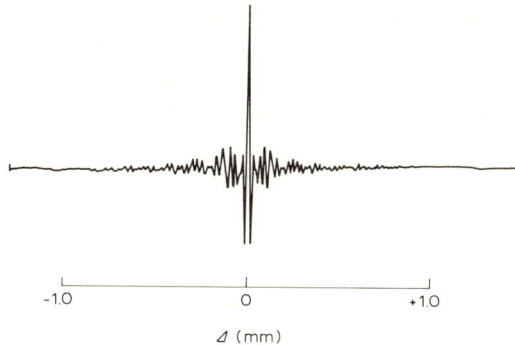

Δ (mm)

Fig. 44. Typical interferogram; recorded during a 2-mm traverse. Reproduced by courtesy of Dr. G.P. McQuillan.

determine $I(\nu)$ for the source and then a second run is made with the sample. It is then a simple matter to include in the computer programme a calculation of the absorbance over the spectral region covered.

One difficulty with eqn. (45) as it stands is that it requires integration over an infinite change in path length, which is obviously unrealistic. Because the scan is limited to some maximum value (say ± 5.0 cm) an approximation is made to allow integration over this shortened range and the resolution is limited correspondingly. As a rough guide, the resolution is given as the reciprocal of the path difference scanned, so that a 10 cm travel of the mirror should enable a resolution of 0.1 cm^{-1} to be obtained. In the spectrum of

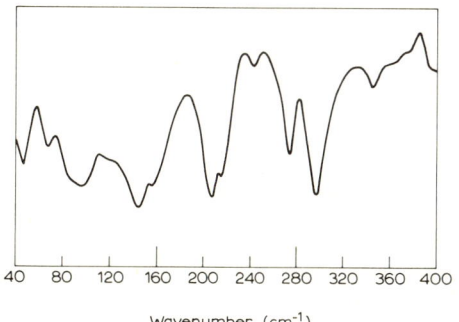

Wavenumber (cm^{-1})

Fig. 45. Far-IR absorption spectrum of an organomercury compound computed from the interferogram in Fig. 44. Reproduced by courtesy of Dr. G.P. McQuillan.

Fig. 45 which was computed from the interferogram in Fig. 44, fine structure was not expected and was therefore not looked for. Hence a relatively low resolution was sufficient and the scan over about 3 mm gave sufficient data with economies in both running time and computing time.

The earlier instruments employed a continuous drive for the mirror with sampling of the intensity at regular intervals using, for example, a Moiré fringe system to follow the movement of the mirror accurately over short distances. The chance of a measurement being made exactly at the position corresponding to $\Delta = 0$ was small and ways were developed of computing the spectrum from double scans (such as that in Fig. 44) passing through the zero point. Computing time was considerable and the cost of obtaining a spectrum in high resolution was rather high. Parrett [139], in a short article on Fourier transform spectroscopy, has included a typical programme in Elliot 503 Algol. The use of a stepping drive for the mirror improved the efficiency of data collection greatly since the relatively long travel times between the sampling times in the continuous sweep are almost eliminated. Two more recent approaches to interferometric spectroscopy have opened up new possibilities and caused a great increase in interest in the technique. They will be described briefly here.

(1) Periodic scanning

If the movable mirror is driven relatively fast (say about 1 sec per scan) the intensity at the detector due to the interferogram varies at audio frequencies, and the signal from the detector will have a low-frequency component (that of the sweep) together with components of higher frequencies. Spectrum analysis of these higher frequencies by conventional electronic (tuned-circuit filter) techniques enables any desired frequency to be isolated and its intensity measured, or a complete spectrum to be plotted. Digital computation is not required. Because of the low-frequency response of the common IR detectors, particularly of the Golay cell, such techniques were first applied successfully in the far-IR where the audio frequencies detected are relatively low [140,141].

The introduction of sensitive sources with faster responses — particularly of the bolometer type — has made possible the extension of rapid scanning interferometry to the whole of the IR spectrum

[142]. Low and Coleman [143] have described the application of one such instrument (the Block Model 200 Interference Spectrometer) coupled with a coherent information adder so as to facilitate improvement of signal-to-noise ratio by making multiple scans, to the study of IR emission spectra from cool sources. The resolution was reported to be about 40 cm^{-1}. This work demonstrates the enormous increase in sensitivity achieved by interferometric techniques compared with that of dispersion spectrometers.

(2) Aperiodic scanning with real-time computing

This approach, in spite of the expense of the equipment, would appear to offer the greatest possibilities for the future, and has already been demonstrated as a high resolution technique not only for the far IR (for example the Coderg Interferometer) but also for the medium IR range (Block Engineering). Since increasing the length of a scan (movement of the mirror) simply increases the resolution obtainable, it is theoretically possible to compute the complete spectrum from a very short scan. If this spectrum is stored electronically, it can be refined by continuous computing to improve the resolution during the scan and scanning is terminated when the spectrum, as displayed on a cathode-ray tube, shows as good a resolution as the operator requires. Final presentation is then as a chart-recording.

Double-beam operation is also now possible with interferometer spectrometers, resulting in a big improvement in photometric accuracy. This is particularly important when scans taking up to one hour are recorded for high resolution studies, as the mercury vapour lamp is not sufficiently stable over such long periods.

(3) A comparison with grating spectroscopy

In far-IR spectrophotometry, the very low output powers of the sources reduce the signal level very nearly to the noise level at the detector; moreover, the large slits which must be used to get some measurable signal reduce the resolution obtainable. With the interferometer, the full output of the source over the range covered is allowed to fall on the detector so that the noise is no longer a matter for concern, and practical resolving powers closely approach the theoretical. A further advantage over grating instruments is the relatively large spectral range handled without the need for changing filters or other parts of the optics. Commercial instruments cover the approximate range 500—10 cm^{-1} (20—1000 μm). The polyethylene lenses and quartz windows act as filters to cut off the wavelengths below 50 μm (200 cm^{-1}). Black polyethylene, which contains suspended carbon black, is particularly useful as it removes all the near IR, visible, and UV radiation from the mercury lamp. A diamond-windowed detector will extend the range to about 25 μm. The actual range scanned is then largely determined by the thickness of the beam splitter, thinner films being used for shorter wavelengths.

It should be stated that it is certainly possible to record spectra with grating instruments in the far-IR and to get reasonable resolution. Several instruments have been described in recent years, and for one a resolution of 0.1 cm^{-1} was claimed for the range 30—60 cm^{-1} [144]. Of the commercially available instruments, the Perkin—Elmer FIS-3, for example, scans the region 25—333 μm (30—400 cm^{-1}) with only one change of grating, in about 30 min. Far IR grating spectroscopy has recently been discussed by Wilkinson and Martin [135].

Richards [145] made a comparison of one grating and two interferometer spectrometers for the far IR in 1964. It was found that the resolution obtainable with the latter sort was far superior to that of the grating instrument, which used several echelettes to cover the range 10—150 cm^{-1}. But he also pointed out that this situation could very easily be reversed by the use of improved detectors which he thought would certainly become available in the near future. It would seem that the need for data processing in connection with Fourier transform spectroscopy will always be a disadvantage, mainly because of the cost, but for high-resolution studies it will remain the preferred technique.

(4) Applications

The first spectrum obtained with an interferometer type spectrometer was that of adsorbed water molecules, by Gebbie and Vanasse [146] in 1956 and this was followed by the spectra of many smaller molecules for which the pure rotational transitions may be observed in this region. Considerable interest has also been shown in the far IR spectra of crystalline solids, particularly at low temperatures, when the lines may become quite sharp. Richards [145] has described a spectrometer in which the light from the interferometer is passed

64

through a light-pipe to the sample in liquid helium and is then detected by a doped germanium bolometer, also at liquid helium temperature. Whatever the temperature used, it is standard practice to evacuate the instrument completely, since water vapour absorbs strongly over most of the far-IR. Full advantage has been taken of the fast scanning capability of the Block Model 200 instrument for continuous examination of the effluent from a gas chromatograph in the medium-IR. Though the sensitivity cannot compare with that of other detectors, its specificity offers considerable attractions. This application has been reviewed recently by Low [176].

4. Detectors

(A) INTRODUCTION

Detectors of electromagnetic radiation fall into two very broad groups described as photon detectors and thermal detectors. The latter are applicable over the whole spectrum (the thermocouple has been used as a reference detector for calibrating a phosphor-coated photomultiplier in the vacuum UV [147]) but the former function only when the energy of the individual photons is sufficient to cause emission of an electron or an electronic transition in the detector material, thus exhibiting a threshold wavelength or frequency.

Thermal detectors have been used since the beginning of spectroscopy and, in their present state, achieve sensitivities and noise levels which approach very closely the theoretical values. The amount of research which has been pursued in recent years in connection with the use and detection of IR radiation, particularly with reference to missile-guidance systems and other topics of military interest, has resulted in the development of a variety of sensitive new detectors which also possess very fast responses to changing levels of illumination. These are mostly based on semi-conductors and include the now widely used lead sulphide cell and the indium antimonide detector, both of which are usually used in the photo-conductive mode.

The use of photon detector materials of relatively high ionisation potential results in a device responding only to short wavelength radiation. For example, the so-called "solar-blind" photomultipliers can be used for measurements below 300 nm without the necessity of shielding from daylight.

(1) Factors affecting the choice of a detector

Spectral response. This is the most obvious criterion, but the other properties which may be sought in a detector will depend on the mode of operation of the instrument.

Linearity of output in relation to incident intensity. When measurement of comparative intensity of two light beams is required, as in spectrophotometry with a manual instrument, or in emission spectroscopy, linear dependence of the electrical output on the light intensity is very important. The photomultiplier is probably the best detector from this point of view.

Frequency response. Most double-beam spectrophotometers employ a photocell only as a detector of unbalance between two chopped beams of radiation, the output of balance signal being used to activate a servo-mechanism which restores the balance of the two beams. The speed with which a spectrum may be scanned depends on the speed with which this balance can be maintained and hence on the chopping rate. The requirement for fast response in the detector is obvious. In general, photon detectors are fast (flat response up to 10 MHz for photomultipliers is common, and for lead sulphide detectors up to several hundred Hz), and thermal detectors are slow (10 Hz for a Golay cell or thermocouple).

Electrical impedance. With the exception of the photovoltaic selenium cell which drives a galvanometer without any external power supplies, most detectors require an amplifier to increase the small signal to a level more easily measured. Though thermionic valve circuits are being rapidly replaced by transistor analogues and the even smaller integrated circuits, they have the advantage of being particularly easy to match to high-impedance sources such as the lead sulphide cell or the anode load resistor of a photomultiplier. On the other hand, lower impedance detectors like the bolometer are more suited to a transistor amplifier. Very low impedance detectors such as the thermocouple must be transformer-coupled to an a.c. amplifier and can then only be used to detect chopped radiation.

Auxiliary power supplies. For many detectors, including photoconductive cells and vacuum phototubes, the signal current is also proportional to the applied voltage. In the case of the photomultiplier, however, the gain varies considerably with change of supply voltage the stability of which should be some ten times that required of the output signal.

Operating temperature. Detector noise decreases with decrease in temperature, but sometimes additional advantages may accrue such as an extended spectral response in the case of the lead sulphide cell. For most applications in spectroscopy and spectrophotometry, the advantages offered by cooling of the detector are outweighed by the bother and cost of running a suitable cooling system.

For further considerations and details concerning the design theory and properties of detectors, particularly for the IR, the reader is referred to the excellent book by Kruse, McGlauchlin, and McQuistan [36]. A useful review of IR detectors emphasising the theoretical approach to the calculation of limits of detection is given by Moss [148]. The two papers by Sharpe [149,150] serve as a good introduction to photo-emissive tubes.

(B) PHOTO-EMISSIVE DETECTORS

(1) Vacuum photocells

With these relatively simple photon detectors the light passes through the envelope, which is made of glass or other transparent material, and falls on the light-sensitive cathode. The electrons which are released are picked up by the anode taking the form of a single loop of wire, or a piece of widely spaced wire mesh in front of the cathode. Typical characteristics for such a tube (Mullard 90 CV, with S1 cathode) are shown in Fig. 46. At anode potentials above about 20 V, corresponding to 100% collection efficiency of the emitted electrons, the current is independent of the anode voltage and pro-

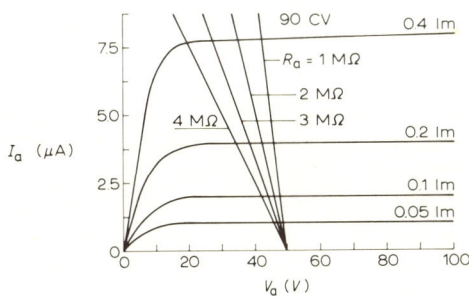

Fig. 46. Characteristics of a vacuum photocell (90CV). Anode current vs. anode voltage as a function of illumination. Reproduced by courtesy of Mullard Ltd.

TABLE 5

Coatings for photo-sensitive cathodes

Code	Composition	λmax	λpeak	Notes
S1	AgOCs	1.2 μm	800 nm	Low quantum efficiency
S4	SbCs$_3$	630 nm	450 nm	
S5	SbCs$_3$	630 nm	350 nm	
S10	BiAgOCs	750 nm	420 nm	
S11	SbCs$_3$O	630 nm	390 nm	Commonly used in fused
S20	Sb(Na$_2$ K)Cs	850 nm	380 nm	silica envelopes for
"S"	SbCs$_3$	630 nm	380 nm	extended UV response

Note: Cut-off for lime-soda glass is about 330 nm and for fused silica about 170 nm.

Data reproduced by courtesy of EMI Electronics Ltd.

portional only to the illumination. Fairly high values of anode-load resistor are common with these tubes.

Many combinations of caesium with other metals, sometimes partly oxidised, have been used as cathodes. A number of the commoner preparations is listed in Table 5, and some of the response curves are reproduced in Fig. 47.

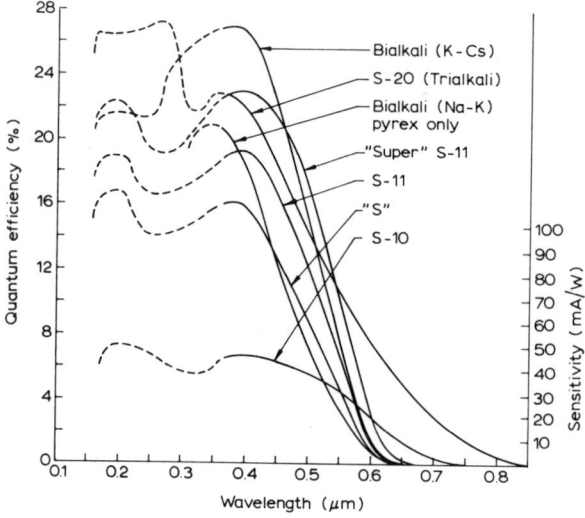

Fig. 47. Spectral response curves of semi-transparent photocathodes. ——, All window materials; - - -, fused silica windows. Reproduced by courtesy of EMI Electronics Ltd.

(2) Photomultiplier tubes

An electron with moderately high energy, say around 100 eV, falling on a suitable electrode material can eject other electrons, the number depending on the energy of the first electron and the nature of the electrode material. The phenomenon is known as secondary emission and the number of emitted electrons due to one incident electron is the secondary emission coefficient, δ. It can reach values around ten for SbCs-coated electrodes but is more commonly around five and may be as low as two for AgMgO surfaces. If several such electrodes, known in this case as dynodes, are arranged to multiply successively the small number of photo-electrons from the cathode, the resulting tube is called a photomultiplier. Some typical geometries are shown in Fig. 48. The number of dynodes is commonly between nine and fifteen but it may be less if small physical size is more important than high sensitivity. Very high gains are obtainable with these tubes — assuming a δ-value of 3, an 11-stage tube would have a gain of 3^{11} or about 18,000.

Strict proportionality between the current and the light intensity is often obeyed to much higher current levels than can be allowed to

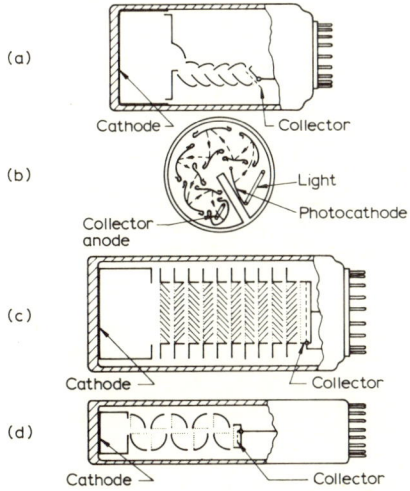

Fig. 48. Electrostatic dynode systems. (a) Focused structure; (b) compact focused structure; (c) venetian-blind structure; (d) box-and-grid structure. Reproduced by courtesy of EMI Electronics Ltd.

References pp. 88—93

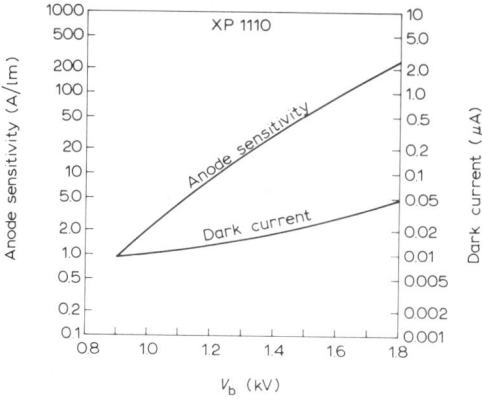

Fig. 49. Variation in sensitivity and dark current with total applied voltage for a 10-stage tube with S11 photocathode (XP1110). Reproduced by courtesy of Mullard Ltd.

flow continuously. A maximum continuous rating of 1 mA is typical but for pulsed signals, outputs up to 100 mA may still be within the limits of linearity if fairly high supply voltages are used. A high-precision spectrophotometer making use of this extremely good linearity (the tube in question being an EMI 9558 B) has been described by Budde [151]. A precision of ± 0.02% was obtained for anode currents not greater than 1 μA.

Dark-currents may vary from 1 to 100 nA depending on the tube type. The variation of sensitivity (i.e. gain) and of dark-current for a 10-stage tube with an SbCs cathode is shown in Fig. 49. The characteristics of the popular RCA tube 931-A have been described by Engstrom [152].

The speed of response of a photomultiplier depends on the spread of transit times of electrons from different parts of the photocathode, and also of the secondary electrons, so that small tubes with focussing dynodes have the fastest response. Typical high-frequency limits are in excess of 10 MHz with the result that very low light intensities may be estimated by photon counting rather than in terms of the total current passed by the photomultiplier [153].

The magnitude of the *dark-current* is of importance when a photomultiplier is to be used for the measurement of low intensities, as for example in Raman spectroscopy. Marrinan has discussed this particular example with reference to the desirability of cooling the tube to

70

reduce the noise [154]. The *noise* in a photomultiplier falls mainly into two categories, thermionic and statistical fluctuation in the emission when illuminated. The signal-to-noise ratio is given by

$$(S/N)^2 = \frac{i_p^2}{2e\Delta f(i_t + i_p)} \tag{46}$$

where i_p is the photocurrent, Δf is the bandwidth ($\sec^{-1}$) and e is the charge on the electron. For very low currents the noise is mainly thermal (i_t), the so-called shot-noise, and can be reduced substantially by cooling the tube. Marrinan observed a hundredfold reduction in dark-current noise by cooling the tube from room temperature to $-174°C$. At even moderate photocurrents, however, the noise was largely dependent on the current and no significant advantage was gained by cooling the tube.

Solar-blind photocells are those with a long-wave cut-off around 300 nm or shorter so that they do not respond to that part of the sun's emission spectrum that is seen on earth. In 1936, Renschler described the responses of cathodes of cadmium, tantalum, titanium, and platinum [155]. Platinum, nickel, and tungsten were used at a much later date [156] in windowless cells for work in the vacuum UV, and two years later Dunkelman [157] reported on the performances of modified 1P28 tubes with cathodes of various metals. A tube with a nickel cathode showed only a 25% increase in dark current when exposed to bright sunshine. A photomultiplier with a tungsten cathode, which responds only below 140 nm, has been used as the detector to measure absorption by water vapour at 121.6 nm for analytical purposes [158], the radiation being the Lyman α line from a hydrogen discharge tube. The response curve for a commercially available tube is shown in Fig. 50. The maximum sensitivity of its CsTe cathode corresponds to a quantum efficiency of only about 0.8% compared with about 20 times that value for most of the common photo-cathodes.

The solar-blind photomultiplier has also been used recently in the construction of *resonance detectors* for atomic absorption spectrophotometry [159]. The detector is a hollow-cathode lamp in which the atomic-fluorescence radiation emitted at right angles to the incident radiation (which suffers atomic absorption) falls on the cathode of a solar-blind cell. The use of these detectors eliminates the need

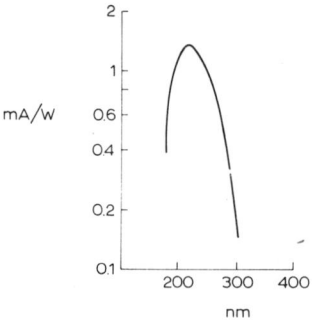

Fig. 50. Spectral response curve of a solar blind photomultiplier (R-166). Reproduced by courtesy of Hamamatsu TV Co. Ltd.

for a monochromator and, of course, stray light from other sources does not interfere.

Another approach to the problem of detecting vacuum UV radiation is to coat a conventional glass-envelope tube with some fluorescent material and to seal it against a rubber O-ring to the vacuum spectrograph. Johnson et al. [160] investigated several phosphors and found that sodium salicylate deposited by evaporation of a methanolic solution on the window of an RCA 1P21 tube gave excellent results. It was later shown that this had a remarkably constant response from 85 nm to 300 nm, when compared against a reference thermocouple, and even at 58 nm the output was only 15% low [147].

A novel application of the image-orthicon television picture tube to spectroscopy has been in the field of fast scanning of emission spectra of flames and explosions [161,162]. The photo-cathode of the tube replaced the photographic plate in a large-aperture spectrograph and the output from a single-line sweep was displayed on a synchronised oscilloscope. The spectral range was 350—1000 nm and the orthicon tube could resolve 20 lines/mm.

An extension of the response of photomultiplier tubes into the near IR has been achieved using an image-converter tube [163]. Thus tubes such as the 931-A could be used up to 1.2 μm in spectrophotometry and good sensitivity obtained for potassium determinations by flame photometry [164].

Overexposure of a photomultiplier to intense radiation will cause a significant increase in the dark-current. When this happens, one must just keep the tube in the dark and wait for the dark-current (and the noise) to decrease to normal. In the event of overexposure when the high-voltage supply is on, serious damage may be caused by the heavy current flowing. Many instruments employ automatic circuit breakers actuated by, for example, a cell-compartment lid, to avoid this danger.

(3) Gas-filled photocells

These cells are similar in design to the vacuum photocells, the only difference being that they contain an inert gas such as argon at low pressure. At low anode potentials, say less than 30 V, these tubes behave as do their vacuum counterparts, but as the anode potential is increased the electrons passing through the gas cause some ionization and a higher current results. The gas amplification factor thus obtained may be of the order of 5—10, but at the expense of severely reducing the speed of response. The action is not unlike that of a Geiger—Müller tube and indeed a photosensitive Geiger—Müller tube sensitive in the range 105—250 nm has been developed [165].

(4) Barrier-layer cells

These devices are also known as photovoltaic cells because they generate a small voltage themselves. They therefore do not need any external power supply or amplifier and can be used to drive a galvanometer directly. A thin layer of a semiconducting material (fre-

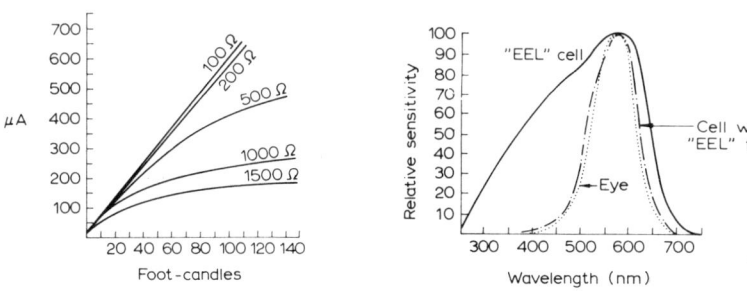

Fig. 51. Current output curves for a 45 mm "EEL" barrier-layer cell. Reproduced by courtesy of Evans Electroselenium Ltd.

Fig. 52. Spectral response curve of an "EEL" barrier-layer cell. Reproduced by courtesy of Evans Electroselenium Ltd.

quently selenium) is applied to a base of iron or aluminium metal. The surface of the selenium is then sputtered with a conducting, but transparent, layer of silver or gold on to which is held by pressure a metallic contact strip. Incident photons can excite electrons in the selenium into a conducting "band" energy level. These are then picked up by the silver layer but do not flow into the iron because of the so-called barrier between the selenium P-type semiconductor and the iron. Provided that a low resistance galvanometer is used (less than 100—200 Ω) the current output is proportional to the radiation intensity and currents of a few hundred microamps may be taken from the cell (Fig. 51). The spectral response extends down to around 300 nm, but can easily be matched to that of the human eye by the use of a suitable filter (Fig. 52).

(C) PHOTO-CONDUCTIVE DETECTORS

Lead sulphide, selenide, and telluride, deposited as thin films on a substrate, show very high dark-resistance, but become electrically conducting when radiation of certain wavelengths falls on them. They are known as *intrinsic* photon detectors. The simplest explanation may be in terms of the semiconductor band theory. Two more likely proposals are: the single-crystal recombination theory, in which holes and electrons are produced by the interaction of the photons with the number of carriers being dependent on the intensity of the radiation; and the barrier theory, according to which the barriers (in the thin layer) are disturbed by the incident photons and allow conduction through increased mobility of a nearly constant number of carriers. Moss, discussing these detectors in detail [148], considers that the evidence is weighted in support of the "numbers" theory. These detectors all exhibit a long-wave cut-off (though not a sharp one) beyond which the photons do not possess sufficient energy to cause an electronic transition. The spectral response of the lead sulphide cell extends from about 300 nm to the near IR. The response curves taken at different temperatures are shown in Fig. 53. Lowering the temperature, facilitated by the cells being mounted in a small glass Dewar, does extend the long-wave limit somewhat, to around 4 μm for lead sulphide and to about 8 μm for the telluride. The dark-resistance of a typical cell (Mullard 61SV) is approximately 1—2 MΩ, which, in series with a load of 1 MΩ, would be supplied from a stabilised 200 V supply. Chopped radiation up to about

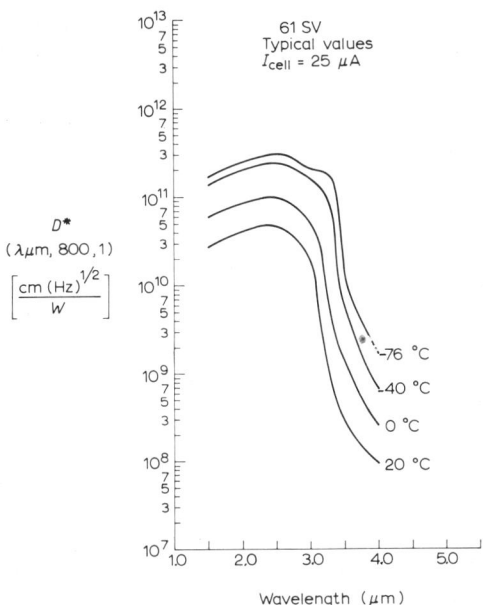

D^*
$(\lambda\mu m, 800, 1)$

$\left[\dfrac{cm\,(Hz)^{1/2}}{W}\right]$

61 SV
Typical values
$I_{cell} = 25\ \mu A$

−76 °C
−40 °C
0 °C
20 °C

Wavelength (μm)

Fig. 53. Spectral response curves of a lead sulphide cell (61SV) at different temperatures. Reproduced by courtesy of Mullard Ltd. D^* is a figure of merit for a detector. $D^* = [(S/N)\,(A.\Delta f)^{1/2}/W]$. S/N is the signal-to-noise ratio for a given power, W, incident on the detector area, A, at bandwidth Δf.

1 kHz may be detected with little loss in response. Lead sulphide cells are widely used in spectrophotometers covering the near IR down to the visible, a change being made to a photomultiplier at about 600 nm.

Semiconductors such as indium antimonide also show photo-conductive properties, but as yet have not been widely used in spec-trophotometers. The response has a long-wave cut-off at about 7 μm, corresponding to a transition energy of 0.18 eV [166].

(D) THERMAL DETECTORS

This group of detectors, all working as small heat engines, are particularly suited to work in the IR where the energy of the indi-vidual photons is not sufficient to cause electronic transitions and only the heating effect (which is proportional to the number of photons) can be measured. Moss has given a very thorough treatment of the theoretical sensitivities of thermocouples and bolometers [148] and has shown that many commercial models have achieved sensitivities and noise levels not far removed from the theoretical. Infrared radiation was discovered by Herschell [21] in 1800 using as a thermal detector a thermometer with a blackened bulb.

(1) Thermocouples

When two wires of dissimilar metals are joined to complete an electrical circuit and the junctions are maintained at different temperatures, a current will flow in the circuit. The potential is proportional to the temperature difference over a wide range of temperature. When such a thermocouple is used as a detector for weak beams of radiation the heat capacity must be kept to a minimum and the receiver area must have uniform response to as wide a range of wavelengths as possible (in practice, for IR work, to about 30 μm, beyond which blackening becomes less efficient). In some modern thermocouple detectors a thin gold foil, blackened to make it an efficient absorber, is attached to the two dissimilar wires which act as support pins. The foil itself may then be made as thin as possible in order to minimise its heat capacity.

Sometimes two such thermocouples are mounted in the same envelope, one acting as a reference to eliminate the effect of ambient temperature fluctuations. In modern scanning IR spectrophotometers, a single thermocouple is sufficient because it is only the chopped signal which is amplified to drive the servo-mechanism. As the a.c. output voltage may be typically only some nanovolts, transformer step-up of the signal followed by high amplification is necessary. Most modern thermocouple detectors will respond adequately to radiation chopped at 5—10 Hz. The minimum detectable energy for a number of commercial devices lies around 10^{-10} W [167] which is only some ten times the theoretical value. It must be remembered, when calculating the efficiency of a thermocouple, that the Peltier effect (whereby the junction which is heated externally will also be cooled internally when a current flows) is operative, and creates a small back-EMF in the circuit.

(2) Bolometers

Julius [1], in the 1890's, measured the IR spectra of various organic compounds using a bolometer as detector. In its general form this consists of a thin film of an electrically conducting material with a high temperature coefficient of resistance, on which the light falls. The change of resistance is measured electrically in a bridge circuit. The thin film is supported in an evacuated envelope along with a second film which is screened from the radiation and is balanced against the detecting film in the bridge circuit to minimise variations in resistance due to causes other than the signal. A steady current is passed through the film to raise its temperature above room temperature (optimum conditions are for $T/T_0 = 1.4$, where T is the operating temperature and T_0 the temperature of the surroundings, both in °K), and the changing potential across the film caused by the chopped radiation falling on it is stepped up with a suitable transformer and further amplified.

(3) Thermistors

In the 1940's the thermistor was developed [168]. This is a semi-conducting piece of mixed metal oxides possessing a very large temperature coefficient of resistance so that a much bigger voltage change is produced for a given radiant intensity than for a metal bolometer. Its higher electrical resistance also makes it more suitable for direct connection to an amplifier. By preparing the thermistor as a thin film on a piece of glass the response time is kept small. Though the signal is higher than that from a metal bolometer, so is the noise level and the minimum detectable energy is around 10^{-10} to 10^{-9} W which is also slightly worse. The envelope is not usually evacuated for these detectors. A comparison of thermocouples and bolometers has been given by Williams [169].

(4) Superconducting bolometers

Superconductors are materials in a state possessing zero electrical resistance, a state occurring only at very low temperatures typically those of liquid helium. Over a very small range of temperature around the transition temperature, the electrical resistance changes from zero very rapidly with change in temperature and the material makes a very sensitive bolometer. Control of the ambient temperature of the device must be very exact otherwise the signal gets lost in the noise due to temperature fluctuations. Andrews [170] described an early experimental device of niobium nitride operated at 15°K which had a response time of 500 μsec. Martin and Bloor [171] produced a superconducting tin bolometer with a transition temperature of 3.7°K, but the response time of 1.25 sec was rather long. Low [172] developed a low-temperature gallium-doped germanium detector which had a very wide spectral bandwidth and very short time constant. It is marketed along with its liquid-helium cryostat for operation around 4°K.

(5) Pneumatic heat detectors — The Golay cell

Very sensitive pneumatic detectors depending on the expansion of a heated gas were described by Weis [173] and by Zahl and Golay [174] in 1946. Golay himself improved his original model [175] and his name has since become associated with this type of detector.

Energy absorbed by a small volume of enclosed gas causes the latter to expand and the resulting pressure change distorts a diaphragm carrying a mirror. The very small change in the angle of the mirror is detected by passing a parallel beam of light through a grid on to the mirror and back through the same grid to a photocell. A small change in angle of the mirror is sufficient to change the light intensity falling on the photocell from zero to maximum. The principle of the detector can be seen from Fig. 54 and in Fig. 55 a diagram of a commercial model is shown. This particular detector, with a sensitive area of 7 mm^2 will respond, without drop in signal, to radiation chopped at up to 10 Hz. At this frequency, the minimum detectable energy is 5×10^{-11} watt. The detector has a very wide spectral response which in practice is determined largely by the window material. For the far-IR (see section 3.(D)) a diamond window is

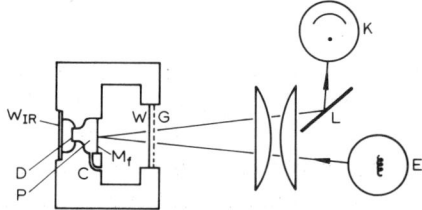

Fig. 54. Golay pneumatic detector (schematic). W_{IR}, IR-transparent window; D, heat-absorbing diaphragm; P, pneumatic chamber; M_f, flexible mirror; C, capillary leak; W, window; G, grid.

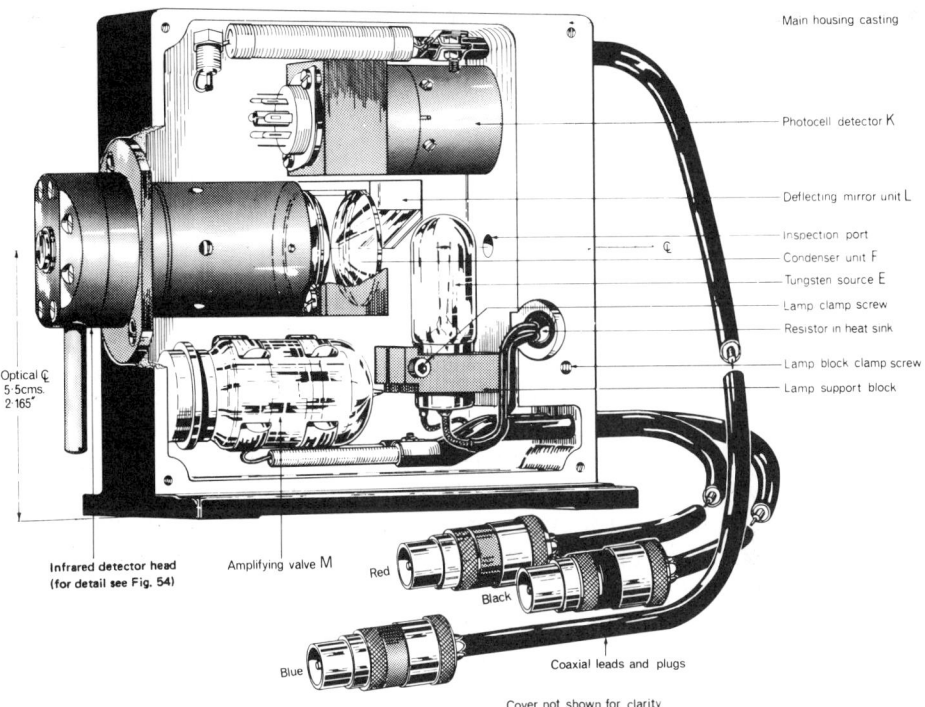

Main housing casting

Photocell detector K

Deflecting mirror unit L

Inspection port
Condenser unit F
Tungsten source E
Lamp clamp screw
Resistor in heat sink

Lamp block clamp screw
Lamp support block

Optical ₵
5·5cms.
2·165″

Infrared detector head
(for detail see Fig. 54)

Amplifying valve M

Red

Black

Blue

Coaxial leads and plugs

Cover not shown for clarity

Fig. 55. Golay pneumatic detector (complete). Reproduced by courtesy of Pye-Unicam Ltd.

used, and good response as far as 20 cm^{-1} is obtained. A disadvantage is that a sudden large pressure change may cause the diaphragm to "pop". A factory replacement has then to be fitted.

(E) THE PHOTOGRAPHIC PLATE

An emulsion of one or more of the water-insoluble silver halides suspended in gelatin is light-sensitive, in that particles of the halide which have been activated by photons of suitable energy (*exposure*) are then easily reduced to metallic silver. Chemical reduction (*development*) of an exposed plate, which is a piece of glass with a uniformly thin coating of emulsion, results in greatest darkening due to the metallic silver in areas subjected to the greatest exposure. The excess of silver halide is removed by complexation of the silver with

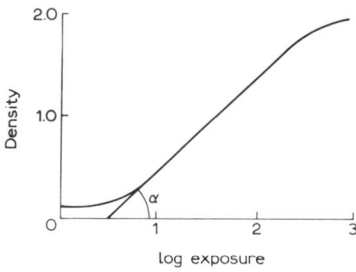

Fig. 56. Hurter—Driffield curve for a photographic emulsion.

thiosulphate (*fixing*). The relationship between the density (i.e. absorbance = log $1/T$) and exposure is best shown diagramatically as in Fig. 56 which shows a typical Hurter—Driffield curve for an emulsion. It can be seen that even with little or no exposure there will be some blackening — the value of the density being referred to as "gross fog" — and also that a linear relationship exists between the density and the exposure over a limited range. But above this the darkening tends to a maximum whatever the increase in exposure.

(1) Contrast

The slope, tan α, of the linear portion of the Hurter—Driffield curve, or *characteristic curve*, is called the gamma (γ), or contrast of the emulsion and is an important factor affecting the choice of such materials for any given purpose. In the case of certain emulsions, the contrast can be varied by the time of development.

The contrast of an untreated emulsion exposed to UV light is commonly around unity, but contrasts as high as 5 or 6 are found for some special-purpose materials at certain wavelengths. At short wavelengths, the UV light is strongly absorbed by the gelatin so that only the surface layer is actually exposed. Longer wavelengths penetrate further and hence cause more darkening with a resulting increase in the contrast (e.g. Spectrum Analysis Plate No. 1 (Kodak) in Fig. 57). The addition to the emulsion of certain dyes which absorb only the near UV and visible, maintains the contrast nearly constant over a very wide range of wavelengths, as in the case of the Type 103-F plate (Kodak) in Fig. 57. In spectrography, an emulsion with a high contrast is useful for quantitative analyses over a restricted range and

80

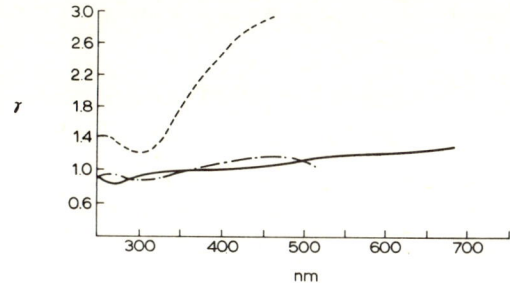

Fig. 57. Variation in contrast (γ) with wavelength for three Kodak emulsions. - - -, Spectrum analysis plate and film, No. 1; —·—, spectrum analysis plate and film No. 3; ——, spectroscopic plate and film, Type 103-F. Reproduced with permission from a copyrighted Kodak publication.

one with a low contrast for qualitative analyses over a wider range of concentrations.

(2) Sensitivity and speed

The exposure required to cause a given density on a plate after carefully standardised development is a measure of the sensitivity of the emulsion, the sensitivity S_A being the reciprocal of the exposure required over a period of 1 sec to give a density of 0.6 above gross fog. The ASA (American Standards Association) speed of an emulsion is also a linear measure of the sensitivity and is widely used for relatively fast (i.e. high-speed) emulsions employed in normal photography. The DIN speed rating, however, is based on a logarithmic scale. The relationship between the two will be evident from the following table.

ASA speed	25	50	100	200	400	800
DIN°	15	18	21	24	27	30

(3) Reciprocity failure

It is found that for very long exposure times used for recording very low levels of light and also for very short exposure times, sensitivities decrease. This point should be remembered when multiple exposures are made in spectrographic detection and determination of trace constituents. Calibration should be carried out with the same exposure times as are used for the analyses.

(4) Spectral sensitivity

The normal range of sensitivity of the photographic plate is up to about 500 nm (Fig. 58), but the addition of suitable sensitising dye-stuffs to the emulsion enables the halide particles to be activated by

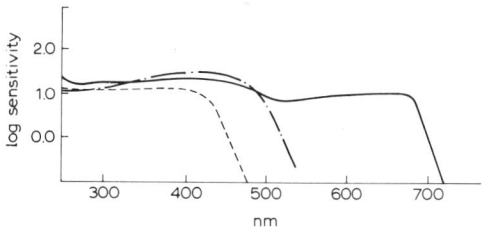

Fig. 58. Spectral sensitivities of three Kodak emulsions. - - -, Spectrum analysis plate and film, No. 1; —·—, Spectrum analysis plate and film No. 3; ——, Spectroscopic plate and film, Type 103-F; (D = 0.6 above gross fog). Reproduced with permission from a copyrighted Kodak publication.

light of longer wavelengths. Orthochromatic plates are sensitive also in the green and panchromatic plates in the yellow and red. IR-sensitive plates are also available, extending the photographic region to around 900 nm and even to over 1.1 μm if the plates are hypersensitised by exposure to ammonia before exposure to the light.

Extension of the short wavelength limit may be achieved by incorporating a fluorescent dye in the emulsion, or, more commonly, by coating a normal plate with a phosphor just before use. In this way the whole of the vacuum UV down to the region of soft X-rays may be detected.

(5) Graininess and Granularity

The impression of graininess obtained by looking at a plate or film after exposure and development is due to the large number of fine particles of silver in the gelatin. The larger these particles, the more obvious is the graininess, and the greater the granularity of the emulsion, which is a measure of the microscopic variations in density and the poorer the resolution obtainable with the emulsion. As larger particles mean a greater sensitivity, since incident photons activate particles rather than individual ions, higher speed will always be achieved at the expense of resolution. In practice, the resolution of fast emulsions is still adequate for spectrographic work, but the limit of detection might not be as good as with a slower plate, since the "noise level" as seen by a microphotometer is determined by the granularity at the gross fog level.

(6) The microdensitometer

The density of an exposed plate can be measured with a densitometer, which is a photometer using white light. For measurement of the density of a line image on a spectrographic plate, a light beam considerably smaller than the line width is used and the spectrum is scanned by slowly passing the plate through the light beam and simultaneously recording the transmittance or the density. Microdensitometers for spectrographic work often have the added facility for projection of the spectrum together with a reference spectrum, on to a ground-glass screen thus enabling the operator to identify lines more easily.

(F) THE HUMAN EYE

Though the human eye is a sensitive detector to light of certain wavelengths when it is dark-adapted, it cannot compare intensities quantitatively. It can, however, distinguish small differences in the intensities of two similarly coloured sources placed side by side. This ability is made use of in the old colorimetric method using Hehner's Cylinders, with which colour densities (and hence concentration multiplied by path length) could be matched to within ± 5% with certain colours.

There are two mechanisms of vision giving rise to two types of response. The dark-adapted eye displays *scotopic* vision depending on the rods of the retina which are not colour sensitive. The wavelength of maximum sensitivity is around 510 nm in this case. The response curve shown in Fig. 52 (p. 73) is that of *photopic* vision, obtained with the cones of the retina, and at higher intensities. The maximum sensitivity occurs at about 560 nm. The limits of visibility are approximately 400 and 700 nm.

5. Basic Instruments

In this section it is intended to discuss very briefly how some of the components described in the preceding sections may be combined to make a complete instrument, with the emphasis on spectrophotometers, designed either for measurements on flames, or on solutions, crystals or gases.

In any but the simplest instruments, radiation is chopped at least once in the course of its passage through the optics, and occasionally, as in the Perkin—Elmer far-IR instrument FIS-3, twice, at two different frequencies. Chopping may simply be a method of passing a beam alternately through two different light paths, but it will in any case give rise to an a.c. electrical signal from the detector, which is easily amplified and may be used to power a servo-mechanism. But in the case of absorption measurements being made on emitting substances such as flames, or any material at room temperature when the far-IR is concerned, chopping offers a convenient method of distinguishing between radiation from two different sources and of allowing the contribution of the unwanted one to be corrected for. Chopping is generally accomplished mechanically by a rotating mirror in the form of a maltese cross, the arms of which spend half of the time in the light path reflecting the beam. Constant speed of rotation is secured by using a synchronous motor. In atomic absorption spectrophotometry, it is more convenient to modulate the hollow-cathode lamp and to dispense with the mechanical chopper.

(A) PHOTOMETERS

The simplest instruments — colorimeters, or filter photometers — follow the basic arrangement of Fig. 59. The detector is usually a barrier-layer cell (see p. 73) which requires only a sensitive galvanometer and a series variable resistor for adjustment of the sensitivity.

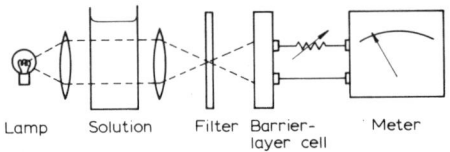

Lamp　　Solution　　Filter　Barrier-　　Meter
　　　　　　　　　　　　　　layer cell

Fig. 59. Simple photometer.

84

With water in the cell, the sensitivity is adjusted so that a reading of 100% is obtained. With the sample in the light beam, the transmittance is then read directly. The absorbance can usually also be read directly, for which purpose the galvanometer has a second scale appropriately calibrated. The lamp should be run from a constant-voltage transformer or from a lead accumulator to maintain the constancy of its output. Such simple apparatus is particularly convenient for photometric titrations.

(1) Filters

A typical set of filters covering the visible spectrum is the Ilford Spectrum series, the transmission characteristics of which are shown in Fig. 60. These filters contain pieces of dyed gelatin between two pieces of glass for support. The transmittance at its maximum is in the region 5—10%, and the bandwidth some 40—50 nm.

A rather narrower spectral bandwidth is exhibited by interference filters which are constructed by evaporating on to an optically flat piece of glass a semi-transparent film of silver followed by a thin layer of transparent material of low refractive index such as magnesium fluoride and finally a second film of silver. Constructive interference allowing transmission takes place for light with wavelength equal to twice the spacing, while all other wavelengths are reflected. The transmittance can be quite high (about 50%) and the bandwidth is typically 10 nm for a single-layer filter. The use of multiple layers results in bandwidths as low as 1 nm. These filters, though more expensive than gelatin filters, find application in flame photometers, where spectral resolution may be quite important.

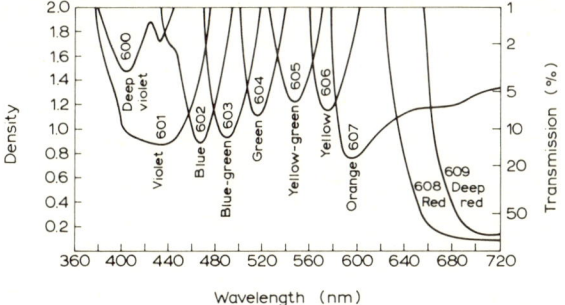

Fig. 60. Transmission spectra of Ilford Spectrum series filters. Reproduced by courtesy of Ilford Ltd.

(B) SPECTROPHOTOMETERS

(1) Manual instruments, single-beam

The basic scheme is shown in Fig. 61. A vacuum photocell with associated power supply and simple amplifier is normally used in this

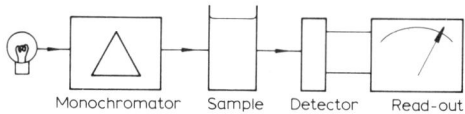

Fig. 61. Single-beam spectrophotometer.

type of instrument [85]. The instrument may be direct-reading (on a meter) or null-balance, the latter being preferred for its higher precision resulting from the improved readability. The luminous energy is controlled by the slit width in the monochromator.

(2) High-precision, single-beam instruments

Reference has already been made to the high-precision spectrophotometer described by Budde [151]. This was, however, a manually operated instrument and as photomultipliers do not pass a steady current until up to a minute after exposure has started, it was not suitable for rapid measurements. Recently a new automatic, single-beam, high-precision spectrophotometer has been described [177] and marketed, in which the photomultiplier is illuminated by chopped radiation from a reference source and the intensity of the monochromatic beam is compared with the intensity of the chopped beam. Absorbance measurements can be made with errors less than 0.2% with this instrument which also presents the absorbance either as a digital read-out to minimise operator fatigue, or on an electric typewriter.

(3) Recording instruments, double-beam

In these instruments light passes through a sample cell or through a reference cell and the intensities of the two beams are then compared. We will consider two variations on this theme.

86

(a) Double-beam in space

In this mode the monochromatic beam falls on two mirrors so that half the beam passes through each cell to its respective detector (Fig. 62). The amplifier is virtually an analogue computer presenting an electrical current proportional to the logarithm of the ratio of the

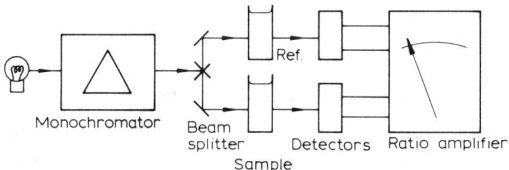

Fig. 62. Double-beam in space spectrophotometer.

two photomultiplier currents, to the galvanometer. For fixed-wavelength work this approach offers high precision because of the exceptionally linear response of the photomultipliers. A suitable circuit using operational amplifiers with transistors as non-linear feedback elements to accomplish the conversion has been described [178]. For spectrum scanning the matching of the two photomultipliers presents a serious problem. In the Cary Model 11 spectrophotometer electrical compensation is adjusted on a large set of preset potentiometers to cover the whole wavelength range.

(b) Double-beam in time

By far the most common arrangement found in recording spectrophotometers is that in which the detector acts as a null-balance device generating a signal to activate a servo-mechanism which drives a comb-shaped wedge into, or out of, one beam to restore balance of intensities. Good precision has also been realised by a pair of vanes, one above and one below the light beam, which move into the beam to cause attenuation. The travel is relatively large and the difficulties associated with cutting a precision comb are avoided. The simplest scheme is shown in Fig. 63. The a.c. component of the photomultiplier current is amplified and, after passing through a phase-sensitive detector locked to the chopping frequency, powers a servo-motor, M, to drive the wedge, W, to which is attached the pen of the recorder or a precision potentiometer for an electrical output. The precision

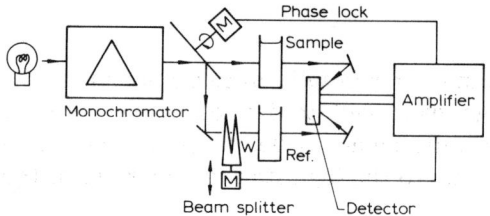

Fig. 63. Double-beam in time spectrophotometer.

obtainable in the measurement of absorbance depends mainly on the accuracy of the wedge, but to a certain extent (e.g. in recording the spectra in Fig. 38 (p. 53)) on the speed of the scan, and on the readability of the graphical presentation.

REFERENCES

1. W.H. Julius, Arch. Neer. Sci. Exactes Natur. 22 (1888) 310.
2. M. Donbrow, Instrumental Methods in Analytical Chemistry, Vol. 2, Pitman, London, 1967.
3. J. Mika and T. Török, Emissziós Szinképelemzés, Akadémiai Kiadó, Budapest, 1968.
4. D.H. Rank, in H.W. Thomson (Ed.), Advances in Spectroscopy, Vol. 1, Interscience, New York, 1952, p. 79.
5. 11th General Conference on Weights and Measures, Paris, 1960.
6. J. Terrien, C.R. Acad. Sci., 246 (1958) 2362.
7. W.F. Meggers and F.O. Westfall, J. Res. Nat. Bur. Stand., 44 (1950) 447.
8. J.R. McNally, J. Opt. Soc. Amer., 47 (1957) 1032.
9. Trans. Int. Astron. Union, 10 (1960) 228.
10. B. Edlén, J. Opt. Soc. Amer., 43 (1953) 339.
11. C.J. Bakker, J. Opt. Soc. Amer., 43 (1953) 410.
12. Joint Commission for Spectroscopy, J. Opt. Soc. Amer., 47 (1957) 1035.
13. IUPAC, Tables of Wavenumbers for Calibration of Spectrometers, Butterworths, London, 1961.
14. W.F. Meggers and C.J. Humphreys, J. Res. Nat. Bur. Stand., 13 (1934) 293.
15. K.S. Gibson, Nat. Bur. Stand. (U.S.) Circ. 484 (1949).
16. R.P. Bauman, Absorption Spectroscopy, Wiley, New York, 1962, Chap. 3, p. 129.
17. S. Walker and H. Straw, Spectroscopy, Vol. 2, Chapman and Hall, London, 1962.
18. B.S. Cooper, A.E. Gillam, G.F. Lothian and R.A. Morton, Analyst, 67 (1942) 164.
19. H.K. Hughes, Anal. Chem., 24 (1952) 1349.

20. G.F. Kirkbright, Talanta, 13 (1966) 1.
21. Sir W. Herschel, Phil. Trans. Roy. Soc. London, Ser. A, (1800) 255.
22. L. Boltzmann, Ann. Phys., 22 (1884) 31.
23. W. Wien, Ann. Phys., 58 (1896) 662.
24. O. Lummer and E. Pringsheim, Verh. Deut. Phys. Ges., 2 (1900) 176.
25. Lord Rayleigh, Phil. Mag., 49 (1900) 539.
26. M. Planck, Ann. Phys., 4 (1901) 553.
27. W. Ritz, Phys. Z., 9 (1908) 521.
28. N. Bohr, Phil. Mag., 26 (1913) 1, 476, 857.
29. T. Maiman, Nature, 187 (1960) 493.
30. A. Javan, W. Bennett and D. Herriott, Phys. Rev. Lett., 6 (1961) 106.
31. D.R. Herriott, J. Opt. Soc. Am., 52 (1962) 31.
32. A. White and J. Rigden, Proc. I. R. E., 50 (1962) 1697.
33. J.C. Evans, Recent advances in raman spectroscopy, in Advances in Analytical Chemistry and Instrumentation, Vol. 7, Interscience, New York, 1968.
34. E.F. Runge, R.W. Minck and F.R. Bryan, Spectrochim. Acta, 20 (1964) 733.
35. W.C. Price, Spectroscopy in the vacuum UV, in Advances in Spectroscopy, Vol. 1, Interscience, New York, 1959, p. 56.
36. P.W. Kruse, L.D. McGlauchlin and R.B. McQuistan, Elements of Infrared Technology, Wiley, New York, 1962.
37. R. Mavrodineanu and H. Boiteux, Flame Spectroscopy, Wiley, New York, 1965.
38. J.A. Dean and T.C. Rains, Flame Emission and Atomic Absorption Spectrometry, Vol. 1, Marcel Dekker, New York, 1969.
39. T. Lyman, Astrophys. J., 60 (1924) 1.
40. T. Lyman, Science, 64 (1926) 89.
41. P. Schulz, Ann. Phys., 1 (1947) 95, 107.
42. W.A. Baum and L. Dunkelman, J. Opt. Soc. Amer., 40 (1950) 782.
43. W.T. Anderson, J. Opt. Soc. Amer., 41 (1951) 385.
44. P.G. Wilkinson and Y. Tanaka, J. Opt. Soc. Amer., 45 (1955) 344.
45. W.R.S. Garton, J. Sci. Instrum., 30 (1953) 119.
46. P.L. Hartman and J.K. Nelson, J. Opt. Soc. Amer., 47 (1957) 646.
47. R.W. Wood, Phil. Mag., 7 (1929) 744.
48. H.L. Welsh, M.F. Crawford, T.R. Thomas and G.R. Love, Can. J. Phys., 30 (1952) 577.
49. B.P. Stoicheff, Can. J. Phys., 32 (1954) 330.
50. T.K. McCubbin and W.M. Sinton, J. Opt. Soc. Amer., 42 (1952) 113.
51. E.K. Plyler, D.J.C. Yates and H.A. Gebbie, J. Opt. Soc. Amer., 52 (1962) 859.
52. F. Paschen, Ann. Phys., 50 (1916) 901.
53. F. Paschen, Ann. Phys., 71 (1923) 142.
54. H. Schüler, Z. Phys., 35 (1926) 323.
55. H. Schüler, Z. Phys., 59 (1930) 149.
56. W.T. Elwell and J.A.F. Gidley, Atomic-Absorption Spectrophotometry, Pergamon, Oxford, 1966, Chap. 3.
57. B.J. Russell, J.P. Shelton and A. Walsh, Spectrochim. Acta, 8 (1957) 317.

58. J.V. Sullivan and A. Walsh, Spectrochim. Acta, 21 (1965) 721.
59. H.M. Crosswhite, G.H. Dieke and C.S. Legagneur, J. Opt. Soc. Amer., 45 (1955) 270.
60. E. Fenner, Spectrochim. Acta, 1 (1941) 164.
61. R.M. Dagnall, K.C. Thompson and T.S. West, Talanta, 15 (1968) 677.
62. A.S. Young, J. Sci. Instrum., 28 (1951) 207.
63. J.H. Taylor, C.S. Rupert and J. Strong, J. Opt. Soc. Amer., 41 (1951) 626.
64. F.J. Studer and R.F. Van Beers, J. Opt. Soc. Amer., 54 (1964) 945.
65. R.C. Lord and T.K. McCubbin, J. Opt. Soc. Amer., 47 (1957) 689.
66. L.G. Smith, Rev. Sci. Instrum., 13 (1942) 63.
67. C.S. Rupert and J. Strong, J. Opt. Soc. Amer., 40 (1950) 455.
68. J.H. Jaffe, J. Opt. Soc. Amer., 43 (1953) 619.
69. G. Kirchhoff and R. Bunsen, Ann. Phys., 110 (1860) 161.
70. Sir I. Newton, Opticks, reissued by Dover Publications, New York, 1952.
71. J.R. Partington, An Advanced Treatise on Physical Chemistry, Vol. 4, Physico-chemical Optics, Longmans, London, 1953, pp. 78—96.
72. R.C. Gore, R.S. McDonald, V.Z. Williams and J.U. White, J. Opt. Soc. Amer., 37 (1947) 23.
73. R. Hilsch and R.W. Pohl, Z. Phys., 59 (1930) 812.
74. A. Matsuishi, Y. Yamada and H. Yoshinaga, J. Opt. Soc. Amer., 52 (1962) 14.
75. R.E. Dodd, Chemical Spectroscopy, Elsevier, Amsterdam, 1962.
76. Landolt—Börnstein Tables, I, 4, Sect. 15091, Springer, Berlin, 1962.
77. S. Walker and H. Straw, Spectroscopy, Vol. 2, Chapman and Hall, London, 1962, Chap. 5.
78. M. Czerny, Z. Phys., 65 (1930) 600.
79. H.W. Hohls, Ann. Phys., 29 (1937) 433.
80. N. Acquista and E.K. Plyler, J. Opt. Soc. Amer., 43 (1953) 977.
81. Y. Yamada, A. Mitsuishi and H. Yoshinaga, J. Opt. Soc. Amer., 52 (1962) 17.
82. Landolt—Börnstein Tables, II, 8, Sect. 28221, Springer, Berlin, 1962.
83. O. Littrow, Amer. J. Sci., 35 (1862) 413 (Abstract).
84. O. Littrow, Sitzungsber. Kaiserl. Akad. Wiss. Wien, Math. Naturwiss. Kl., 47 (1863) 26.
85. H.H. Cary and A.O. Beckman, J. Opt. Soc. Amer., 31 (1941) 682.
86. Pye—Unicam Ltd., Handbooks for SP 90, SP 600, Cambridge.
87. A.H. Pfund, J. Opt. Soc. Amer., 14 (1927) 337.
88. F.L.O. Wadsworth, Astrophys. J., 2 (1895) 264.
89. F.L.O. Wadsworth, Phil. Mag., 38 (1894) 137.
90. C. Féry, C.R. Acad. Sci., 150 (1910) 216.
91. W.C. Miller, G. Hare, D.C. Strain, K.P. George, M.E. Stickney and A.O. Beckman, J. Opt. Soc. Amer., 39 (1949) 377.
92. P. Pellin and A. Broca, J. Phys. (Paris), 8 (1899) 314.
93. N.J. Rumsey, Nature, 195 (1962) 168.
94. A.C. Hardy, J. Opt. Soc. Amer., 25 (1935) 305.
95. J.L. Michaelson, J. Opt. Soc. Amer., 28 (1938) 365.
96. G. Cario and H.D. Schmidt-Ott, Z. Phys., 69 (1931) 719.

97. E.B. Brown, Modern Optics, Reinhold, New York, 1965.
98. R. Wood, J. Opt. Soc. Amer., 34 (1944) 509.
99. R. Wood, Nature, 140 (1937) 723.
100. G.R. Harrison, J. Opt. Soc. Amer., 39 (1949) 413.
101. J.S. Ames, Phil. Mag., 27 (1889) 369.
102. R.W. Wood and A. Trowbridge, Phil. Mag., 20 (1910) 886, 898.
103. A. Michelson, Astrophys. J., 8 (1898) 37.
104. W. Williams, Proc. Phys. Soc. London, 45 (1933) 699.
105. G. Harrison, J. Opt. Soc. Amer., 39 (1949) 522.
106. A.E. Martin, Infra-red Instrumentation and Techniques, Elsevier, Amsterdam, 1966, p. 50.
107. H.A. Rowland, Phil. Mag., 13 (1882) 469.
108. H.A. Rowland, Phil. Mag., 16 (1883) 197.
109. J. Strong, Sci. Amer., 186 (1952) 45.
110. W. de W. Abney, Phil. Trans. Roy. Soc. London, 177 (1886) 457.
111. A. Eagle, Astrophys. J., 31 (1910) 120.
112. G. Runge and F. Paschen, Ann. Phys., 61 (1897) 641.
113. F.L.O. Wadsworth, Astrophys. J., 3 (1896) 54.
114. W.F. Meggers and K. Burns, J. Res. Nat. Bur. Stand. Sect. A, 18 (1922) 185.
115. H. Ebert, Ann. Phys. Chem. (Leipzig), 38 (1889) 489.
116. M. Czerny and A.F. Turner, Z. Phys., 61 (1930) 792.
117. G.R. Rosendahl, J. Opt. Soc. Amer., 52 (1962) 412.
118. W.G. Fastie, J. Opt. Soc. Amer., 42 (1952) 641, 648.
119. J.R. Churchill, Ind. Eng. Chem. Anal. Ed., 16 (1944) 653.
120. P.A. Wilks, Instr. Contr. Syst., 41 (1968) 67.
121. W.W. Sleator, Astrophys. J., 48 (1918) 125.
122. P.M.B. Walker, J. Leonard, D. Gibbs and P.J. Chamberlain, J. Sci. Instrum., 40 (1963) 166.
123. P. Day, A.F. Orchard, A.J. Thomson and R.J.P. Williams, J. Chem. Phys., 42 (1965) 1973.
124. R.F. Stewart and N. Davidson, J. Chem. Phys., 39 (1963) 255.
125. R. Tousey, F.S. Johnson, J. Richardson and N. Toran, J. Opt. Soc. Amer., 41 (1951) 696.
126. G.C. Chandler, J. Opt. Soc. Amer., 58 (1968) 895.
127. A. Ericson and B. Edlén, Z. Phys., 59 (1930) 656.
128. H.J. Babrov and F. Casden, J. Opt. Soc. Amer., 58 (1968) 179.
129. P.H. van Cittert, Z. Phys., 65 (1930) 547.
130. T. Grubb, Proc. Roy. Soc., 22 (1874) 308.
131. H. Roemer and R.A. Oetjen, J. Opt. Soc. Amer., 36 (1946) 47.
132. J.R. Edisbury, Practical Hints on Absorption Spectrometry, Hilger and Watts, London, 1966.
133. A.A. Michelson, Rep. Brit. Assoc., (1892) 170.
134. A.E. Martin, Infra-red Instrumentation and Techniques, Elsevier, Amsterdam, 1966.
135. P.L. Richards, Fourier transform spectroscopy, in D.H. Martin (Ed.), Spectroscopic Techniques for the Far Infra-red, North Holland, Amsterdam, 1967.

136. J. Strong and G.A. Vanasse, J. Opt. Soc. Amer., 49 (1959) 844.
137. F. Twyman, Phil. Mag., 35 (1918) 49.
138. H.A. Gebbie and N.W.B. Stone, Infrared Phys., 4 (1964) 85.
139. F.W. Parrett, Lab. Pract., 19 (1970) 68.
140. L. Genzel and R. Weber, Z. Angew. Phys., 10 (1958) 127.
141. H. Happ and L. Genzel, Infrared Phys., 1 (1961) 39.
142. L.C. Block and A.S. Zachor, Appl. Opt., 3 (1964) 209.
143. M.J.D. Low and I. Coleman, Spectrochim. Acta, 22 (1966) 369.
144. F.K. Kneubühl, J.F. Moser and H. Steffen, J. Opt. Soc. Amer., 56 (1966) 760.
145. P.L. Richards, J. Opt. Soc. Amer., 54 (1964) 1474.
146. H.A. Gebbie and G.A. Vanasse, Nature, 178 (1956) 432.
147. K. Watanabe and C.Y. Inn, J. Opt. Soc. Amer., 43 (1953) 32.
148. T.S. Moss, Modern IR Detectors. Advances in Spectroscopy, Vol. 1, Interscience, New York, 1959, p. 193.
149. J. Sharpe, Electron. Technol., 38 (1961) 196.
150. J. Sharpe, Electron. Technol., 38 (1961) 248.
151. W. Budde, Appl. Opt., 3 (1964) 69.
152. R.W. Engstrom, J. Opt. Soc. Amer., 37 (1947) 420.
153. E.H. Piepmeier, D.E. Braun and R.R. Rhodes, Anal. Chem., 40 (1968) 1667.
154. H.J. Marrinan, J. Opt. Soc. Amer., 43 (1953) 1211.
155. H. Renschler, J. Amer. Inst. Elec. Eng., 49 (1936) 576.
156. H.E. Hinteregger and K. Watanabe, J. Opt. Soc. Amer., 43 (1953) 604.
157. L. Dunkelman, J. Opt. Soc. Amer., 45 (1955) 134.
158. W.R.S. Garton, M.S.W. Webb and P.C. Wildy, J. Sci. Instrum., 34 (1957) 496.
159. J.V. Sullivan and A. Walsh, Spectrochim. Acta, 21 (1965) 727.
160. F.S. Johnson, K. Watanabe and R. Tousey, J. Opt. Soc. Amer., 41 (1951) 702.
161. J.T. Agnew, R.G. Franklin, R.E. Benn and A. Bazarian, J. Opt. Soc. Amer., 39 (1949) 409.
162. R.E. Benn, W.S. Foote and C.T. Chase, J. Opt. Soc. Amer., 39 (1949) 529.
163. E.R. Holiday and W. Wild, J. Sci. Instrum., 28 (1951) 282.
164. R.L. Mitchell, Spectrochim. Acta, 4 (1950) 62.
165. T.A. Chubb and H. Friedman, Rev. Sci. Instrum., 26 (1955) 493.
166. P. Bratt, W. Engeler, H. Levinstein, A. Macrae and J. Pehek, Infrared Phys., 1 (1961) 27.
167. D.A.H. Brown, R.P. Chasmar and P.B. Fellgett, J. Sci. Instrum., 30 (1953) 195.
168. W.H. Brattain and J.A. Becker, J. Opt. Soc. Amer., 36 (1946) 354.
169. V.Z. Williams, Rev. Sci. Instrum., 19 (1948) 135.
170. D.R. Andrews, R.M. Milton and W.J. Desorbo, J. Opt. Soc. Amer., 36 (1946) 518.
171. D.H. Martin and D. Bloor, Cryogenics, 1 (1961) 159.
172. F.J. Low, J. Opt. Soc. Amer., 51 (1961) 1300.
173. R.A. Weis, J. Opt. Soc. Amer., 36 (1946) 356.

174. H.A. Zahl and M.J.E. Golay, Rev. Sci. Instrum., 17 (1946) 511.
175. M.J.E. Golay, Rev. Sci. Instrum., 18 (1947) 347; 20 (1949) 816.
176. M.J.D. Low, in W. Lodding (Ed.), Gas Effluent Analysis, Edward Arnold, London, 1967, Chap. V.
177. D.D. Shrewsbury, Spectrovision, 19 (1968) 2.
178. M.D. Morris and J.B. Orenberg, Talanta, 16 (1969) 539.
179. P.J. Hendra and C.J. Vear, Analyst, 95 (1970) 321.
180. J.B. Headridge and J. Richardson, Lab. Pract., 19 (1970) 372.
181. A.B. Shafer, L.R. Megill and L. Droppleman, J. Opt. Soc. Amer., 54 (1964) 879.
182. J. Reader, J. Opt. Soc. Amer., 59 (1969) 1189.

Chapter 2

Analytical atomic absorption and fluorescence spectroscopy

G.F. KIRKBRIGHT and M. SARGENT

1. Introduction

The dark bands observed in the continuum spectrum of the sun by Wollaston [1] (1802) were thoroughly studied by Frauenhofer (1814) and ascribed by Brewster [2] (1820) to absorption in the sun's atmosphere. Kirchoff and Bunsen [3—5] (1860) gave a precise explanation of the Frauenhofer lines and showed that the yellow emission from sodium salts introduced into a flame corresponded exactly to the dark D line from the spectrum of the sun. Kirchoff discovered the fundamental relation between emission and absorption spectra and showed that *all matter absorbs light at the wavelength at which it emits.* The study of atomic absorption in stellar spectra has long provided astronomers with information concerning the elements present in the atmosphere of stars. Early difficulties with the quantitative interpretation of atomic absorption spectra were caused by inadequate data on oscillator strengths, atomic line absorption coefficients and thermal ionisation. Apart from the determination of mercury in air [6], even when the requisite data were available, the development of the technique as a method of analysis was restricted by the lack of suitable sources and means of production of atoms of the elements to be analysed. In 1953 Walsh [7] proposed the general application of atomic absorption spectroscopy to chemical analysis. A hollow-cathode lamp emitting a sharp line spectrum of the element to be determined, was recommended as the source; in this way the atomic absorption coefficient at the absorption line centre was measured rather than the integrated absorbance over the complete absorption profile using a continuum source. Walsh proposed the use of a flame for the production of atoms of the analyte element from solution samples introduced as a fine mist in air. Alkemade and Milatz [8,9] independently proposed an atomic absorption system utilising a flame as radiation source and a second flame as the sample cell. Although the first applications to chemical analysis were not published until 1958, in the decade since then the technique of atomic absorption spectroscopy (AAS) has experienced extremely rapid growth. It has become an established technique of analysis and is applied intensively in many laboratories wherever the sensitive and highly selective determination of any one of about 65 elements is required.

The *atomic fluorescence* of metal vapours in quartz cells was studied by several workers at the beginning of the century. The atomic fluorescence of metal atoms in flames stimulated by an external source of radiation was first observed by Nichols and Howes [10] in 1924. Alkemade [11] described the use of atomic fluorescence spectroscopy (AFS) in flames for the measurement of quantum efficiencies, and his investigations suggested the use of the technique for practical chemical analysis. Robinson [11a] had previously observed the fluorescence of magnesium at 285.2 nm in an oxyhydrogen flame, and Winefordner and co-workers [12—14] pioneered the use of AFS flame methods in analysis. At the time of writing, several groups of workers have explored the potential of AFS, and new commercial instrumentation specifically designed for AFS as well as attachments to permit use of the technique with existing flame spectrophotometers are being developed. To date, only relatively few reports of the use of AFS for the analysis of practical samples have appeared.

Monographs devoted to atomic absorption spectroscopy have been written by Elwell and Gidley [15], Ivanov [16], L'Vov [17], Robinson [18], Ramirez-Munoz [19], Angino and Billings [20], Slavin [21], Rubeska and Moldan [22] and Rousselet [23]. Many general reviews of AAS have been written; amongst these are reviews by Herrman [24,25], Hulanicki [26], Massman [27], Mavrodineanu [28], Petrakev [29], Prugger [30], Rubeska [31], Slavin [32], Suzuki and Takeuchi [33], Walsh [34] and Walsh and Willis [35]. Kahn [36,37] has reviewed instrumentation, while applications of AAS to biology, agriculture, metals analysis and forensic science have been reviewed by Girard and Rousselet [38,39], Schueller [40], Kahn [41] and Ramirez-Munoz [42], respectively. This list of reviews is not exhaustive; many other workers wrote early reviews of

96

AAS before 1966. The technique has assumed such importance and progress in its world-wide application is currently so rapid that it is apparently necessary for annual reviews in several languages. The development of the technique of atomic fluorescence spectroscopy has been reviewed by Winefordner and Mansfield [43,44] and by West [45]. No monographs specifically devoted to AFS have yet been published.

At its present stage of development AAS is primarily useful for the serial determination of trace concentrations of elements in solution. It is becoming one of the most popular methods for the reliable routine determination of elements in solution samples. Although molecular absorption spectroscopy in solution is still widely employed for this type of inorganic trace analysis, AAS is increasingly finding favour over this technique for the determination of elements such as copper, iron, magnesium, etc.. The recent developments in the application of AFS to chemical analysis, and the renaissance which is occurring in analytical flame emission spectroscopy, show such promise that in the near future the analytical chemist will be faced with the choice of three complementary flame techniques for the determination of many elements. It should be emphasised, however, that the techniques are likely to remain complementary. It is improbable that one or other of the techniques will offer superior sensitivity, precision and selectivity for all elements regardless of the nature of the sample matrix. Owing to the fact that each of the three techniques is at present best suited to deal with the determination of a single element in a large number of solution samples, AAS, FES and AFS have not hindered the development and use of techniques such as arc and spark spectrography and X-ray fluorescence spectroscopy which may provide direct multi-element analysis on large numbers of solid samples. Much research and development work is in progress, however, concerned with the use of non-flame cells for AAS and AFS which may be used to produce atomic vapour directly from solid samples.

2. Theory

It is proposed here to discuss the fundamental theory of atomic absorption and fluorescence spectroscopy necessary for a proper understanding of their application as analytical techniques. In view of the nature of the work and the existence elsewhere of comprehensive theoretical treatment only a brief description of the theory involved is presented here. Most of this chapter is concerned with those factors which govern the widths of spectral lines and the relationships between sample concentration and absorption or fluorescence signals. These topics have not been widely discussed in reference to AAS and AFS in the text-books mentioned in Section 1, but are essential if experimental observations are to be interpreted correctly.

(A) ESSENTIAL SPECTROSCOPIC THEORY

It is assumed that the reader has a basic knowledge of the simpler theories of atomic structure and atomic spectra and is aware that spectral lines are the result of the emission or absorption of energy by the atom during a transition from one energy state to another. A clear discussion of the origin of atomic spectra may be found in Chapter 16 of the excellent treatise by Mavrodineanu and Boiteux [46].

The intensity of a spectral line produced by a gas which comprises a population of emitting or absorbing atoms depends on two factors. First, with a given number of atoms in each energy state which is permitted to atoms of that element, some transitions between states will occur more frequently than others. This probability of a transition occurring is a fundamental property of an atom as expressed by the quantum mechanics. It is discussed below as it involves the introduction of a number of quantities which are an essential part of AAS and AFS expressions to be derived subsequently. Secondly, the number of times a particular transition occurs will depend on the number of atoms existing at any time with the energy of the initial level from which the transition occurs. As the population of any level will depend on the means employed to obtain the free atoms (or to excite emission lines) this subject will be discussed in Section 2 (E).

(1) Einstein Transition Probability

The probability of transitions from given energy levels of a fixed population of atoms was expressed by Einstein in the form of three coefficients. These are usually written A_{ij}, B_{ji} and B_{ij} and are termed the *transition probabilities* of spontaneous emission, absorption and stimulated emission, respectively. These are defined as follows.

Let a unit volume of emitting gas contain N_i atoms in an excited state of energy E_i above the ground state, and let a number of these atoms, $dN_{i \to j}$, per unit time, undergo spontaneous transitions to a lower energy state E_j. Then

$$dN_{i \to j} = A_{ij} N_i \tag{1}$$

The intensity of the resulting emission line is then

$$I_{em} = h\nu \, dN_{i \to j} = A_{ij} h\nu N_i \tag{2}$$

where ν is the frequency of the line.

Conversely, if radiation of frequency ν and spectral volume density $\rho(\nu)$ is sent through the gas, let $dN_{j \to i}$ atoms undergo the transition from E_j to E_i in unit time and the population of E_j be N_j atoms per unit volume. Then

$$dN_{j \to i} = B_{ij} \rho(\nu) N_j \tag{3}$$

The observed absorption is then given by

$$\frac{dI}{I_0} = B_{ji} \frac{h\nu}{c} N_j \tag{4}$$

where I_0 is the initial light intensity, dI is the amount of light absorbed, and c is the velocity of light. Einstein also postulated, however, that this incident radiation could induce a number of atoms $dN'_{i \to j}$ to undergo the transition back from E_i to E_j. Then

$$dN'_{i \to j} = B_{ij} \rho(\nu) N_i \tag{5}$$

In this case the absorption observed is no longer given by eqn. (4) but by the expression

$$\frac{dI}{I_0} = \frac{h\nu}{c} (B_{ji} N_j - B_{ij} N_i) \tag{6}$$

If the statistical weights * of the levels E_i and E_j are denoted as g_i

---

* The statistical weight of each level is defined as $g = 2J + 1$, where J is the resultant magnetic quantum number for the level.

and g_j, it may be shown that the following useful expressions hold true.

$$g_i B_{ij} = g_j B_{ji} \tag{7}$$

$$\frac{A_{ij}}{B_{ji}} = \frac{8\pi h \nu^3}{c^3} \frac{g_j}{g_i} \tag{8}$$

$$\frac{A_{ij}}{B_{ij}} = \frac{8\pi h \nu^3}{c^3} \tag{9}$$

These equations are presented here as they are useful for substitutions into formulae defined only in terms of the transition probabilities. For example, in practical spectroscopy it is frequently advantageous to replace the transition probability by a second coefficient, termed the *oscillator strength*, f. The emission oscillator strength, f_{ij} is defined by

$$A_{ij} = \frac{8\pi^2 e^2}{\lambda^2 mc} f_{ij} \tag{10}$$

where e and m are the electron charge and mass and λ is the wavelength corresponding to the frequency ν.

Substituting the values of the constants into eqn. (10) gives

$$A_{ij} = \frac{0.667 \times 10^8}{\lambda^2} f_{ij} \tag{11}$$

where A_{ij} is in $\sec^{-1}$ and λ is in microns.

The absorption oscillator strength, f_{ji}, is related to f_{ij} by

$$g_i f_{ij} = g_j f_{ji} \tag{12}$$

(B) THE WIDTHS OF SPECTRAL LINES

Even the sharpest spectral line that can be produced has a finite width, so that a line may never be considered as represented by a unique frequency (or wavelength). This fact has important consequences on the applications of both AAS and AFS. In view of this,

100

the factors contributing to the final line width observed from sources will be discussed in some detail.

(1) Definition of Line Width

The intensity distribution of an emission line may be represented by a curve of the type shown in Fig. 1.

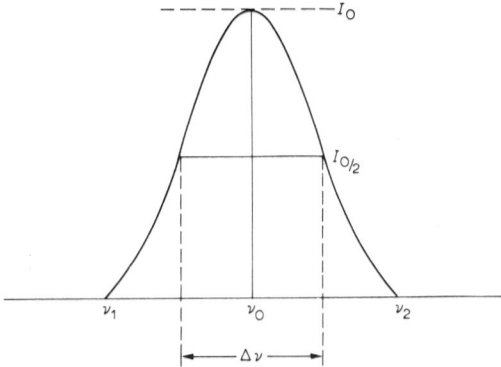

Fig. 1. Shape of a spectral line.

Here the intensity is written as I_ν which is a function of the frequency ν. The maximum intensity I_0 occurs at frequency ν_0. As the values of ν_1 and ν_2 cannot be measured accurately, it is customary to define the breadth of any line in terms of its *half-width*, $\Delta\nu$, i.e. the distance between the two points on the curve where I_ν = - $I_0/2$.

The same nomenclature is used for absorption lines with I_ν replaced by the *absorption coefficient*, k_ν, which attains a maximum value of k_0 at frequency ν_0. All of the factors considered below which act to broaden emission lines apply equally to absorption lines (except for self-reversal broadening). Similarly, the formulae used to represent line broadening apply to both emission and absorption lines.

References pp. 253—262

(3) Phenomena responsible for Spectral Line Widths

(a) Natural Broadening

Natural broadening is the result of the finite lifetime of any atom in an excited state. The finite lifetime leads, by the Heisenberg uncertainty principle, to an uncertainty in the energy of the level. Thus the frequency ν_0 for a transition between levels E_i and E_j is no longer given uniquely by $\nu_0 = (E_i - E_j)/h$ and the line has a finite width, $\Delta\nu_N$. It may be shown that $\Delta\nu_N$ is of the order of 10^8 sec^{-1}. In terms of wavelength this represents a value of 10^{-5} nm at 300 nm. The natural line width is usually quite negligible in comparison with that caused by other broadening factors.

(b) Doppler Broadening

Doppler broadening is caused by the thermal motion of the emitting or absorbing atoms. According to the Doppler effect, when an atom moving towards the observer with velocity v emits light of frequency ν the radiation received by the observer is displaced to frequency $(\nu + d\nu)$ where $d\nu = \nu v/c$. Similarly, when the atoms are moving away from the observer the frequency received is $(\nu - d\nu)$. As, in a gas, the atoms move in different directions with differing velocities, the observed line is broadened and has the shape of a probability distribution curve (Gaussian). The half-width of the line, denoted by $\Delta\nu_D$, is directly proportional to ν_0 and to the square root of the absolute temperature and inversely proportional to the square root of the atomic weight. Thus, for a given atom, $\Delta\nu_D$ increases with increasing temperature and becomes smaller at short wavelengths. In flames commonly used in analytical flame spectroscopy, $\Delta\nu_D$ is usually in the range 5×10^{-4} to 50×10^{-4} nm.

(c) Lorentz Broadening

This is also known as collisional or pressure broadening. Whereas the Doppler width depends on the temperature and mass of the individual atoms in a population, Lorentz broadening depends on the temperature, pressure and nature of the gas in which the emitting or absorbing atoms are situated. This type of broadening is the result of collisions by the atoms with other neutral atoms of a different kind (foreign gas perturbation). The relationships between the Lorentz broadening half-width, $\Delta\nu_L$, and the collisions which cause it are complex. It may be shown, however, that $\Delta\nu_L$ is usually inversely proportional to the square root of the temperature. In flames at atmospheric pressure, there is a high population of foreign atoms and molecules and Lorentz broadening is quite important, with $\Delta\nu_L$ of the same order of magnitude as $\Delta\nu_D$.

In most sharp line spectral sources used in AAS and AFS, the foreign gas pressure is usually very low (less than 10 Torr) and the Lorentz broadening is generally negligible. This also applies to the simultaneous shift of the line maximum and the assymetry of the distribution curve produced by the collisions. This shift may not be negligible for the absorption line in the flame, however, so that some loss of peak absorption sensitivity may occur.

(d) Holtsmark Broadening

This is also known as *resonance broadening* and, like (c), as collisional or pressure broadening. Holtsmark broadening is produced by collisions with atoms of the same kind as those responsible for the emission or absorption of the radiation. The line width, $\Delta\nu_H$, produced by collisions with like atoms at resonance lines is considerably greater than that produced by foreign atoms because the effect is the result of a strong electrostatic interaction between an atom in an excited state and another in the ground state rather than a simple kinetic collision process. In view of this, accurate values of $\Delta\nu_H$ can only be calculated using a complex theoretical treatment to evaluate a correction factor for each particular type of transition. The number of results available at present is quite small. Actual practical measurements also prove difficult because conditions within a source which produce sufficient resonance broadening for study also produce self-absorption effects. It seems, however, that in flames $\Delta\nu_H$ is negligible compared with $\Delta\nu_L$ unless very high sample concentrations (in the molar range) are introduced. In the spectral sources used for AAS and AFS the value of $\Delta\nu_H$ will usually be fairly small compared with $\Delta\nu_D$ but may begin to reach significant values when the atom concentration is high enough to produce serious self-absorption effects. It should be noted that a line shift is not predicted by the theories used to describe pure resonance broadening, so that any shift or assymetry observed can be only of the same order of magnitude as that predicted for the simple Lorentz effect.

(e) Stark and Zeeman Broadening

These effects are also known as *field broadening* effects. Stark broadening is observed as the splitting of an atomic line due to the effect of a strong electric field on the spectral terms of the atom. It is negligible in flames and is also considered small in hollow-cathode discharge lamps and electrodeless discharge tubes under the conditions used for AAS and AFS. However, it becomes serious in arc and spark sources. Zeeman broadening is a similar splitting effect obtained in the presence of a strong magnetic field. As this environment is not encountered in analytical work in flames it will not be considered here.

(f) Hyperfine Structure

When many spectral lines are examined with an ultra-high resolution monochromator, they are found to be composed of a number of closely spaced lines. These hyperfine structure (hfs) components are the result of the presence of a number of isotopes and/or the interaction of the nuclear spin with the spins of the electrons. These components are therefore characteristic of the atoms and will always be present, whatever the excitation conditions, but fortunately the practical effect on AAS and AFS measurements is quite small. It should be appreciated, however, that the hfs components have a variable and complex effect on the line-width and may considerably affect the validity of any calculations made in connection with AAS and AFS which ignore their existence.

Each hfs line is a single spectral line and will be broadened by all the above broadening factors. If this results in a line-width much greater than the hfs separation, so that complete overlap occurs, the line may still be treated as a single line for most purposes. In cases where the hfs separation is much wider (e.g. the Hg 546.1 nm line has its extreme components ca. 0.045 nm apart), any calculations will be extremely complex and depend on the amount of overlap between components (i.e. the extent of the broadening). A conventional monochromator of the type used in AAS will not resolve any of the hfs components; unless they are completely overlapped, how-

104

ever, they will still have an effect on the light absorbed. This arises because the equations derived in Section 2(C) must be applied to each hfs component individually and the results summed. The value obtained will differ from that derived using a 'total' value of the absorption coefficient, k_ν, for the whole line in a single equation and will depend on the separation of the components. Although, in many cases, the difference between the two values may be quite small it has been shown to be appreciable for some elements (e.g. the resonance lines of copper and gold).

(g) Self-Reversal Broadening

As the lines used for AAS and AFS are usually those which give strong absorption (i.e. the resonance lines or others with a lower level at or near the ground state) it follows that they are also the lines which suffer the greatest self-absorption under suitable conditions. This is particularly serious from the point of view of the emission sources used for AAS because the self-absorption is accompanied by a strong broadening effect. This may increase the half-width several-fold (depending on the type of the profile distribution) and this multiplication factor applies to the sum of all the other broadening factors. Thus self-reversal broadening may be more important than any one of the other causes of broadening alone. For many elements self-reversal is the most serious problem which is encountered when attempts are made to obtain a sharp line source.

In order to explain the effects of self-absorption it is necessary to utilise the relationships derived later in Section 2(C). However, it is sufficient here to make use of two results which follow from these relationships.

(i) The maximum absorption occurs at the centre of the line (i.e. at ν_0 or λ_0). This means that when self-absorption occurs, a greater proportional amount of radiation is lost from the centre of the emission line than from the edges, i.e. the shape of the line profile changes as well as its intensity. This effect is shown in Fig. 2. The original emitted line is shown with the profile ABCDE and has a half-width BD. After self-absorption has occurred, the line assumes the shape AXYE. The half-width of the line is then XY and is greater than BD.

(ii) At vapour pressures which are high enough to produce sufficient self-absorption to affect the line breadth, the degree of self-

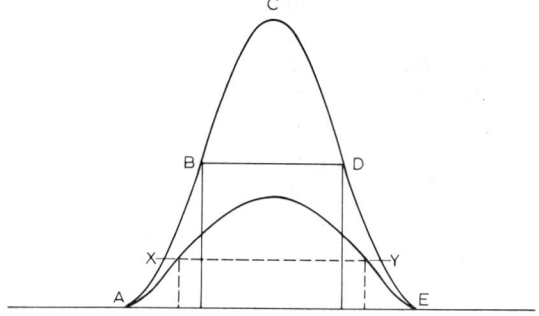

Fig. 2. Effect of self-absorption on half-width.

absorption is proportional to $N^{1/2}$ where N is the number of absorbing atoms in the path length considered. As N is also proportional to the vapour pressure, the self-reversal broadening may be minimised by using a small path length and low vapour pressure.

Obviously if the self-absorption continues to increase (by increasing N) more and more of the line centre is lost. Eventually the stage is reached where the centre of the line is actually less intense than the wings. This is the well-known case of *self-reversal* and is shown in various stages in Fig. 3.

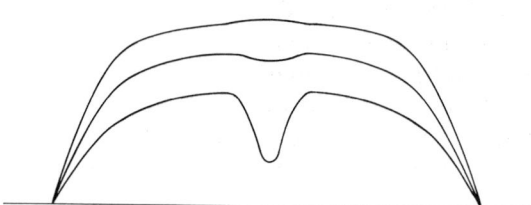

Fig. 3. Self-reversal of a spectral line.

In practice, self-reversed lines are not seen emitted from a source unless it contains a temperature gradient. If the source volume has a uniform temperature, light emitted from a layer at the edge of the volume will have the same initial intensity as light from a layer at the centre of the source. However, a line from the edge of the source is much less self-absorbed (because the optical path length through the source atoms is shorter). This emission therefore augments that at the centre of the self-reversed line from the centre of the source and

106

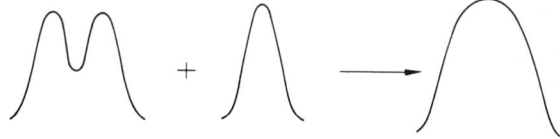

Fig. 4. Loss of self-reversal in a source of uniform temperature.

an observer will only view a rather broad, self-absorbed line. This is illustrated schematically in Fig. 4.

When the source is actually cooler at the edge than at the centre, however, light emitted from the outer part of the source does not have sufficient intensity to compensate completely for the radiation lost by self-absorption en route from the source centre. In this case, which occurs frequently in practice, a degree of self-reversal is actually observed in the line.

There is a second mechanism giving rise to self-reversal which may be significant in sources such as hollow-cathode lamps when Doppler broadening is important and a sharp temperature gradient is present. As Doppler broadening decreases with decreasing temperature, the absorption line at the cooler edge of the source is narrower than the emission line emitted at the hotter centre of the source. Thus, as the light from the hot centre passes through the cooler outer region, the proportion of the radiation removed only from the centre of the line is even higher than usual. Self-absorption by these cooler atoms then itself appears as self-reversal even though the amount of self-absorption is lower than that normally needed for the onset of self-reversal.

(3) The Total Width of a Spectral Line

The combined effect of all the broadening factors on the intensity, I_ν, or the absorption coefficient, k_ν, of a line can be expressed in rigorous fashion utilising many different equations. In the discussion of sharp line sources for AAS and AFS, however, it may be necessary to make use of the actual width of the source line. For example, when measuring the peak absorption the narrower the source line the higher will be the absorption and the better the linearity of the calibration graphs. For most purposes it is sufficient to denote this source width as, say, $\Delta\nu_s$ or $\Delta\lambda_s$ and assume that it can be deduced from other factors where necessary. $\Delta\lambda_s$ is a complex value, as the shape of the profile resulting from, for example, collision broadening is Lorentzian whereas that resulting from Doppler broadening is Gaussian. A useful approximation is available, however, which gives an indication of the relative effects of the different broadening factors [47]

$$\Delta\lambda_s = [(\Delta\lambda_D)^2 + (\Delta\lambda_N + \Delta\lambda_L + \Delta\lambda_H)^2]^{1/2} \tag{13}$$

It should be remembered that eqn. (13) takes no account of the effects of hyperfine structure or self-reversal broadening.

(C) ATOMIC ABSORPTION SPECTROSCOPY

(1) Introduction

The fact that all absorption or emission lines have a finite width introduces a degree of complexity into the derivation of expressions for use in AAS; integration with respect to ν or λ must be performed over the width of the line. In view of this, the theory of AAS is best described in terms of two quantities: the absorption coefficient and the total absorption factor. Care must be taken not to confuse these two quantities.

The absorption coefficient, k_ν. The absorption coefficient, k_ν, describes the shape of the absorption line. Its measurement, therefore, requires a record of the actual profile of the spectral line. The use of a continuum background source would be necessary to accomplish this experimentally; k_ν is independent of an outside source and is a function only of the conditions prevailing within the absorption cell.

The total absorption factor, A_T. This is a measure of how much radiation is actually absorbed during experimental atomic absorption measurements. It depends, therefore, on both the emission line from the source and the absorption line of the corresponding atom in the cell. It may be calculated simply from a knowledge of the radiant power of the light entering and leaving the cell. For the reasons described below, this means that nearly all analytical AAS measurements are directly concerned with A_T rather than k_ν. The parameters normally measured in AAS (percentage absorption and absorbance) are both closely related to A_T.

108

(2) The Absorption Coefficient, k_ν

The absorption coefficient at a frequency ν is defined by

$$I_\nu = I_\nu^0 \exp(-k_\nu L) \tag{14}$$

where I_ν^0 and I_ν are the incident and transmitted intensities of radiation of frequency ν passed through an absorption cell of path length L.

Normally, in order to make use of k_ν it is necessary to replace it with an expression which shows its dependence on frequency and which can be integrated with respect to ν. This will be dealt with in Section 2(C)(4). However, we may consider here the one special case where this is not necessary. If k_ν is integrated with respect to ν *over the whole line*, the result is independent of ν.

(3) The Integrated Absorption Coefficient, K

This special case may be used as follows. It may be shown (ref. 48, p. 95) that K is given by dI/I_0 of eqn. (6), i.e.

$$K = \int k_{\nu_0} \, d\nu = \frac{h\nu_0}{c} [B_{ji} N_j - B_{ij} N_i] \tag{15}$$

where the small variation in ν over the width of the line is neglected and ν_0 is the frequency of the centre of the line.

Substitution into eqn. (15) of the relationships between the Einstein coefficients (eqns. (7)—(9)) then gives

$$K = \frac{c^2}{8\pi\nu_0^2} A_{ij} [\frac{g_i}{g_j} N_j - N_i] \tag{16}$$

which may be written

$$K = \frac{\lambda_0^2}{8\pi} \frac{g_i}{g_j} A_{ij} N_j \left[1 - \frac{g_j}{g_i} \frac{N_i}{N_j} \right] \tag{17}$$

Now in a number of absorption cells, including flames, the value of N_i/N_j is very small. It is therefore possible to approximate eqn. (17) to

$$K = \frac{\lambda_0^2}{8\pi} A_{ij} \frac{g_i}{g_j} N_j \tag{18}$$

Alternatively, substituting for A_{ij} using eqns. (10) and (12)

$$K = \frac{\pi e^2}{mc} f_{ij} \frac{g_i}{g_j} N_j = \frac{\pi e^2}{mc} f_{ji} N_j \tag{19}$$

Equations (18) and (19) are most important because they show not only that K is represented by a simple linear relationship with N_j, but also that this relationship is quite independent of whatever physical processes are responsible for the formation of the absorption line.

Thus the evaluation of K would appear to provide a very simple means of measuring atomic populations by atomic absorption. Unfortunately, in order to make use of it, it is necessary to plot k_ν against ν for the line and take the area under the curve which is obtained. This is only possible by the measurement of the actual profile of the absorption line (using a continuum background source) when k_ν may be calculated from the value of I_ν at each point on the profile using eqn. (14). The experimental difficulties in making this measurement are considerable; the method would certainly prove too complex to permit its routine use for analytical AAS. Thus, the integrated absorption coefficient, K, is seldom used in this way.

It is important to realise that the value of the integrated absorption coefficient, K, which is obtained by the above procedure with a continuum source and high resolution monochromator *does not* correspond to the absorption measured using a continuum source and a conventional medium resolution monochromator. In this latter case, measurements are merely being made of the absorption factor, A_T, for the whole line. As will be shown in Section 2(C)(6), this only approximates to the value of K at low absorption levels and is otherwise quite different.

(4) The Voigt Expression for the Absorption Coefficient, k_ν

As already stated, k_ν is a function of the frequency. Any expression used to represent it, therefore, for a given absorption line must take into account the factors affecting the shape of the line, i.e. the broadening parameters. The most commonly used expression for k_ν

is due to Voigt and accounts only for natural, Doppler and Lorentz broadening. It is usually written in the form

$$k_\nu = \frac{k_0 a}{\pi} \int_{-\infty}^{\infty} \frac{\exp(-y^2)\,dy}{a^2 + (w-y)^2} \tag{20}$$

where

$$k_0 = \frac{2\sqrt{\ln 2}}{\Delta\nu_D \sqrt{\pi}} K \tag{21}$$

$$a = \frac{(\Delta\nu_N + \Delta\nu_L)\sqrt{\ln 2}}{\Delta\nu_D} \tag{22}$$

$$w = (\nu - \nu_0) \frac{2\sqrt{\ln 2}}{\Delta\nu_D} \tag{23}$$

$$y = \frac{2\delta\sqrt{\ln 2}}{\Delta\nu_D} \tag{24}$$

k_0 represents the *maximum absorption coefficient* which would be obtained (at ν_0) if it were possible to have a line with only Doppler broadening. a is known as the *damping ratio*. As $\Delta\nu_N$ is very small, eqn. (22) is sometimes approximated to $\Delta\nu_L\sqrt{\ln 2}/\Delta\nu_D$ and is then referred to as the *collision damping ratio*. δ represents a variable distance from a fixed point which is itself at a distance $(\nu-\nu_0)$ from the centre of a line showing only natural broadening. Further details are given in ref. 48, p. 100.

It can be seen that the expression for k_ν is quite complex. In fact its evaluation is only possible by the introduction of simplifying assumptions. As the nature of these assumptions must depend on the prevailing experimental conditions, a number of different approximations for k_ν (and hence A_T, which depends on it) may be obtained. The most useful of these will be given below as equations for A_T.

(5) The Total Absorption Factor, A_T

The total absorption factor is defined in terms of the total power of the radiation passing through the absorption cell at the frequency of the line. The intensities used are, therefore, integrated with respect to the whole range of frequencies of the line at which absorption can occur. It is given by

$$A_T = \frac{\text{intensity of absorbed radiation}}{\text{intensity of incident radiation}} = \frac{\Delta I}{I_0} \tag{25}$$

i.e. $\Delta I = I_0 - I$ where I is the intensity of the radiation transmitted by the cell.

The *percentage absorption* commonly employed in analytical AAS is defined as

$$\frac{\Delta I}{I_0} \times 100 \tag{26}$$

so that for most purposes it may be considered as identical to A_T.

The *absorbance*, A, which is also widely used in AAS (see Section 2(C)(8)) is defined as

$$A = \log_{10}(I_0/I) \tag{27}$$

Thus

$$A = \log_{10} 1/(1-A_T) \tag{28}$$

A is thus related to A_T in quite complex fashion (except for values of A_T less than about 0.1 when A is, to a good approximation, related linearly to A_T).

Now, by definition

$$I_0 = \int I_\nu^0 \, d\nu \qquad I = \int I_\nu \, d\nu$$

Therefore writing eqn. (25) as

$$A_T = \frac{I_0 - I}{I_0}$$

and substituting eqn. (14) gives

$$A_T = \frac{\int I_\nu^0 \, d\nu - \int I_\nu^0 \exp(-k_\nu L) \, d\nu}{\int I_\nu^0 \, d\nu} \tag{29}$$

112

If the assumption is now made that I_ν^0 will always be integrated over the same limits as k_ν, i.e. imposing the condition that the integration will be made over the source line width if it is narrower than the absorption line, or the absorption line width if it is narrower than the source line, eqn. (29) may be written

$$A_T = \frac{\int I_\nu^0 \, d\nu \int [1 - \exp(-k_\nu L)] \, d\nu}{\int I_\nu^0 \, d\nu} \tag{30}$$

$$= \frac{I_0 \int [1 - \exp(-k_\nu L)] \, d\nu}{I_0}$$

i.e.

$$A_T = \int [1 - \exp(-k_\nu L)] \, d\nu \tag{31}$$

The limits over which the integration is performed will depend on the width of the narrower line and cannot be generalised. It is therefore now necessary to consider a number of special cases.

(6) The Total Absorption Factor where the Absorption Line Half-width is Narrow Compared to the Source Line Half-width

This situation is most commonly encountered when a continuum source is employed in AAS. It also applies when an arc or flame source giving a very broad, self-absorbed line is used. As the source line is so much wider than the absorption line, the assumption may be made that in all cases the total absorbed radiation may be found by integrating across the entire absorption line, i.e. from zero to infinity. Thus eqn. (31) becomes

$$A_T = \int_0^\infty [1 - \exp(-k_\nu L)] \, d\nu \tag{32}$$

As explained earlier, this may only be evaluated by making approximations in the expression for k_ν. The two most convenient approximations are for small and large absorbance values.

(a) Small $k_\nu L$

In this case the well known approximation $[1 - \exp(-x)] \to x$ as

$x \to 0$ may be applied, and eqn. (32) becomes

$$A_T = \int_0^\infty k_\nu L \, d\nu = L \int_0^\infty k_\nu \, d\nu \tag{33}$$

But $\int_0^\infty k_\nu \, d\nu$ is K, the integrated absorption coefficient. Thus substituting the expression for K from eqn. (19)

$$A_T = \frac{\pi e^2}{mc} f_{ji} N_j L \tag{34}$$

i.e. A_T, and hence the percentage absorption increases linearly with N_j and thus usually with sample concentration.

(b) Large $k_\nu L$

The approximation needed here is rather more complex than that used above; it may be shown [49] that A_T is given by

$$A_T = \left(\frac{k_0 \Delta\nu_D^2 L a \pi^{1/2}}{\ln 2} \right)^{1/2} \tag{35}$$

Substituting for k_0 from eqns. (19) and (21)

$$A_T = \left(\frac{2\pi e^2 \Delta\nu_D a f_{ji} N_j L}{mc\sqrt{\ln 2}} \right)^{1/2} \tag{36}$$

i.e., in this case, A_T increases linearly with $N_j^{1/2}$.

Thus if AAS measurements are made with a continuum source, a plot of $\log (\Delta I/I_0)$ against log concentration should result in a straight line graph with a slope of 1 at low concentrations and a slope of 0.5 at higher concentrations.

In practice it is found that at very high concentrations the square root relationship does not hold, and $\Delta I/I_0$ increases more slowly than expected, eventually reaching a limiting value. This results from the fact that as k_ν increases (i.e. at higher absorption) its value at the edge of the line becomes more important (since the centre of the line quickly reaches 100% absorption). This is shown in Fig. 5. As a narrow monochromator band-pass is always used with a continuum source (so that the detector receives the minimum amount of radia-

114

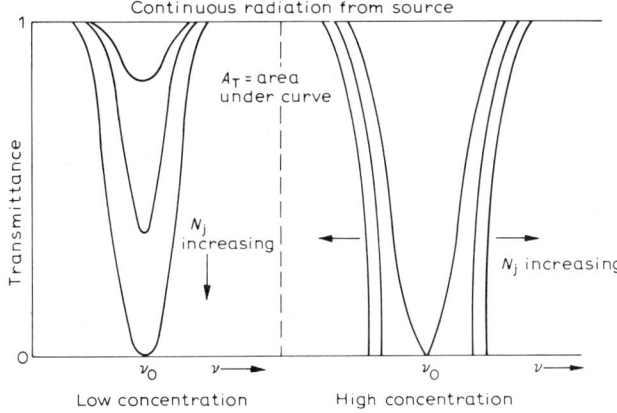

Fig. 5. Effect of increasing concentration on absorption line width.

tion which has not passed through the absorption line profile), the stage is eventually reached where increases in absorption fall outside the band-pass of the monochromator and are not detected. The problem is easily solved by increasing the spectral band-pass.

It will be seen, therefore, that with the one exception mentioned above, a plot of log $(\Delta I/I_0)$ versus concentration will produce linear calibration graphs. This is not true, however, for the absorbance, A, except when A is related linearly to A_T at very low concentrations. Thus unlike in AAS with a sharp line source, when a continuum source is employed calibration graphs of absorbance vs. concentration should not be used.

(7) The Total Absorption Factor where the Absorption Line Half-Width is Wide Compared to the Source Line Half-Width

This situation is encountered when a sharp line source such as a hollow-cathode lamp is employed in AAS.

As the source line is narrow it may be assumed that absorption will occur over the entire width, $\Delta \nu_s$, of the source line, i.e. the integration in eqn. (31) must be from 0 to $\Delta \nu_s$. The calculation is conveniently made after converting the Gaussian function of the line to a triangular one. This is done by multiplying $\Delta \nu_s$ by the factor $\sqrt{\pi}/(2\sqrt{\ln 2})$. Equation (31) then becomes

$$A_T = \int_0^{\frac{\Delta\nu_s\sqrt{\pi}}{2\sqrt{\ln 2}}} [1 - \exp(-k_\nu L)]\, d\nu \tag{37}$$

As in the case mentioned in Section 2(C)(6) this equation is most easily evaluated for small and large optical densities.

(a) Small $k_\nu L$

Using the approximation $[1-\exp(-x)] \to x$ as $x \to 0$, eqn. (37) becomes

$$A_T = \int_0^{\frac{\Delta\nu_s\sqrt{\pi}}{2\sqrt{\ln 2}}} k_\nu L\, d\nu \tag{38}$$

As the interval of the integration is very narrow for a sharp line source, it is a valid assumption that k_ν is a constant and equal to the peak absorption coefficient of the line centre. This coefficient is given by bk_0, where b is a factor correcting for the fact that, in practice, the absorption line is broadened by both Doppler and other processes. Hence eqn. (38) becomes

$$A_T = bk_0 L \int_0^{\frac{\Delta\nu_s\sqrt{\pi}}{2\sqrt{\ln 2}}} d\nu$$

i.e.

$$A_T = \frac{bk_0 L \Delta\nu_s \sqrt{\pi}}{2\sqrt{\ln 2}} \tag{39}$$

Substituting for k_0 from eqns. (19) and (21)

$$A_T = b\, \frac{\Delta\nu_s}{\Delta\nu_D}\, \frac{\pi e^2}{mc}\, f_{ji} N_j L \tag{40}$$

i.e. A_T varies linearly with N_j.

116

(b) Large $k_\nu L$

As the absorption can only be measured near the line centre (due to the narrow source width) it may be assumed that A_T is the result of complete absorption of the incident light, i.e. $e^{-k_\nu L} = 0$. This is illustrated in Fig. 6. Thus eqn. (37) becomes

$$A_T = \int_0^{\frac{\Delta\nu_s\sqrt{\pi}}{2\sqrt{\ln 2}}} d\nu \qquad (41)$$

i.e.

$$A_T = \frac{\Delta\nu_s\sqrt{\pi}}{2\sqrt{\ln 2}} \qquad (42)$$

This means that at high optical density (high sample concentrations in AAS) the absorption will no longer increase with concentration. Thus analytical AAS measurements are not possible. It should be

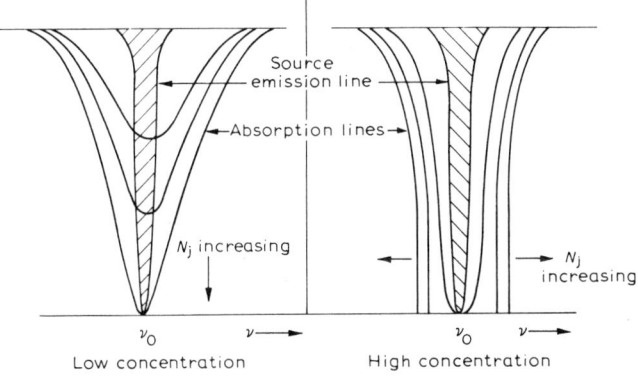

Fig. 6. Effect of concentration on absorption with a narrow line source.

noted that this situation cannot be remedied simply by increasing the spectral band-pass of the monochromator as was the case with a continuum source.

(8) The Absorbance, A, with a Sharp Line Source

In practice it is found that eqns. (40) and (42) hold true only at the extreme ends of the concentration range. Unlike the case for a continuum source, the equations provide no simple relationship between A_T and concentration in the range most important for AAS. This situation arises because, although eqns. (34) and (40) are only valid at very low values of $k_\nu L$ (owing to the use of the approximation $(1-e^{-k\nu L}) = k_\nu L$), eqn. (42) for a sharp line source only takes over at much higher values of $k_\nu L$ than does eqn. (36) for continuum sources.

The problem is solved by making use of the absorbance, A, which may be evaluated as follows for a sharp line source without making use of the approximation $(1-e^{-k\nu L}) = k_\nu L$.

Substitution into eqn. (27) of

$$I_0 = \int_0^{\Delta\nu_s} I_\nu^0 \, d\nu \quad \text{and} \quad I = \int_0^{\Delta\nu_s} I_\nu \, d\nu$$

gives

$$A = \log_{10}\left[\frac{\int_0^{\Delta\nu_s} I_\nu^0 \, d\nu}{\int_0^{\Delta\nu_s} I_\nu \, d\nu} \right]$$

Use of eqn. (14) transforms this to

$$A = \log_{10}\left[\frac{\int_0^{\Delta\nu_s} I_\nu^0 \, d\nu}{\int_0^{\Delta\nu_s} I_\nu^0 \exp(-k_\nu L) \, d\nu} \right] \quad (43)$$

As a sharp line source is being used, the assumption that k_ν is a constant and equal to bk_0 is valid as before. Hence eqn. (43) becomes

$$A = \log_{10}\left[\dfrac{\displaystyle\int_0^{\Delta\nu_s} I_\nu^0 \, d\nu}{\exp(-bk_0L)\displaystyle\int_0^{\Delta\nu_s} I_\nu^0 \, d\nu}\right]$$

$$= \log_{10}\left[\dfrac{1}{\exp(-bk_0L)}\right]$$

i.e.

$$A = \log_{10}\exp(bk_0L) = 0.434\, bk_0L \tag{44}$$

Substituting for k_0 from eqns. (19) and (21)

$$A = 0.434\, b\, \frac{2\sqrt{\ln 2}}{\Delta\nu_D\sqrt{\pi}}\, \frac{\pi e^2}{mc}\, f_{ij}\, N_j\, L \tag{45}$$

Thus a plot of absorbance, A, vs. concentration should be linear and should extend to high concentrations (as no limit has been placed on $k_\nu L$ during the evaluation of eqn. (35)). Indeed, if the line width of the source was very narrow, eqn. (35) would hold true until the onset of Holtsmark broadening in the flame (i.e. up to molar solutions) when the damping ratio, a, of the absorption coefficient would change. In practice, however, the line width from a hollow-cathode lamp is never negligible with respect to the width of the absorption line. In this case, the assumption that $k_\nu = bk_0$ is no longer valid. Fortunately, the error introduced is quite small except at large values of $k_\nu L$, and it is often found that graphs of absorbance vs. concentration are linear up to about $A = 0.7$. At higher values of A the graphs may curve towards the concentration axis.

(D) ATOMIC FLUORESCENCE SPECTROSCOPY

The principal types of atomic fluorescence effect which may be observed are detailed below. Not all of these types are encountered in analytical AFS, mainly because, so far, the technique has employed flames as the fluorescence cells.

(a) Resonance Fluorescence

When the atom re-emits radiation at the same spectral line as that used to excite it, the effect is termed resonance fluorescence. Pure resonance fluorescence (i.e. with no other emissions occurring simultaneously from the same excited state) is only observed when there is no other possible transition by which a return to the ground state can occur, so that the ground state must consist of a single level. However, pure resonance fluorescence can still be observed even if the excited level is not the first one, as long as the selection rules forbid transitions into the intermediate level(s). Examples of pure resonance fluorescence with, and without, an intermediate level are given by the zinc 213.86 nm and 307.59 nm lines respectively. This is illustrated in Fig. 7.

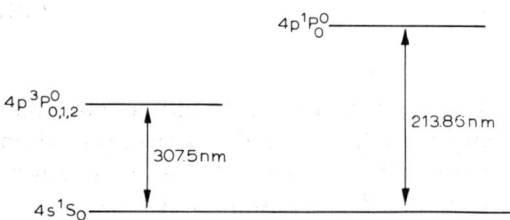

Fig. 7. Pure resonance fluorescence for zinc.

An example of resonance fluorescence accompanied by a second type of fluorescence from the same excited level is encountered for thallium at the 377.6 nm line (see Fig. 8).

(b) Direct Line Fluorescence

Direct line fluorescence is observed when transitions between the excited state of the resonance line and a lower intermediate level are not forbidden by the selection rules. Thus the emission of resonance fluorescence is accompanied by the emission of a second line at a longer wavelength corresponding to the transition between the excited level and this metastable level. This situation will arise if the

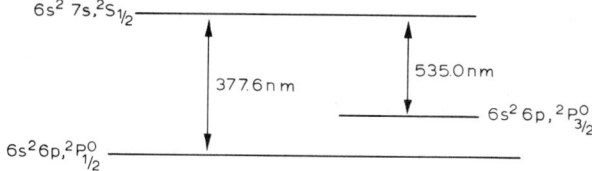

Fig. 8. Fluorescence from the lower thallium excited state.

ground state is a multiplet. For example, after excitation of thallium by the Tl 377.6 nm line, emission at the Tl 535.0 nm line (direct line fluorescence) is observed as well as at the Tl 377.6 nm line (resonance fluorescence). This is shown in Fig. 8.

If the excited state of the resonance line lies above another excited state, this may also act as the intermediate level. For example, absorption of the antimony 217.6 nm line also produces direct line fluorescence at 267.1 nm and 277.0 nm. This is illustrated in Fig. 9.

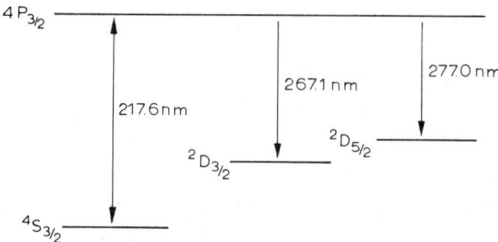

Fig. 9. Example of resonance and direct line fluorescence by antimony.

(c) Stepwise Line Fluorescence

Stepwise line fluorescence results when the atom is initially excited to a higher excited state by absorption at a resonance line and then undergoes deactivation (often by a collision or other radiationless process) to a lower excited state. If a transition is permitted from this state to a still lower state (which is often the ground state) stepwise fluorescence may be observed. For example, when sodium atoms are excited by only the resonance lines at 330.2 and 330.3 nm, the resonance fluorescence at this wavelength is accompanied by fluorescence from the sodium D lines. This is illustrated in Fig. 10. The deactivation from 4^2P_0 to 3^2P_0 levels probably occurs

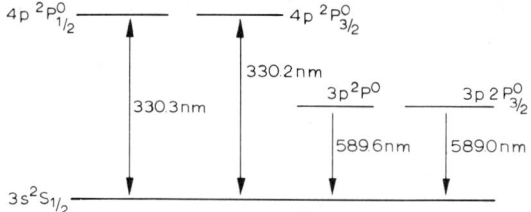

Fig. 10. Stepwise line fluorescence by sodium.

via intermediate energy levels (3d ^{2}D and 4s ^{2}S) and may also give other fluorescence lines at long wavelengths.

(d) Stepwise Excitation Fluorescence

Stepwise excitation fluorescence occurs as a result of a stepwise absorption of energy by the atom in order to reach a particular excited state. A number of fluorescence lines might then be obtained during the transition back to the ground state. A simple example of this occurs when the first state reached above the ground state is populated thermally due to the temperature of the flame. Absorption of a line having this state as its ground state can then result in the observation of fluorescence. For example, at sufficiently high temperatures, absorption of the thallium 535.0 nm line could result in stepwise excited fluorescence at both 535.0 and 377.6 nm lines. This type of fluorescence is of low intensity in most flames.

Another well-known mechanism for stepwise excited fluorescence involves the use of a very intense source. An excited state, reached by absorption from the ground state, may then itself act as the lower level for a further absorption of a line to reach a higher excited state, e.g. absorption of the mercury 253.7 nm resonance line may be followed by absorption of the 435.8 nm line. Again, fluorescence intensities resulting from this type of process are likely to be low in most flames.

(e) Sensitised Fluorescence

Sensitised fluorescence may be obtained when an atom is excited by collisional excitation by a foreign (donor) atom which has itself been previously excited by the absorption of resonance radiation.

122

For example, irradiation with the mercury 253.7 nm line of a mixture of mercury and thallium atoms at sufficiently high vapour pressure may result in the observation of fluorescence at the thallium 377.6 and 535.0 nm lines.

It is extremely improbable that this type of fluorescence will ever be of analytical use except in non-flame cells. In flames the concentration of donor atoms may not be made sufficiently high. In flame systems the donor atoms are also much more likely to lose their energy by collisional deactivation with the molecules of the flame gases than by energy transfer of the type required for sensitised fluorescence.

(1) Intensity of Atomic Fluorescence

The theoretical relationships between atomic fluorescence emission intensity and atomic concentration are generally somewhat more complex than the corresponding relationships for AAS. This is a result of the simultaneous absorption and re-emission of energy involved in the fluorescence process. In order to avoid the unnecessary complication of the expressions for the observed fluorescence intensity, and to make the understanding of the subject rather easier, it will be divided into two parts. First, the ideal case is considered where the expression for the intensity of the fluorescence is derived assuming that all of the absorption of the source radiation occurs within the region of the fluorescence cell* used for measurements, and that none of the fluorescence intensity is lost by re-absorption before it can be measured. Secondly, factors are introduced which allow for experimental deviations from the ideal case. These factors are discussed from a qualitative viewpoint, with emphasis on experimental techniques by which they may be reduced to unity. They must allow for the possibilities that

(i) Some of the incident radiation is lost by absorption before it reaches the region of the flame where measurements are made.

(ii) Some of the fluorescence radiation produced is lost by re-absorption before it can leave this region of the flame.

(iii) There is a further region near the edge of the flame which,

---

* The fluorescence cell most widely used in AFS to date has been the premixed or unpremixed flame; the text refers to flames repeatedly, although most of the theory given is quite generally applicable to other cells.

although not irradiated by the source, also reduces the fluorescence intensity by absorption.

These possibilities can obviously be expressed mathematically in terms of absorption coefficients, but the resultant expressions for the fluorescence intensity reveal extremely complex relationships between intensity and atomic concentration. Such a detailed treatment serves no useful purpose for almost all analytical AFS applications and is therefore not given here.

(2) The Ideal Fluorescence Intensity

When no fluorescence emission is lost by reabsorption, the entire cell is within the solid angle over which excitation occurs, and the entire cell is within the solid angle of radiation received by the monochromator, the integrated atomic fluorescence intensity is given by

$$I_F = \frac{I_A \phi}{4\pi} \tag{46}$$

where I_A is the total absorbed intensity of the radiation which is responsible for the excitation of the fluorescence, and ϕ is a factor known as the fluorescence yield or efficiency, i.e. it is the ratio of the amount of energy emitted by the fluorescence process to the amount of energy causing it.

The value of ϕ depends on the quenching of the fluorescence by collisions with foreign atoms or molecules in the flame. As it depends mainly on the type of flame used, therefore, and has no effect on the actual shape of the calibration graphs obtained in analytical AFS, it will not be discussed in detail here. It is sufficient to state that in order to obtain high sensitivity via maximum signal strength in AFS the flame is frequently chosen to make ϕ as near unity as possible for the particular atomic species. This is accomplished by the use of flames containing low concentrations of atomic and molecular species which have high quenching cross-sections.

Thus in order to establish a relationship between I_F and the number of fluorescing atoms (i.e. the sample concentration) it is necessary only to consider the value of I_A. In general this is given by

$$I_A = \Omega_A \sum I_{O_i} A_{T_i} \tag{47}$$

124

where Ω_A is the solid angle over which excitation occurs, I_{0_i} represents the incident intensity for each of the absorption lines contributing to the excitation, and A_{T_i} represents the total absorption factor for each of these lines. Equation (47) is most easily evaluated when there is only one absorption line to consider. This applies in pure resonance fluorescence and also in direct line and stepwise line fluorescence for many of the elements whose spectra are simple. This case therefore applies to many analytically useful fluorescence lines and is easily extended to more complex line spectra (as replacement of the summation in eqn. (47) by the summation for several lines does not affect the shape of calibration graphs of fluorescence vs. concentration).

Equation (47) may be re-written

$$I_A = \Omega_A I_0 A_T \tag{48}$$

where I_0 and A_T now apply to the single absorption line considered. For a continuum source

$$I_o = I_c \tag{49}$$

where I_c is the power emitted by the source per unit frequency interval (per cm^2/steradian).

For a sharp line source in order to obtain the intensity per unit frequency interval the integrated line intensity must be divided by the line width. It was stated in Section 2(C) (7) that A_T could be evaluated conveniently in AAS in this case after converting the line width to a triangular function. The same operation applied in AFS gives

$$I_0 = I_L \frac{2\sqrt{\ln 2}}{\sqrt{\pi}\,\Delta\nu_s} \tag{50}$$

where I_L is the integrated intensity for the source line in question.

The value of I_F may now be evaluated using eqns. (46) and (48)—(50) and the expressions obtained for A_T in Section 2(C) for certain limiting cases.

(3) Ideal Fluorescence Intensity when the Absorption Line Half-Width is Narrow compared to the Source Line Half-Width

This situation applies when a continuum source of excitation is employed.

Substituting eqns. (48) and (49) into eqn. (46)

$$I_F = \frac{\phi \Omega_A I_c A_T}{4\pi} \tag{51}$$

(a) For Small Optical Density

Substituting eqn. (34) for A_T into eqn. (51)

$$I_F = \frac{\phi \Omega_A}{4\pi} I_c \frac{\pi e^2}{mc} f_{ji} N_j L \tag{52}$$

i.e. I_F increases linearly with N_j and hence with the concentration of the sample.

(b) For Large Optical Density

Substituting for A_T in eqn. (51) from eqn. (36)

$$I_F = \frac{\phi \Omega_A}{4\pi} I_c \left[\frac{2\pi e^2 \Delta\nu_D a f_{ji} N_j L}{mc\sqrt{\ln 2}} \right]^{1/2} \tag{53}$$

i.e. the fluorescence intensity, I_F, increases linearly with $\sqrt{N_j}$.

(4) Ideal Fluorescence Intensity when the Absorption Line Half-Width is Wide compared to the Source Line Half-Width

This situation applies when sharp line sources such as hollow cathode lamps and electrodeless discharge tubes are employed as excitation sources.

From eqns. (48), (50) and (46)

$$I_F = \frac{\phi \Omega_A 2\sqrt{\ln 2} A_T I_L}{4\pi\sqrt{\pi} \Delta\nu_s} = \frac{\phi \Omega_A I_L \sqrt{\ln 2} A_T}{2\pi^{3/2} \Delta\nu_s} \tag{54}$$

(a) Small Optical Density

Substituting for A_T in eqn. (54) from eqn. (40)

$$I_F = \frac{\phi\sqrt{\ln 2}\,\Omega_A I_L}{2\pi^{3/2}\,\Delta\nu_s}\, b\, \frac{\Delta\nu_s}{\Delta\nu_D}\, \frac{\pi e^2}{mc}\, f_{ji} N_j L$$

$$= \frac{\phi\Omega_A I_L \sqrt{\ln 2}}{2\sqrt{\pi}}\, \frac{be^2 f_{ji} N_j L}{\Delta\nu_D mc} \tag{55}$$

i.e. I_F increases linearly with N_j.

(b) Large Optical Density

Substituting for A_T in eqn. (54) from eqn. (42)

$$I_F = \frac{\phi\Omega_A I_L \sqrt{\ln 2}}{\Delta\nu_s\, 2\pi^{3/2}}\, \frac{\Delta\nu_s \sqrt{\pi}}{2\sqrt{\ln 2}}$$

$$= \frac{\phi\Omega_A I_L}{4\pi} \tag{56}$$

i.e. at sufficiently high sample concentrations the fluorescence intensity, I_F, is independent of increases in N_j. A calibration graph of I_F vs. sample concentration will take the form of a line parallel to the concentration axis at these high optical densities.

As stated earlier (Section 2(C)(8)) in AAS with sharp line sources the linear relationship between A_T and concentration only holds at the very low end of the concentration range on a calibration graph, making it necessary to introduce the absorbance, A. The introduction of a corresponding extra quantity in AFS is quite unnecessary. For those elements determined by AFS the sensitivity is usually very much greater than the corresponding AAS sensitivity. Also the absorption path length in the flame is usually smaller. Thus the *high* concentration end of a linear calibration range in AFS may correspond to an absorption (i.e. $A_T \times 100$) of only 2—3%. Such an absorption value would lie at the *lower* concentration end of an AAS calibration range where the absorption is indeed related linearly to concentration. It is shown in Section 2(D)(5)), concerning practical factors which affect the ideal fluorescence intensity, that these factors may curtail the range of linearity of I_F with N_j before the absorption becomes too large for eqn. (40) to hold true.

Finally it should be repeated that even when absorption of radiation occurs at more than one wavelength, the relationships obtained above between I_F and N_j still hold. For example, when I_F is proportional to $\sqrt{N_j}$ (continuum source, large optical density), this still holds when summation over all absorption lines is performed. Essentially, this summation may be seen as an evaluation of I_F for a given fluorescence line individually for each absorption line contributing to its excitation, and the summation of the I_F values to obtain a resultant. A similar procedure may also be applied when fluorescence emission is received simultaneously from more than one line (for example from a doublet when a wide spectral band-pass is employed).

(5) The Actual Fluorescence Intensity

The ideal case outlined in Sections 2(D)(2), (3) and (4) is seldom achieved in practice. The simplifying assumptions made for the ideal case may not be justified. From the nature of these assumptions the actual fluorescence intensity obtained will obviously depend on the geometry of the fluorescence cell, as this affects the extent of the overlap volume of the excitation and analysing solid angles. In Fig. 11 the cell has been drawn as a cylinder, as the most commonly used cell in AFS has been an approximately cylindrical flame.

The departures from the ideal situation described in Section

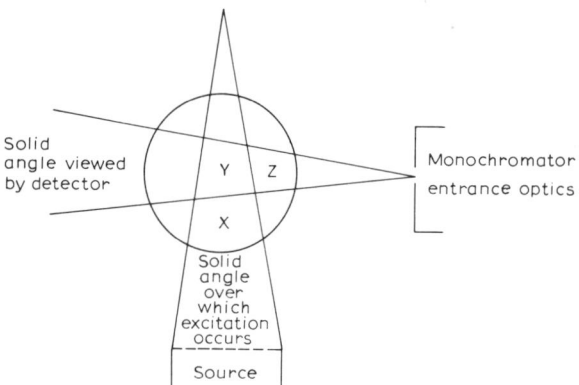

Fig. 11. Schematic diagram of AFS cell geometry.

2(D)(2) may be represented by three factors, each of which is defined as unity in the ideal case.

F_X represents the absorption of the incident radiation by the atoms of the sample element in the region X in Fig. 11, i.e. in the volume irradiated but not viewed.

F_Y represents the re-absorption of the fluorescence in the region Y where the fluorescence is both excited and received by the detector.

F_Z represents the re-absorption of fluorescence in the region Z, where it has left the zone of excitation.

As stated earlier, no attempt is made here to derive expressions for F_X, F_Y and F_Z and incorporate them into the expressions for I_F. Several simple cases will be discussed in a qualitative fashion.

(i) When the flame is entirely within the solid angles of excitation and viewing, F_X and F_Z will both be equal to unity. This was one of the assumptions made to evaluate the ideal fluorescence intensity. The situation is conveniently achieved in practice with a small circular flame.

(ii) For resonance fluorescence, F_Y will depend directly on the atom concentration so that except at low concentrations it will be considerably less than unity. As a result, the analytical curves expected from eqns. (52), (53), (55) and (56) are further bent towards the concentration axis. In practice, it is found that at high concentrations the fluorescence excited by a continuum source remains constant with increasing concentration whereas that excited by a sharp line source actually decreases.

(iii) For direct line fluorescence (and for other types of non-resonance fluorescence where the fluorescence line does not involve a direct transition to the ground state) re-absorption will be negligible in most flames; i.e. F_Y will be unity and F_z will be unity regardless of condition (i).

(iv) Thus the "ideal" expressions for fluorescence intensity derived here apply generally only at very low concentrations, when I_F is directly proportional to N_j under all conditions, and to direct line fluorescence over a much wider concentration range provided the experimental arrangements are such that no absorption of incident radiation occurs before the region in which the fluorescent signal is detected.

(E) LAMBERT—BEER LAW

The fundamental and well-known equation of molecular absorption spectrophotometry in solution may be written

$$\log_{10}(I/I_0) = -\epsilon LC \qquad \text{or} \qquad I = I_0\, 10^{-\epsilon L C} \tag{57}$$

where I_0, I and L are defined as in atomic absorption spectroscopy, ϵ is a constant of proportionality (termed the molar absorptivity) and C is the concentration of the absorbing species in the solution.

The equation is known as the Lambert—Beer Law. In recent years it has been widely quoted not only as the fundamental law of solution spectrophotometry but also in atomic absorption spectroscopy.

It has, however, been possible, as shown in the preceding sections, to derive all of the equations required to relate atomic concentration to atomic absorption without any reference to the Lambert—Beer Law. The justification of its use in discussion of AAS apparently lies in the fact that it allows the absorbance, A, as defined by eqn. (27) to be expressed neatly in the form

$$A = \epsilon CL \tag{58}$$

Although this equation is ideally applicable to molecular solution spectrophotometry, it provides a deceptively simple, and possibly misleading, explanation of the situation when applied to AAS. This stems partly from the fact that in AAS C no longer represents both the concentration of the sample solution and the concentration of the absorbing species in the absorption cell, it only represents the atomic concentration of the absorption cell. Even more important is the fact that the constant ϵ may take many different forms in AAS depending on the conditions in the source and the absorption cell. Thus, it has already been shown that the AAS absorbance is linearly related to the atomic concentration only under carefully controlled conditions involving a number of parameters.

In view of this, it is much more satisfactory to discuss AAS relationships in terms of the more complex equations derived earlier rather than by invoking the Lambert—Beer Law. The Law is presented here merely because its incorrect interpretation can lead to a number of wrong conclusions. Perhaps the most common of these is the misconception that the absorbance is in all cases directly related

130

to the length of a flame used as the absorption cell, just as its counterpart in solution spectrophotometry depends on the length of the sample cell. That this is not, in general, true is shown in Section 2(F) below.

(F) EFFECT OF FLAME LENGTH ON ABSORBANCE

When the Lambert—Beer Law is applied to a change of path length in solution spectrophotometry, it is implicitly assumed that the solution concentration remains constant. This is perfectly logical, as the larger cell needed for a longer path-length will be filled by the addition of more sample solution. Thus doubling the path length doubles the absorbance.

In AAS, the most widely used absorption cell has been a flame burning at a long slot burner. It should be evident that the maintenance of a constant concentration of atoms in the flame per unit volume on doubling the flame length is a much more difficult matter. Unless the rate at which sample is supplied to the flame is simultaneously doubled, increasing the flame volume merely lowers the concentration per unit volume by dilution with the additional gases employed. It is difficult to increase the rate of supply of sample to the flame as required without at the same time altering the efficiency of atomisation and nebulisation.

This is allowed for in all of the expressions for A_T and A given in Section 2(C) A_T and A are always expressed as factors of N_j; this was defined (Section 2(A)) as the number of atoms *per unit volume*. Thus if the length of the flame is changed, both N_j and L are varied and the terms $N_j L$ or $(N_j L)^{1/2}$ which appear in the expressions must be used to calculate the change in absorbance.

Similarly, if eqn. (58) is rewritten as

$$A = \epsilon LN \tag{59}$$

where N is the number of atoms per unit volume, no confusion should arise when the Lambert—Beer Law is used to calculate the effect of change in path length. It must still be realised, however, that the relationship between N and the sample solution concentration is complex, and that for a given value of C the value of N actually obtained can vary widely, depending on the experimental conditions. This is discussed further in Section 2(G).

(G) RELATIONSHIP BETWEEN SAMPLE CONCENTRATION, C, AND THE CONCENTRATION, N_j, OF ATOMS FOR ABSORPTION

In general the following statements hold true.

(i) Not all of the free atoms of the sample element present in the absorption cell are capable of absorbing the incident radiation at the wavelength of measurement.

(ii) Not all of the sample element is present in the absorption cell as free atoms.

These facts have a pronounced effect on the relationship between C and N_j and indicate that each particular combination of an element and an absorption cell must be treated individually.

(i) The proportion of the free atoms capable of absorbing radiation depends on the electronic spectrum of the atoms of the sample element and on the absorption line chosen. It is denoted by N_j/N, where N_j is the number of absorbing atoms per unit volume and N is the total number of free atoms per unit volume. The value of N_j/N depends on the number of low lying energy levels, their relative energies (particularly that of E_j, the energy of the lower level of the absorption line to be employed) and the temperature. It may be shown (see ref. 46, p. 499) that for any energy level E_x,

$$\frac{N_x}{N} = \frac{g_x \exp(-E_x/kT)}{\sum_n g \exp(-E_n/kT)} \tag{60}$$

where N_x is the number of atoms in level E_x, N represents the total number of atoms in *all* of the energy levels of the atom, and there are n levels each with its own energy value of E_n; g is the statistical weight of the level concerned, k the Boltzmann Constant and T the absolute temperature in K.

It is evident from eqn. (60) that at normal flame temperatures (not greater than about $3000°$K) the value of N_x is very small if E_x is more than $1–2$ eV. For example, if T is 3000 K, which is close to the highest flame temperature usually used in AAS and $E_x = 2$ eV, we have $\exp(-E_x/kT)$, $\approx 4 \times 10^{-4}$, so that the value of N_x/N is extremely small. Thus in analytical AAS in flames the population of the excited state atoms is negligible compared with the total number of free atoms present.

In view of the above conclusion, it has occasionally been stated that, assuming a constant population of free atoms, the number of absorbing atoms will always be independent of the temperature. In fact this is only true for atoms which have simple spectra where the only low lying level is the ground state. Thus mercury, for example, has a $6s^2\ {}^1S_0$ ground state and the next level is $6p\ {}^3P_1^0$ at 4.88 eV. When there are a number of levels which lie within about 1 eV above the ground state, however, their relative populations will vary with temperature to produce a similar variation in the measured absorption at a particular line. Although the variations observed are usually quite small, sometimes the fortuitous combination of a large statistical weight and low lying energy level can give rise to an appreciable change with temperature for certain elements. An example of this effect occurs for tin, the data for which are shown in Table 1. Tin has three energy levels below 1 eV, which comprise the three levels of the $5p^2\ {}^3P_{0,1,2}$ triplet state. It is evident from Table 1 that the population of the true ground state level decreases rapidly as the temperature is increased, so that at 2500 K the next level at 0.210 eV actually has a higher population. As a result of this the relative sensitivities of different tin lines when the element is determined by AAS depend quite significantly on the flame chosen as the free atom source.

TABLE I

Relative populations of ^3P levels of tin at different temperatures

J	E_j (eV)	g_j	N_j/N				
			1000 K	1500 K	2000 K	2500 K	3000 K
0	0.000	1	0.772	0.563	0.433	0.354	0.302
1	0.210	3	0.200	0.332	0.384	0.400	0.407
2	0.425	5	0.028	0.105	0.184	0.246	0.291

Two other points are of importance. First, if the ground state and the next level above it are very close, the latter may have the higher population even at quite low temperatures. This is shown in Table 2 for the example of aluminium, in which the two levels of the ground state doublet ($3p\ {}^2P_{\frac{1}{2},1\frac{1}{2}}$) are separated by only 0.0139 eV. Secondly, even if the ground state of a particular atom is the most highly

TABLE 2

Relative populations of ^2P levels of aluminium at different temperatures

J	E_j	g_j	N_j/N				
	(eV)		1000 K	1500 K	2000 K	2500 K	3000 K
1/2	0.0000	2	0.370	0.358	0.351	0.348	0.345
3/2	0.0139	4	0.630	0.642	0.648	0.652	0.655

populated, it may not give rise to the most sensitive absorption lines. This arises because a line whose lower energy level lies somewhat above the ground state may have a much greater transition probability than any of the ground state lines, thus giving it the highest absorption coefficient. An example of this effect is found in the zirconium 360.12 nm line (transition between 0.154 and 3.600 eV) which gives greater absorption in a nitrous oxide-acetylene flame than the ground state line which has the highest gA value i.e. 298.54 nm (transition between 0 and 4.155 eV).

(ii) The fraction of the sample element which is actually present in the cell as free neutral atoms can vary greatly, depending on the element and environment in the cell. The major factors affecting the fraction present are the degree of ionisation and the volatility and stability of the compounds formed by the element. These factors are discussed in Section 4.

3. Instrumentation

(A) INSTRUMENTAL SYSTEMS

The general requirements and arrangement of the instrumentation used in AAS and AFS will be summarised briefly here; the individual components of the systems employed are described in more detail in later sections.

The early work in AAS was conducted with a variety of different instrumental arrangements frequently based on existing flame spectrophotometers or solution spectrophotometers. A standard instrumental assembly has now found favour, however, and is used in most AAS instruments. This is illustrated diagrammatically in Fig. 12.

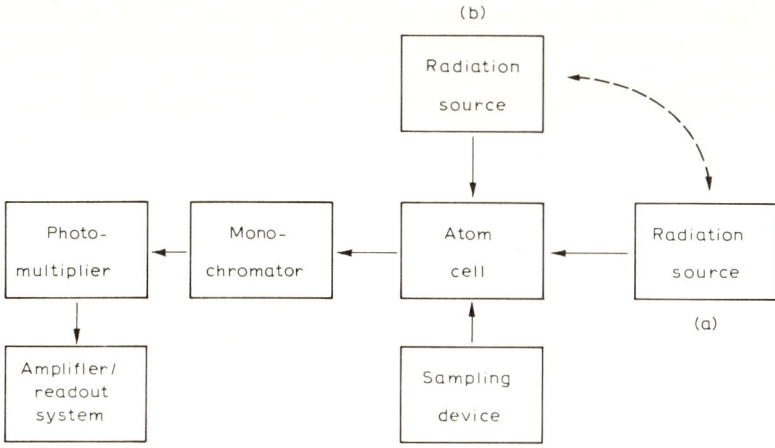

Fig. 12. Instrumental arrangement for atomic absorption spectroscopy (radiation source position (a)) or atomic fluorescence spectroscopy (radiation source position (b)).

With the majority of commercial systems the monochromator is a grating device of moderate resolution, the atom cell takes the form of a flame at a long slot burner and the radiation source is a hollow cathode lamp. Most of the published work in AFS has employed similar instrumentation with the radiation source placed at 90° to the atom-cell/monochromator axis; the atom cell most frequently employed has been a cylindrical, low- or moderate-temperature flame. In view of the similar instrumental arrangement, several commercial AAS instruments are supplied with facilities for changing the position of the radiation source and an accessory burner suitable for AFS work.

(1) Special requirements for AAS

The development of AAS instrumentation has been governed by three major requirements: large signals (i.e. high sensitivity); low signal noise levels (i.e. low detection limits and high precision); and simple and rapid processing of analytical samples. The first of these requirements has been met by extensive development of the atom cells and sources employed. The former are designed to achieve the most efficient atomisation of the sample and to place the maximum number of atoms within the optical path of the instrument. In view of the design of the monochromators and sources used, this has most frequently resulted in a long, narrow atom cell placed end-on to the beam of radiation. Sources have been developed to produce small source line widths, as these result in the maximum absorption signal and also assist the achievement of linear calibration graphs. For linear calibration graphs to be obtained it is also important to ensure that no emission signal received from the atom cell is amplified at the detector; this has resulted in the widespread use of modulated systems.

The maximum absorption signal which may be obtained by these aspects of the instrumental design is limited by theoretical factors, but the attainable detection limits are governed mainly by the signal noise levels. Thus improvements in the intensity and stability of the sharp line sources have also been of importance. These have been accompanied by reductions in noise levels and instability at both the atom cell and photomultiplier-amplifier device. Double-beam operation has also been important in increasing the instrumental stability. These improvements have made possible much greater amplification of signals, with consequent determination of much lower absorption signals than was possible with early instrumentation. Higher precision is, of course, also obtained for samples of higher concentration which produce large absorbance signals.

(2) Special Requirements for AFS

At the time of writing, the instrumentation in use for AFS differs in many respects from laboratory to laboratory. Many workers have employed modified atomic absorption spectrophotometers or flame photometers for AFS, while several attachments for AAS instruments are available commercially which permit their use for AFS.

Much development work has been concerned with the production of suitable AFS sources. In contrast to AAS, very sharp line sources are not essential, and in fact sources which produce lines of a similar width to the absorption lines in the atom cell should result in the most efficient transfer of energy and greatest analytical sensitivity. It is essential, however, that the source employed should emit lines at high intensity; this intensity should also be obtained without self-reversal of the emitted lines.

The atom cells required for AFS are also somewhat different to those used in AAS. The atoms within the cell must lie within the

136

solid angles subtended by both the radiation source and the detector; the long narrow cells used in AAS are therefore not normally satisfactory. The use of a long-path length cell of this type in AFS would result in reabsorption of part of the fluorescence signal before it emerged from the cell and would also require an extremely large solid angle to be subtended from the source. Quite apart from optical geometry requirements the cells used in AFS have two other requirements different to those of cells used in AAS. First, for maximum sensitivity the quenching of the fluorescence must be minimised. This is achieved by careful selection of the other species present. It is particularly important to avoid the presence of high concentrations of molecular species (such as nitrogen, carbon monoxide, carbon dioxide or oxygen) combined with high temperatures. Thus the best results are frequently obtained with cool cells containing mainly monatomic species (such as argon or helium). Secondly, it is important that scatter of the incident beam of radiation should be as small as possible; when resonance fluorescence is measured, scattered source radiation is recorded as a fluorescence signal at the detector. Unfortunately the most serious problems from scatter are often encountered from the cool cells required to produce low quenching of fluorescence, owing to the presence of unvaporised solvent droplets or solid particles.

Many of the instrumental requirements for AFS are, however, similar to those of AAS. Thus high stability and large amplification of the signal are important requirements. The monochromator used for AFS should give high light transmission. The use of a monochromator from an AAS instrument for AFS may not be satisfactory in this respect. Good resolution is of secondary importance as exclusion of non-resonance lines from the detector (as required in AAS) is not of importance.

(B) RADIATION SOURCES

The sources of radiation used in AAS and AFS may be classified according to the type of spectrum emitted and the manner in which they are prepared and used.

(1) Spectral Classification

The source may emit radiation continuously with respect to time or it may be pulsed so that the radiation is emitted as periodic bursts of energy. Sources which are only modulated (so as to provide an alternating signal for amplification by a tuned detector) may generally be referred to as continuous. Either of these types of source may emit a line spectrum or a continuum spectrum. The former is associated with quantised changes of electronic energy, while in a continuum spectrum the energy distribution is unquantised and distributed in an uninterrupted manner between all wavelengths within a given domain.

(2) Practical Classification

The most common types of source used in AAS and AFS are permanently sealed-off. These offer considerable advantages for ease of operation, good reproducibility and simplicity, especially when the source is at very low or high pressure. Alternatively, demountable sources are sometimes used. These usually have similar forms to the sealed-off types but allow alterations to the contents to be made at any time after their manufacture. This is most commonly achieved by simply fitting a replaceable electrode assembly and/or a tap assembly to permit the outgassing and evacuation of the system. Such demountable sources are most useful for research purposes, but are more difficult to handle on a routine practical basis as sources for AAS and AFS. A class of consumable sources must also be recognised. Sources such as flames, arcs, plasmas and flow-through microwave discharges fall into this category.

(C) CONTINUUM SOURCES

In theory, continuum sources appear to offer one outstanding advantage for use in AAS and AFS over line sources, i.e. one source would suffice for any element whose analysis may be required. In practice, there are some difficulties which have prevented the widespread use of continuum sources in this way. As described in Section 2(C)(6), the sensitivity obtained with a continuum source in AAS will only approach that attainable with a sharp-line source when a high resolution monochromator is employed. Even under these conditions calibration curves of percentage absorption vs. concentration show two distinct forms over the normal concentration ranges required. Apart from the expense of suitable monochromators, other problems are introduced by the low light transmittance associated with the use of a continuum source with a high resolution mono-chromator used at narrow slit width.

Although similar problems do not occur in AFS, where a narrow source line and monochromator band-pass are not necessary to obtain high sensitivity, it is difficult to obtain sufficient intensity with available continuum sources to achieve this sensitivity. With most elements the best absorption lines lie in the ultraviolet and when the continuum source radiates as a black body the output in this region of the spectrum is low unless a very high effective temperature is achieved. It is the output of the source integrated over the width of the absorption spectral line which is important in AFS, *not* the intensity integrated over the band-pass of the monochromator. Low-resolution monochromators of high light transmission characteristics may be used for AFS; when such instruments are used in the comparison of line and continuum source intensities the comparison is misleading if the above fact is not taken into consideration. The band-pass of the monochromator may well be one hundred times greater than the line width of a sharp-line source. When fluorescence measurements are made under conditions producing light scatter, the scatter is greatly enhanced relative to the fluorescence intensity when changing from a line to a continuum source as the scatter may be detected over the whole band-pass of the monochromator. Considerable care must thus be taken when continuum sources are employed in AFS or AAS.

In spite of the difficulties mentioned above, a quite considerable amount of research has been undertaken into the use of continuum sources for both techniques. The conventional tungsten filament lamp is generally insufficiently intense for use with either technique, although tungsten iodide lamps with quartz envelopes have been tested for AAS [50]. Similarly, several workers [51—53] have attempted to use hydrogen discharge lamps as sources in AAS; their intensity is somewhat too low, however, to permit the use of a sufficiently narrow band-pass at the monochromator. More satisfactory results have been obtained using the high pressure xenon arc, which has been applied to both AAS [54—56] and AFS [57,58]. The most interesting results have been obtained in the use of this source in the determination of those elements for which line sources are not completely satisfactory, such as the rare earths [59], and in its use for ancillary purposes such as correction for background absorption [60] or the preliminary investigation of absorption spectra [61]. At the present time, the sensitivities attained with the xenon arc source have not been as high as those routinely achieved with sharp-line sources in AAS or AFS. When sensitivity is not the most important factor, the xenon arc source might be employed to exploit its advantages of cost and time saving on lamp changing, warm-up, alignment and current optimisation compared with a hollow-cathode lamp. The status of xenon arc and other continuum sources for use in AAS and AFS may be improved by several recent developments which appear most promising. Thus Svoboda [62] has suggested the modulation of continuum sources with respect to wavelength by filtering the radiation using a device such as a Fabry—Perot interferometer. This in theory allows selection of the band-pass at any desired half-width with a tuning capability which would permit exact coincidence with any desired absorption line, just as is obtained with sharp-line sources. This device has not yet been applied routinely to AAS. A second promising development is the use of pulsed continuum sources; a much greater power input may be used for each pulse of radiation than would be possible with continuous operation. Klein [63], for example, has reported a xenon arc lamp which may be operated at 2.2 kW in the continuous mode but 10 kW when pulsed. The application of a pulsed xenon lamp to AAS has already been reported by Sheklein and Popov [64]. Even greater intensity should be attainable using true flash tubes of the Garton or Lyman types. Such sources have already been used for absorption studies in the vacuum ultraviolet (see for example, ref. 65), and their high intensity should prove useful for AFS.

(D) HOLLOW-CATHODE LAMPS

The hollow-cathode lamp is possibly the most important source of intense sharp-line spectra and has been widely used for this purpose since its invention in 1916. Owing to the great interest in AAS, probably more hollow-cathode lamps have been manufactured in the past five years than in the previous fifty. All commercial AAS instruments available at the time of writing are designed for use with these sources.

Many different types of HCL have been described in the literature. For fundamental applications, such as ultra-high resolution studies of hyperfine structure, the cathode has usually been cooled with water or liquid air to achieve narrow line emission. When line-width requirements have been less critical, or for studies of high energy lines, a hot hollow-cathode has generally proved more useful. In most cases the line-widths obtained with this arrangement are still sufficiently narrow for peak absorption measurements and all commercial lamps used for AAS are of this type. The considerable importance of these sources for AAS has stimulated development studies which have resulted in good combinations of intensity of output, line sharpness, stability and operating lifetime.

(1) Construction and Operation of Hollow-Cathode Lamps

A hollow-cathode lamp consists of a sealed envelope which contains an anode, a hollow cathode made of, or lined with, the element whose spectrum is required, and an inert gas at a pressure of ca. 1—3 Torr. Commercial lamps usually also contain a number of other components designed to improve the performance (see below). When a voltage of several hundred volts is applied between the electrodes a glow discharge is obtained and concentrates within the cathode, giving higher current densities than those normally obtained. Electrons leave the cathode due to this applied field and collide with atoms of the inert filler gas. The atoms are ionised by this collision process and the positive ions produced are accelerated by the field within the cathode. Bombardment of the cathode surface by these ions displaces (or 'sputters') atoms from the surface. The sputtered atoms accumulate inside the cathode and are excited to give their atomic line spectrum; excitation most probably proceeds by electron impact or by second order collisions with the excited atoms and ions of the inert gas.

A typical modern hollow-cathode lamp is shown in Fig. 13, and its dimensions are given on the diagram of Fig. 14. Modern hollow-cathode lamp sources are very compact; until quite recently most were of considerably greater volume. The larger volume was intended to increase the life of the lamp, as the three main factors limiting useful lifetime are: leakage of air into the tube; clean-up of the filler gas by absorption by the metal or the glass envelope; and deposition of the sputtered metal atoms onto the envelope. It has become possible to avoid the first of these problems by careful lamp construction and the use of graded glass—silica seals for the lamp window—envelope interface. A getter may also be added to the lamp during con-

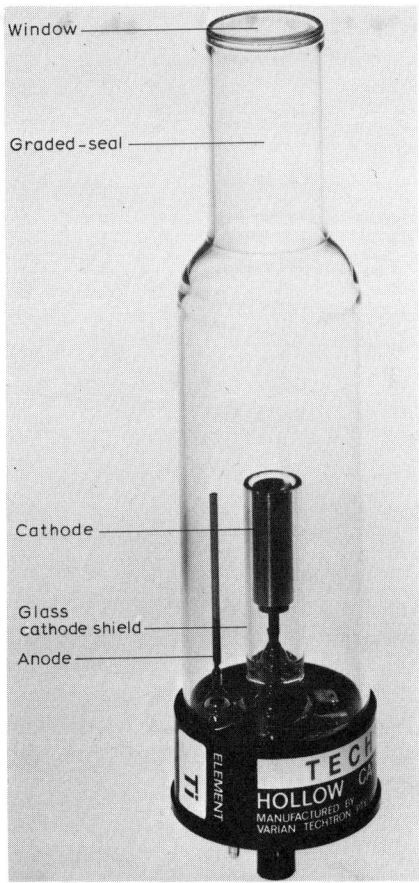

Window ———

Graded-seal ———

Cathode ———

Glass
cathode shield ———
Anode ———

Fig. 13. A modern design of hollow-cathode lamp. (Reproduced by courtesy of Varian Associates.)

struction. This is a chemical, such as activated uranium and zirconium, added to the lamp to remove any impurities present in the filler gas and envelope. Similarly, loss of filler gas is not a problem with modern lamp manufacturing techniques. In order to avoid the deposition of material on the window of the envelope it is necessary to place the cathode a considerable distance from the window. The envelope may be fabricated from Pyrex glass and a similar glass of good optical quality (such as Corning 7740) may be employed for the window when the resonance lines required from the lamp lie in

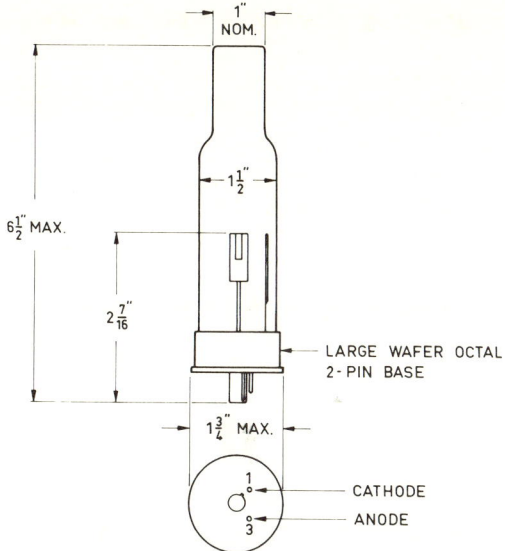

Fig. 14. Dimensions of the hollow-cathode lamp shown in Fig. 13. (Reproduced by courtesy of Varian Associates.)

the visible or near ultraviolet region. With the majority of elements, however, the most useful absorption lines lie in the ultraviolet region and a silica window must be used.

The filler gas normally employed is argon or neon. Helium has proved less successful, as it is too light to produce efficient sputtering for many elements, and its high excitation potential gives rise to

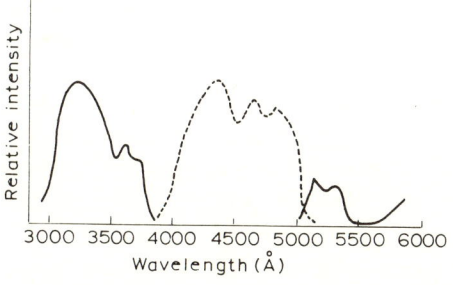

Fig. 15. Major regions of possible spectral interference from filler gas lines. ——, Neon; -----, argon. (Reproduced by courtesy of Westinghouse Electric Corp., Electronic Tube Division.)

References pp. 253—262

complex spectra. Until recently argon was used in most hollow-cathode lamp sources as it was considered to clean-up more slowly than neon. It shows a complex emission spectrum, however, and causes spectral interference problems in AAS at the resonance lines of a number of elements. As shown in Fig. 15, neon exhibits its main emission lines in a different region of the spectrum and is now used for many hollow-cathode lamps. Modern manufacturing techniques have minimised the "clean-up" problem. The use of neon also gives increased intensity for some elements, apparently without adverse effect on the line width. An excellent example of the improvement produced with neon filler gas compared to argon is in the calcium hollow-cathode lamp, as illustrated in Fig. 16. The selection of the optimum pressure of the filler gas in the source is extremely important in order to obtain high intensity, low background emission, sharp-line emission and long operating lifetime.

Apart from improvements in the quality of lamp construction and the optimisation of the filler gas characteristics, considerable development work has been undertaken in the design of the electrodes and their shields. Ideally, the cathode should be shaped such that most of the energy from the discharge is concentrated within the cathode and so that the cloud of metal vapour is maintained within a restricted area at a uniform level of excitation. When a large cloud of weakly-excited atoms is formed at the front of the cathode or just outside it, self-absorption of the resonance lines becomes a serious problem, and the lamp life may be reduced. The exact shape of the hollow cathode and its success in reducing these problems depends on the element concerned. For elements such as tin, indium, and gallium, for example, which have low melting points but are relatively involatile, unshielded lamps may be used provided that the molten element obtained during operation is retained within a suitable cup-shaped cathode. Many other elements are used with a simple cathode shield such as that shown in Fig. 13. More elaborate ceramic shields are employed in some commercially available hollow-cathode lamp sources; a typical design is shown in Fig. 17. The restrictive shield is mostly employed for hollow-cathode lamp sources of those elements which are volatile or molten, such as bismuth or lead, whereas the open shield is used for less volatile elements such as titanium or germanium. Manning and Volmer [66] report that both types of shield help to restrict the metal vapour cloud to a region in front of the

144

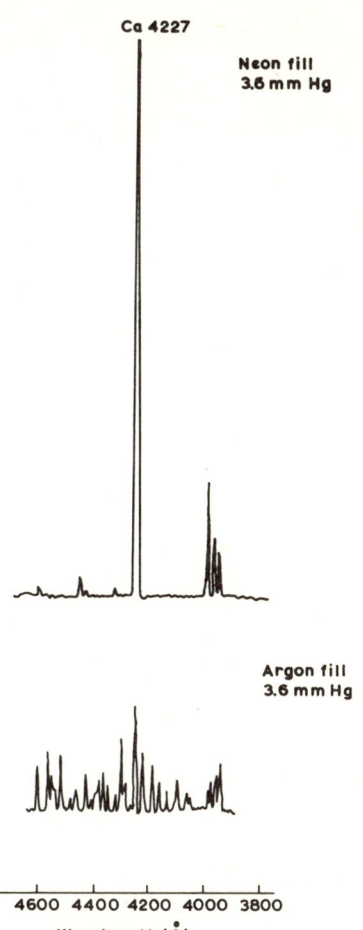

Fig. 16. Effect of type of filler gas on the spectrum of calcium hollow-cathode lamps. Cathode Ca—Mg—Al. Current 20 mA in both lamps. (Reproduced by courtesy of Westinghouse Electric Corp., Electronic Tube Division.)

cathode, concentrate the discharge in the cathode cavity and reduce filler gas clean-up.

The material used for the cathode also has an important effect on the working of the source. With many of the common metals, the cathode may be made entirely of the pure metal itself; when the metal is expensive or simply unsuitable for cathode construction

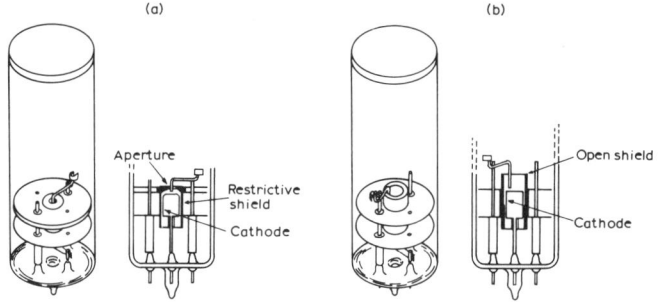

Fig. 17. Shielded hollow-cathode lamps. (a) Restrictive shield; (b) open shield. (Reproduced by courtesy of Perkin—Elmer Corp.)

copper, brass, aluminium or iron cathodes lined with a foil or electrolytically deposited layer of the required element may be employed. In some cases it is possible to form the cathode of a suitable alloy of the required element. Recent improvements to a number of "difficult" hollow-cathode lamp sources have involved careful selection in this way of the matrix material used for the cathode, so that the rate of sputtering of the emitting element may be controlled. This results in better source performance and a longer useful life.

(2) Special Types of Hollow-Cathode Lamp

Demountable hollow-cathode lamp sources are not usually supplied by manufacturers of commercial AAS instrumentation but have been described by a number of workers [67—69]. The performance claimed for these sources compares quite favourably with that of sealed hollow-cathode lamp sources. Their use provides for economical operation when a large number of different elements is to be determined, but their operation is less convenient for routine work.

Multi-element hollow-cathode lamp sources have long been of some interest to analysts engaged in routine AAS work, as they offer the possibility of combining the advantages of the hollow-cathode lamp as a source with a reduction of source cost per element and more convenient operation for routine applications. The source may contain several separate cathodes within one envelope [70,71] or a single cathode formed from different elements. This latter type has been most widely used. The cathode of the multi-element source may be made of a suitable alloy, but the performance often deterio-

146

rates quickly owing to preferential sputtering of one element of the alloy (e.g. zinc when brass is used to obtain a copper-zinc lamp [72]). A second type, described by Massman [73] and Butler and Strasheim [74], consists of a cathode in which rings of the various metals are pressed together. The metal rings may be arranged in order of their volatility, and the operating current is then chosen to give a stable output for all of the elements. A six-element hollow-cathode lamp has been obtained by using a sintered cathode of mixed metal oxide powders [75]. The same workers have reported a number of multi-element hollow-cathode lamp sources containing up to seven elements and using intermetallic compounds to overcome the problem of selective sputtering [76—78].

The emission intensities of the resonance lines from the elements in multi-element hollow-cathode lamp sources are usually lower than the corresponding lines for the same element obtained from a single cathode lamp. This is to be expected when the discharge energy must be shared by all of the elements of the cathode. The choice of the six or seven elements to be used in a single lamp is difficult. It is important that the materials used in the cathode construction should result in a similar useful lifetime for each element, but it is also necessary to select the elements in order to minimise any spectral interferences, and as far as possible to choose elements whose analysis is likely to be required in a particular type of sample or laboratory.

High intensity hollow-cathode lamps have been produced for a number of elements by using a special design with an auxiliary discharge. In the conventional lamps described above, the discharge within the cathode is used both to produce a population of atoms by sputtering and to excite them to emit radiation. Thus, when the discharge current is increased to give stronger excitation and a more intense resonance line, there is also an increase in the vapour density within the cathode. This may result in self-absorption becoming serious at high cathode currents. Simultaneous changes in the intensity of ion line emission for the element concerned also occur. Sullivan and Walsh [79] used the conventional discharge in the cathode to produce the atoms at the optimum vapour pressure but arranged to excite these atoms independently in front of the cathode with a second, low-voltage discharge at relatively high current between two auxiliary electrodes. A considerable increase in resonance line radiation intensity is achieved in this way without increased self-absorption. The intensity of ion line emission is also suppressed due to the

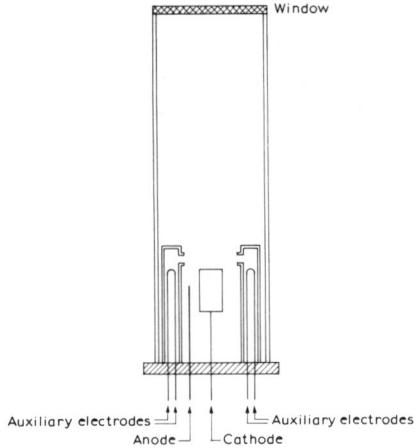

Fig. 18. Electrode assembly of a high-intensity hollow-cathode lamp.

abundance of electrons in the secondary discharge. Improved sensitivity and range of linearity in AAS has been observed with these sources for a range of elements. The design of the Sullivan and Walsh type of high intensity hollow-cathode lamp is illustrated in Fig. 18. The auxiliary electrodes are shielded to insulate them from the primary anode and cathode. The shields are shaped so that the low-voltage discharge is concentrated across the mouth of the cathode. As a result of the small voltage drop the auxiliary discharge uses a high current (200—500 mA).

An improved type of high intensity lamp has recently been described by van Gelder [80] who suggests that its performance is considerably better than the Sullivan and Walsh type. The design of this lamp, which is not actually of the hollow-cathode type, is shown in Fig. 19. A discharge in the filler gas is maintained between the cathode and a perforated anode through which radiation may be viewed. The discharge is maintained as a narrow column by a glass capillary and for part of its length is surrounded by a cylindrical probe. This is made of the required metal, or lined with it, as in a hollow cathode. The probe is maintained at a negative potential so that sputtering occurs and a cloud of metal atoms is obtained. The design of the lamp is such that the vapour density of metal atoms drops very rapidly outside the probe space and all the metal atoms present are electronically excited. This minimises self-absorption. A

148

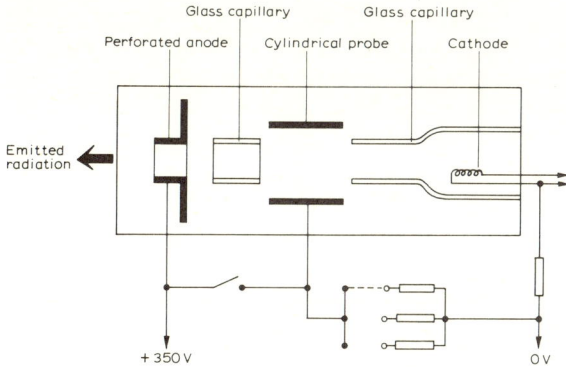

Fig. 19. Schematic construction of van Gelder high-intensity lamp. (Reproduced by courtesy of Appl. Spectrosc.)

further improvement results from the fact that the greatest excitation occurs at the centre of the discharge where the vapour density is lowest (due to the effect of the capillary). Hence it is possible to obtain an intense line source which is practically unaffected by self-absorption; the source line width is then much less dependent on the lamp current.

Pulsed operation of hollow-cathode sources for the isolation of resonance lines is described in Section 3(D)(5). Several workers have described the use of this technique, however, to increase the intensity of hollow-cathode lamp sources. Dawson and Ellis [81] found that pulsed excitation increases the line intensities obtainable from a conventional hollow-cathode lamp source by several orders of magnitude; their absorption measurements suggested that no significant increase in line width or self-reversal was obtained. Similarly, Katskuv et al. [82] have shown that at the same average current pulsed operation gives 20—130 times greater excitation than d.c. operation and only a small increase in line width. They also point out that pulsed operation removes variations of intensity due to overheating of the lamp and results in longer lamp lifetimes. Mitchell and Johannson [83] have recently reported the use of pulsed hollow-cathode lamp sources for atomic fluorescence spectroscopy. They found that the much greater current which could be applied to the lamps during a pulse allowed fluorescence measurements of very short duration to be made without any loss of fluorescence intensity compared with measurements made in the normal way. For example, a pulse current of 190 mA corresponded to a mean current of only 8.5 mA, so that overheating of the lamp was not a problem. In fact, these authors suggested that even greater pulse currents could be used in AFS where narrow source line width is less important than in AAS.

(3) Operation of Hollow-Cathode Sources

One of the principal reasons for the popularity of hollow-cathode lamp sources is the ease with which a stable and reproducible emission intensity may be produced. A voltage-stabilised power supply is usually employed and many units are current-stabilised to provide stable output. The sources are most frequently operated at between 5 and 30 mA and most power supply units are capable of supplying a continuously variable stabilised current of up to 50 mA (d.c.) at between 300 and 800 V. In some commercial instruments this basic facility is supplemented by a number of others. In some cases it is possible to select d.c. operation of the lamp (i.e. fully rectified output), a.c. operation (i.e. modulation at mains frequency), or a modulated output whose frequency is governed by the amplifier of the instrument (see Section 3(E)). A non-stabilised or partly stabilised auxiliary output is also often supplied so that several sources may be maintained on 'warm-up' and are ready for use when required. The power requirements of the auxiliary discharge in high intensity sources are similar, but a much higher current (up to 500 mA) at lower voltage is required. A 'boost' voltage is also supplied to initiate the discharge. As this type of high intensity source is often only employed for a limited number of elements, the power requirements are usually met by a separate accessory unit rather than by the principal lamp power supply unit used for the conventional lamps.

(E) ELECTRODELESS DISCHARGE SOURCES

The high-frequency electrodeless discharge tube has been used for many years as a source of sharp-line spectra. These sources have several useful properties for high resolution studies and other fundamental work, namely ease of construction, absence of electrode contamination, maintenance of the discharge at extremely low vapour pressures, and the requirement of only a small amount of material whose spectrum is required. In comparison with hollow-cathode sources, however, these sources have found relatively little applica-

150

tion in AAS. This is undoubtedly due at least in part to the choice of the hollow-cathode lamp as the most suitable source by Walsh and his co-workers in their original publications on analytical AAS, but at the time of writing their original reasons for this decision still hold; it is more difficult to obtain a stable and reproducible output of radiation of the required frequency from an electrodeless discharge tube source than from a hollow-cathode lamp. The construction of suitable power supplies and tuning cavities is also more demanding and relatively expensive.

Electrodeless discharges may be excited at both radio frequencies (100 kHz to 100 MHz) and microwave frequencies (> 100 MHz). The former have been most widely used, usually at frequencies above 30 MHz, especially for fundamental spectroscopic work, and the mode of construction and operation of tubes and power supplies is well-documented (see, for example, references 84—88). Radio frequency tubes also have the advantage of the skin-effect if the discharge is carefully controlled. This concentrates the discharge near the walls of the tube and helps to reduce self-absorption and self-reversal of the resonance lines. Current stabilised power supplies for this type of tube may also be produced with no more difficulty than for hollow-cathode lamps [85,89,90]. For microwave excited discharges the mode of operation of the discharge and the conditions under which a true skin effect may be obtained appear to be less clearly understood. Development of power supplies and tuning cavities to give an output stability comparable to that of hollow-cathode lamp supplies is currently receiving much attention. In spite of these difficulties microwave frequencies have been used by both groups of workers who have been concerned with the development of electrodeless discharge tube sources for AAS and AFS; Winefordner and co-workers [91,92,14] in the U.S.A. and West and co-workers [93—96] in the U.K. It has been reported that microwave frequencies are more efficient than radio frequencies with respect to the power transferred to the discharge [97,98] and that the lifetime of the sources increases with discharge frequency [98,99]. Convenient medical diathermy power supply units operating at 2450 MHz are the most commonly employed sources of microwave power. These may be used in conjunction with either tuned microwave cavities of the type shown in Fig. 20 or with antennae which focus the microwaves on to the discharge tube.

The construction of electrodeless discharge tube sources for use

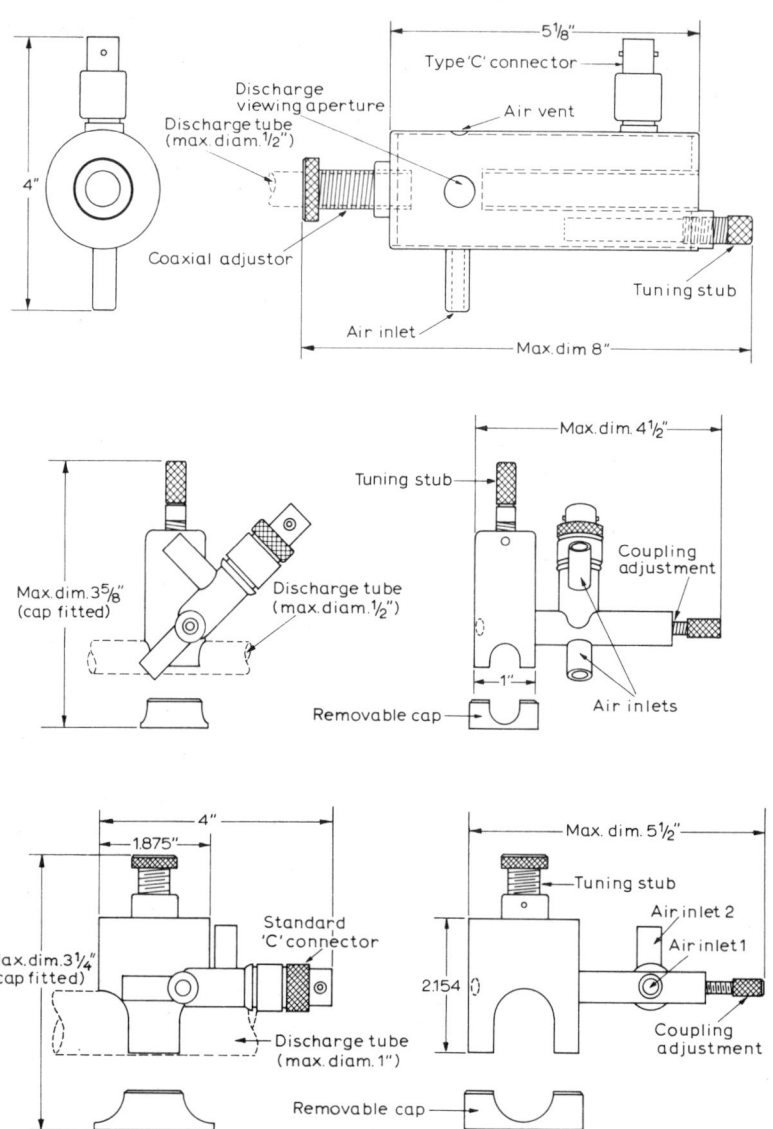

Fig. 20. Microwave discharge cavities for use with electrodeless discharge tubes. (a) No. 210L; (b) No. 214L; (c) No. 216L. (Reproduced by courtesy of Electromedical Supplies Ltd., Wantage, England.)

152

with microwave cavities of the type shown in Fig. 20 has been described in detail [94]. Generally the tubes are constructed from transparent silica tube 8 to 10 mm in diameter which is filled with the appropriate element or compound (< 10 mg) and sealed at an inert gas filler pressure of less than 10 Torr to form a discharge tube 1—10 cm in length. The successful operation of such a tube as a source depends on obtaining a suitable vapour pressure of the element at the operating temperature of the discharge (usually 200—400°C). Discharge tubes for some elements have been prepared using the pure metal (e.g. mercury and zinc) or gas (e.g. hydrogen or neon) but in many cases the iodide or chloride of the element is most satisfactory (e.g. for aluminium, lead, thallium). The elements for which electrodeless discharge tube sources of this type have been prepared are shown in Fig. 21. Most of these sources are operated at power inputs in the range 20—80 W. Some tubes of this type are also available commercially.

Microwave excited electrodeless discharge tubes of the general design outlined above are capable of producing an intense spectral line output, often many times greater than that produced by the corresponding hollow-cathode lamp. Provided that problems of self-reversal can be avoided, (e.g. by optimisation of optical path length and vapour pressure) these sources may be extremely useful as sources for AFS. This is particularly true as many of the elements for which AFS is most useful also give the most intense electrodeless discharge tube sources (e.g. zinc, cadmium, mercury, silver and lead). It seems, however, that considerable development will be necessary

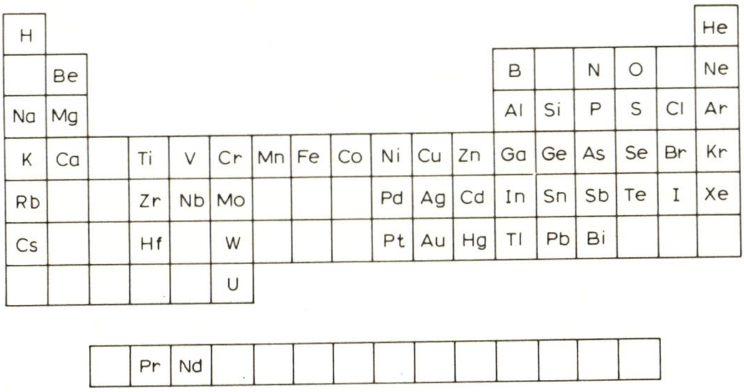

H																	He
	Be											B		N	O		Ne
Na	Mg											Al	Si	P	S	Cl	Ar
K	Ca	Ti	V	Cr	Mn	Fe	Co	Ni	Cu	Zn		Ga	Ge	As	Se	Br	Kr
Rb		Zr	Nb	Mo				Pd	Ag	Cd		In	Sn	Sb	Te	I	Xe
Cs		Hf		W				Pt	Au	Hg		Tl	Pb	Bi			
				U													

	Pr	Nd												

Fig. 21. Elements for which electrodeless discharge tubes have been prepared.

before these sources are able to compete with hollow-cathode lamps for use in AAS; it has been shown [100] that the resonance lines emitted from an electrodeless discharge tube source in the high intensity mode are much broader than those from the corresponding hollow-cathode lamp and may suffer considerable self-absorption. This is largely a result of the high vapour pressure of the element, but when the discharge tube is operated so as to reduce this, it becomes far more difficult to obtain a stable and reproducible discharge. In spite of these difficulties, the advantages of electrodeless discharge tube sources as intense line sources for analytical flame spectroscopy warrant further development work. Studies concerned with the vacuum jacketing of electrodeless discharge tube sources have been undertaken [101], and other work has concerned the use of solutions to obtain reproducible charges in the sources [102], statistical design for optimisation of parameters [103], and the construction of multi--element electrodeless discharge tube sources [104].

(F) VAPOUR DISCHARGE LAMP SOURCES

Vapour discharge lamps have been widely used as sources of line spectra from the volatile elements cadmium, caesium, mercury, potassium, sodium, rubidium, thallium and zinc. These sources give intense line emission with very low background emission and are ideal for purposes such as calibration of monochromators. Unfortunately, when these sources are operated at the currents recommended by the manufacturer (1.0—1.5 A) the line emission obtained consists of broad lines which are strongly self-absorbed or even self-reversed [105,86,100]. Several groups of workers have used these sources for AFS, however, and reduced the degree of self-absorption by operating the lamps at lower currents, (see, for example, references 106,91,92,107—110). Similarly operated lamps have found some application in atomic absorption spectroscopy [111—113].

The construction of a commercial vapour discharge lamp is shown in Fig. 22. The inner discharge envelope is usually made of silica (although a protective inner coating may be required to prevent corrosion by alkali metal vapours) and is surrounded by a glass envelope. This protects the inner envelope from damage and draughts. It is necessary to remove part of the outer glass envelope in order to use the emission from those elements with lines in the ultraviolet region. In addition to the two electrodes which maintain the discharge, the

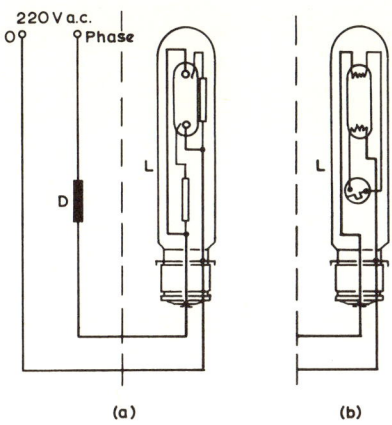

Fig. 22. Construction of a commercial vapour discharge lamp. (a) With one or two auxiliary electrodes; (b) with glow switch. D = choke, L = lamp (Reproduced by courtesy of AEG (Great Britain) Ltd., London.)

lamp design incorporates a starter electrode which operates via a thermistor; it is therefore automatically switched off when the lamp reaches its operating temperature. All the electrodes are normally made of tungsten and are sealed into the silica envelope with a small amount of the required element and an inert filler gas. The lamps may be operated either a.c. or d.c. although it is usually necessary to initiate the discharge on a.c. and then to switch over to d.c.

(G) OTHER SOURCES

Although a number of other types of spectral source can, in principle, be applied to AAS and AFS, they have seen little application in routine analytical work. Strasheim and coworkers have made comparative studies of several of these sources and described the use for AAS of a time-resolved spark [114,115] and gas stabilised d.c. arcs [116,117]. They reported that good precision and the ease of multielement analysis often compensated for the lower sensitivity obtained compared to when a hollow-cathode lamp source was employed. The use of a high frequency plasma torch as a source for AAS has also been described [118].

Flame sources may also be used for AAS, although the attainable sensitivity, due to the relatively broad line emission, is usually considerably lower than when a sharp-line source is used. Flame sources were employed by Alkemade and Milatz [9] in their original work in AAS, and since that time they have been used by several workers for special purposes [119—122].

(H) ABSORPTION AND FLUORESCENCE CELLS

Some of the basic requirements for atom cells in AAS and AFS have already been mentioned in Section 3(A). These requirements may be summarised by stating that the cell must provide for as high a fraction as possible of the sample to be converted to free, neutral atoms which may absorb the incident radiation efficiently. In addition, in AFS the emitted radiation lost by quenching or re-absorption processes must be minimised. These general requirements are fulfilled by a number of devices, although many of them have seen little application to routine analytical practice.

(1) Flames

Flames have been the most widely used of any of the available atom cells. All commercially available AAS instruments are equipped with flame facilities, and it seems likely that this situation will remain for some time. Although the popularity of flames may be due in part to the fact that they were inherited in a useful working form from flame emission spectroscopy, they do possess a number of advantages for analytical AAS and AFS;

(i) Flame systems are convenient to use, reliable and suffer little from memory effects. The burner systems employed are small, durable and inexpensive. Adequate precision is usually attained.

(ii) A wide variety of flames is available to allow the selection of optimum conditions for many different analytical purposes.

(iii) Sample solutions are easily and rapidly handled via the use of relatively simple nebuliser assemblies.

(iv) The signal: background and signal: noise ratios obtainable are quite large, even in flames with sufficient energy to decompose extremely stable compounds.

(v) Most flames in common use are quiet and safe to operate. Flame systems also possess some disadvantages compared to other atom cells. In particular, the sample solution volume required is usually greater than that required by other systems, and flame cells are not able to deal with solid samples directly. It is sometimes inconvenient that gas cylinders be used. There may also be difficulties in the use of flame cells in closed, automated systems when no operator is in attendance.

In view of the importance of flames as atom cells for AAS and AFS at the present time, they are discussed in a separate section (Section 4), where a description of some of the flame processes encountered by analytical chemists is presented. Devices of long path length, where a flame is used to introduce the sample, are also discussed in Section 4.

(2) Furnaces

The high-temperature furnace should provide a useful atom source for AAS and AFS as it offers the possibility of obtaining a high concentration of absorbing atoms within a well-defined volume with very low background emission and noise. Several workers have used conventional furnaces, with the sample in a silica or glass tube, to obtain a long path cell for AAS. Unfortunately the memory effect between samples is often large, chemical interferences may be more serious than in flames, and with many elements the atomisation efficiency is low (as the furnace must rely solely on its temperature to decompose compounds in the sample).

Vidale [123] has used a modification of the well-known King furnace to make AAS measurements on several metallic vapours. Mislan [124] obtained excellent detection limits for cadmium using a long silica tube heated to temperatures up to 1250°C by a wire resistance furnace; solutions were transferred to the cell via a conventional spray chamber. Hudson [125] has made AAS measurements on sodium vapour obtained in a stainless steel absorption cell heated by a resistance wire. Studies of absorption spectra using furnaces have also been made by Choong and Loong-Seng [126] and Tomkins and Ercoli [127].

The most satisfactory analytical systems employing furnaces have been conducted with graphite furnaces which may be heated to 3000°C. Excellent detection limits have been reported by Woodriff et al. [128—130] for 15 elements in this type of system; sample introduction was accomplished by nebulising solutions or placing solid samples directly into the furnace. Many AAS analyses have been carried out by L'Vov and coworkers using a sophisticated graphite furnace [131]; samples may be introduced by striking an arc between the furnace tube and a sample holder. This work has been reviewed by L'Vov. A graphite tube 5—10 cm in length and 2—3 mm in diameter is employed. The tube is lined with tantalum foil and the whole tube is maintained at a high temperature by electrical heating. A small amount of the solid sample is placed on the auxiliary graphite electrode which is positioned in the centre of the tube adjacent to an orifice in the tube wall. A high-current d.c. arc is then used to vaporise the sample, and the vapour collects within the light path in the graphite tube where absorption measurements are made.

Massman [132] has employed a small furnace as the atom cell for AAS and AFS measurements of elements such as zinc, cadmium and silver. The sample is placed in a small graphite crucible and vaporised at 2600°C by heating the crucible in a specially designed electric furnace in an inert atmosphere.

(3) Hot Filaments

Hot filament cells allow the use of very small samples but reduce the problems such as memory effect which are associated with graphite cuvettes and crucibles. With these 'open' devices, on the other hand, it becomes difficult to obtain a high concentration of sample atoms within a given volume. Ulfvarson [133] and Brandenburger and Bader [134] were able to detect nanogram amounts of mercury by collecting it as an amalgam on a wire and then heating the wire to vaporise the mercury into the optical path of an atomic absorption spectrophotometer. Brandenburger has applied a similar technique to the AAS determination of cadmium, zinc, lead, tellurium, copper, silver, gold and platinum.

West and coworkers [135—137] have described a heated carbon rod which may be employed to vaporise samples for AAS and AFS. One form of this 'filament atom reservoir' is shown in Fig. 23. A small drop of the sample solution is placed directly on the rod. The rod is then preheated to remove most of the solvent. The filament may be heated to 2000—2500°C by passage of a large, low-voltage current; this is sufficient to atomise many samples. The vaporisation is carried out in an inert atmosphere of argon. The 'open' design of the cell and its low background emission should prove valuable in AFS.

158

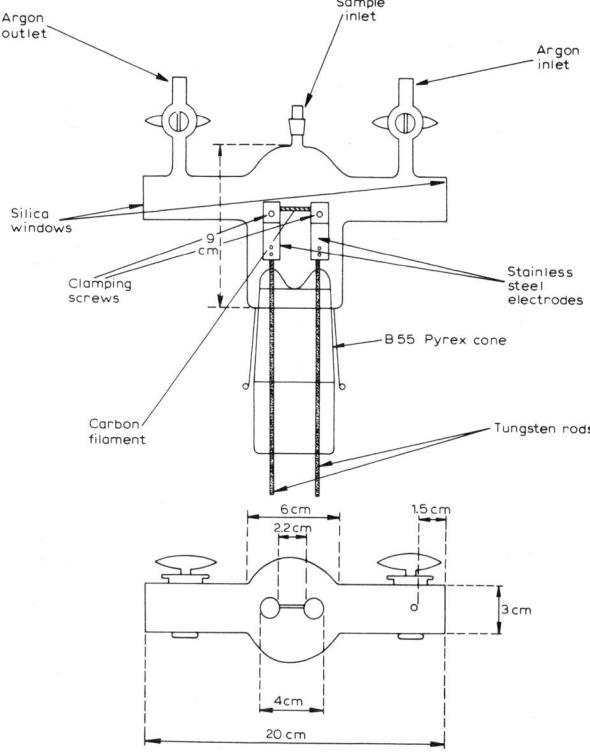

Fig. 23. The Carbon Filament (filament atom reservoir) Cell. (Reproduced by courtesy of Anal. Chim. Acta.)

(4) Hollow-cathode Cells

The sputtering process which produces atomic vapour within a hollow-cathode lamp may be used for the atomisation of samples for AAS or AFS. Walsh and coworkers [138,139] designed a special sputtering chamber in which metal samples, machined to the usual shape of a hollow-cathode, were clamped. The chamber was then evacuated, filled with argon at the required pressure, and a discharge initiated as in a conventional hollow-cathode source. The sputtering chamber was fitted with silica windows at each end and could be placed in the light path of an atomic absorption spectrophotometer. The application of this device to the determination of silver and phosphorus in copper, and silicon in aluminium and steel has been reported [140].

In order to avoid the necessity to machine a solid sample into the shape of a hollow-cathode, Goleb and Brody [141] evaporated sample solutions on to the inner wall of an aluminium hollow-cathode. This was then used as part of a water-cooled demountable hollow-cathode lamp with a continuous flow vacuum system. Only a small sample volume was required and 1 μg quantities of sodium, calcium, magnesium, silicon and beryllium were detectable; considerable inter-element interferences were reported. Goleb has also used the device as an atom source for the isotopic analysis of uranium [142]. Some of the advantages of the graphite cuvette were obtained by Ivanov et al. [143] who used a graphite hollow-cathode in a sputtering cell. Sample solutions were evaporated directly on to the walls or on to a fine molybdenum wire which was then placed along the central axis of the cathode.

(5) Other Cells

Several workers have described the use of the d.c. arc for sample atomisation [144—146]. Robinson has reported that twenty percent absorption was obtained for aluminium using a spark to atomise the sample [147,148]; the solution was nebulised in the usual way and the mist passed between two electrodes used to obtain the spark. There have been several reports of the use of radio frequency or microwave plasma to atomise samples for AAS. Wendt and Fassel [149,150] employed an induction coupled plasma together with a triple-pass optical system to obtain sufficient path-length. Friend and Diefenderfer [151] investigated the possibility of using a plasma jet for the determination of the refractory elements. The radio frequency plasma has recently been applied to AAS by three groups of workers [152—154]. The use of lasers offers promise as a means of free atom formation; they should permit the atomisation of small samples while maintaining low emission background and signal noise. There is also the possibility of sampling small areas on the surface of solid samples; some examples of the use of lasers in this way have already been described in the literature [155—157].

(I) ISOLATION OF SPECTRAL LINES

(1) Introduction

Atomic absorption and atomic fluorescence spectrocopic techniques require that only a signal from the resonance line(s) of interest is detected and amplified at the spectrophotometer. This may be achieved optically using a conventional filter or monochromator device or a resonance detector so that only radiation of the required wavelength is received at the detector, or by selective modulation of the resonance line(s) so that unwanted signals are rejected electronically. Most instrumentation in widespread use at the present time employs conventional monochromators of the prism or grating type.

When a filter or prism or grating monochromator is to be used for the isolation of resonance lines there are two main properties of the device which must be considered, the *transmission* and the *band-pass*. The transmission is represented by the percentage of the incident radiation at the wavelength selected which is passed by the monochromator. For most purposes the filter is chosen or the monochromator is adjusted so that the peak transmission coincides with the required resonance line. For AFS it is important that the peak transmission of the device employed is as high as possible in order to permit the attainment of high sensitivity. A high transmission factor is not necessary for AAS with most line sources, but sufficient light must be transmitted to avoid the necessity for large amplification of signals. The *spectral half-band pass* is usually defined as the width of the transmitted peak, in wavelength units, between the two points at which the transmitted intensity is equal to half that at the maximum. A narrow band-pass is not usually necessary for AFS work, since it serves little purpose with the types of flame source most commonly employed and simply reduces the signal intensity received. A narrow band-pass is also not necessary for many purposes in AAS. Many instrumental AAS systems employ monochromators of good resolution, however, as when used at narrow slit-width a useful gain in sensitivity may be obtained when the resonance line of the element determined lies close to non-absorbing lines. A monochromator of high resolution, which permits a narrow band-pass to be obtained, is also necessary when a continuum source is employed for AAS.

(2) Filters

A filter is simply a device which transmits light over a required narrow range of wavelengths. The transmission curves for some typical glass filters are shown in Fig. 24. This type of filter is quite unsuitable for almost all AAS applications; even for a transmittance of only ca. 15% the half-intensity band-pass is about 50 nm. Filters of this type can only be used for the isolation of the resonance lines of a few elements, and the use of several filters in combination produces a transmission factor which is inadequate for AFS measurements.

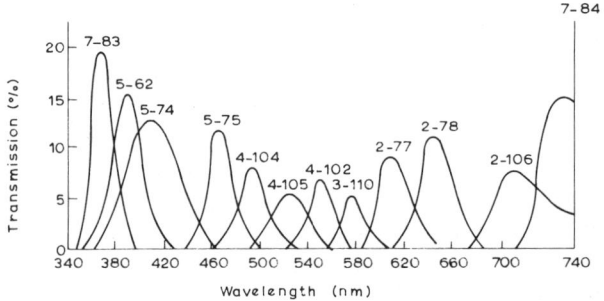

Fig. 24. Transmission curves of some glass absorption filters. (Corning Glass Works).

More satisfactory results may be obtained by the use of interference filters. In these devices a glass or silica plate is coated with a semi-reflecting layer of a material which has a high reflection coefficient but a low absorption factor. This may be silver or aluminium, but modern interference filters use multi-layers of dielectric materials such as zinc sulphide, magnesium fluoride and lead chloride. This layer is then covered by a spacing layer, which is about one-half of a wavelength in thickness, and which may also be a dielectric material (but without a reflecting surface). This spacer layer is itself covered by a second semi-reflecting layer, the reflecting surfaces of the two outer layers being placed face-to-face. Multiple reflections, and hence interference of the light beams, occur when light is passed through the filter. An interference transmission maximum may be obtained at any required wavelength by careful selection of the thickness of the layer between the two reflecting films. Secondary transmission max-

162

ima may occur within the wavelength range detected by the particular instrument employed; these are removed by the use of auxiliary glass blocking filters. The method of manufacture of interference filters allows a wide range of combinations of band-pass and peak transmission. For example, filters may be obtained which give a half-intensity band-pass of 6—8 nm with a peak transmission of about 65%, or a band-pass of 0.3—0.4 nm with a peak transmission of about 25%. The cost of such filters depends on the required combination of high transmission and narrow band-pass and the tolerance in the wavelength of peak transmission. These filters are obtainable for wavelengths as low as 210 nm, but the best quality is available in the visible region.

In general filters are only useful in AAS for the simpler type of instrument to be used for one particular application to the determination of elements giving simple spectra, e.g. sodium or potassium. In AFS the requirement of high transmission may mean that a filter assembly is a suitable alternative to a monochromator for a wider range of elements.

(3) Monochromators

Two basic types of monochromator are available; the distinction depends on whether a prism or a diffraction grating is used as the dispersing medium. In either case the design of the device is similar. Radiation from the source is passed through an entrance slit and collimated by a lens or mirror before passing to the dispersing element (prism or transmission or reflection grating). The dispersed radiation is then brought to a focus at an exit slit by a second lens or mirror. A detector is placed at the exit slit to receive the dispersed radiation; the wavelength of the radiation detected is selected by alteration of the angle of incidence of the incident radiation at the prism or grating.

The quality of a monochromator is governed to a large extent by the nature of the dispersing element. Generally good light transmission (for a given band-pass) is obtained by using larger components, i.e. the ratio of focal length to diameter (the f number) for optical components must be fairly small. A typical instrument for use in AAS requires an aperture of only about $f/8$ to $f/11$, whereas one intended for use in AFS requires greater aperture for high performance; monochromators of $f/2$ to $f/5$ are readily available. The *re-*

solving power of a monochromator defines the least separation of two spectral lines which can be shown to be separate. For a prism monochromator it depends on the size of the prism and the material from which it is constructed★. The use of glass as the prism material results in greater dispersion than silica, but it cannot be employed at wavelengths below ca. 370 nm. With grating instruments the resolving power depends on the number of grooves (or lines) on the grating surface and on the order of diffraction which is employed. The smallest band-pass is obtained using a large number of lines and a high order of diffraction. Most monochromators used in AAS or AFS are designed for use in the first order, as the transmitted spectral intensity decreases when higher orders are employed. A typical grating monochromator used for AAS may have a grating 5 × 5 cm in size with 500—1000 lines/mm, depending on the resolution required. Most modern gratings are also 'blazed' by ruling specially shaped grooves so that for any given wavelength region as much light as possible is gathered into the required order. The wavelength range over which the blazing technique is effective is quite large, but it is usual to employ two differently blazed gratings, one for the visible region and a second for the ultraviolet region.

Although the limit of resolution and the transmission efficiency are governed in any monochromator by its components and design, the resolution and the radiant intensity received at the detector may be altered considerably by use of a variable slit width. The best resolution is obtained at narrow slit width, whereas the greatest intensity at the radiation detector is obtained by using a wide slit. It is most important in both AAS and AFS to optimise the slit width employed. The monochromators used in many AAS instruments allow the selection of only three or four fixed slit widths. This limited adjustment facility may be quite adequate for many AAS applications but does not permit the alternative use of the instrument for AFS in efficient fashion.

The majority of commercial AAS instruments are fitted with grating, rather than prism, monochromators. This results partly from the

---

★ The theoretical resolving power, P, may be shown for a prism instrument to be $P = \lambda/\Delta\lambda = b\mathrm{d}\eta/\mathrm{d}\lambda$, where $\Delta\lambda$ is the separation of two lines at wavelength λ, b is the base length of the prism and η is the refractive index of the prism material. $\mathrm{d}\eta/\mathrm{d}\lambda$ is termed the dispersion, and represents the dependence of the refractive index upon the wavelength. (See also Chapter 1, p.31.)

successful production of low cost, high quality replicate gratings. The use of gratings also has the practical advantage of producing a virtually constant resolving power over the whole wavelength range; the resolving power of a prism monochromator varies with wavelength. It is also easier to obtain a linear wavelength scale with a grating device than with a prism.

(4) Resonance Detectors

The phenomenon of resonance fluorescence has already been described in Section 2(D)(1). Russell and Walsh [158] suggested that the resonance fluorescence principle might be used for the isolation of the resonance absorption line used in AAS. When the radiation which has passed through the flame absorption cell containing the sample atoms is allowed to fall on a cloud of free atoms of the same element, resonance fluorescence is stimulated from these atoms. A detector placed at 90° to the incident light path then responds to this fluorescence radiation and registers changes in intensity of the incident resonance line. In practice this technique has been applied to the isolation of the resonance lines of several elements. The elements studied and the spectral lines isolated are shown in Table 3. Resonance detection is most successful for those elements which have simple atomic spectra. The source used for AAS with resonance detection must produce an intense and stable sharp-line output. Modulation of the source is necessary in order to allow detection and amplification of the resonance fluorescence signal in the presence of the emission from the resonance detector itself and any stray radiation. The atomic population in the resonance detector may be produced thermally for some elements, but Sullivan and Walsh [159] report that the use of the sputtering technique (as in a hollow-cathode lamp) requires less warm-up time, results in a much longer useful lifetime for the detector, and requires less accurate adjustment of the operating current to the detector.

The main advantage of the resonance detector is that it provides at relatively low cost a spectral isolation system which is automatically and invariably centred exactly on the absorption line(s) of the element. At the same time, the band width of the system is extremely narrow, so that very little trouble from, for example, flame emission background is encountered. The use of resonance detection systems

TABLE 3

Atomic lines isolated by resonance monochromators[*]

Element	Wavelengths of lines isolated (nm)
Li	670.8
Na	589.0, 589.6
K	766.5, 769.9
Rb	780.0, 794.8
Cs	852.1, 894.4
Cu	324.7, 327.4
Ag	328.1, 338.3
Au	242.8, 267.6
Be	234.9
Mg	285.2
Ca	422.7
Sr	460.7
Ba	535.5
Zn	213.8
Cd	228.8
Hg	253.7
Al	308.2, 309.27, 309.28, 394.4, 396.2
Ga	287.4, 294.36, 294.42, 403.3, 417.2
In	303.9, 325.5 , 325.9 , 410.2, 451.1
Tl	276.8, 377.6

[*] From ref. 159.

also simplifies the design of multichannel AAS systems without the use of several monochromators or a polychromator.

(5) Selective Modulation

Selective modulation as a means of isolation of spectral resonance lines is closely related to resonance detection in so far as a secondary cloud of atoms of the element concerned is required. Instead of utilising the resonance radiation emitted by this second atom vapour, however, selective modulation requires the detection of the original light beam after it has been transmitted by the secondary vapour. During this passage through the second atom cloud, only the resonance lines of the element are modified by absorption and the non-resonance lines are transmitted unchanged in intensity. If the secondary cloud can be produced periodically, therefore, it is possible to

166

modulate the radiation at the resonance lines 'selectively' without affecting the other wavelengths. A tuned amplifier operating at the same frequency responds only to intensity changes in the resonance lines. As with resonance detectors, the application of selective modulation to AAS has been investigated by workers at CSIRO in Australia [160,161]. They have reported several different methods of producing the selective modulation effect: (i) the pulse of a pulsed hollow-cathode lamp is adjusted in height and width in order to use the cloud of atoms produced to absorb selectively the resonance emission resulting from a d.c. continuous discharge; (ii) by using a hollow-cathode lamp with a built-in selective modulator working on the sputtering principle in the a.c. mode to absorb radiation from a primary cathode operated d.c.; and (iii) by using a normal hollow-cathode lamp with the selective modulator in a separate envelope. An extensive investigation of all three methods has been reported by Sebestyen [162]. It was found that it was necessary to use a filter in conjunction with the selective modulator or a solar blind photomultiplier which does not respond to extraneous radiation received at longer wavelengths from the flame and surroundings.

(J) DETECTOR AND READ-OUT SYSTEMS

(1) The Detector

The device most widely used as the radiation detector in AAS and AFS is the multiplier phototube, commonly referred to as the photomultiplier. As its name implies, this device not only detects the radiation but effects an amplification of the signal which results from it. The gain of these devices is extremely high, amplification factors of greater than 10^6 are readily obtained, and consequently very low light levels may be detected and measured. The principal disadvantage of the photomultiplier is its requirement for a power supply which produces a stabilised high voltage (500—1500 V, depending on type and the required gain). The amplification effect of the photomultiplier is obtained via the process of secondary electron emission. The operating principle and construction of the photomultiplier is shown diagrammatically in Fig. 25. The tube contains a photosensitive cathode, S, an anode, A, and a number of target electrodes known as dynodes, D. A potential difference of about 100 V is applied between each pair of electrodes by the resistance chain, R. The

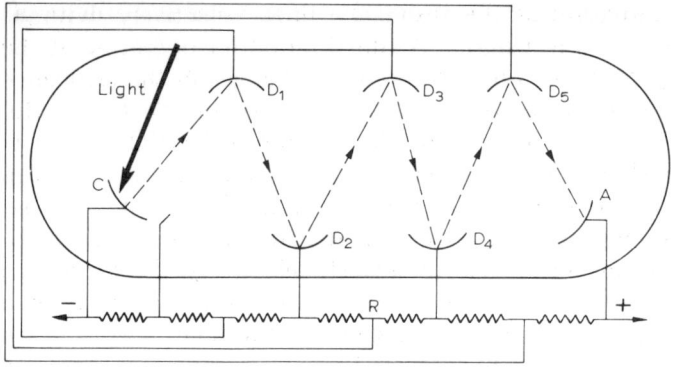

Fig. 25. Schematic construction of a photomultiplier tube.

incident radiation enters through a silica or glass window and falls onto the cathode C. This produces a number of photo-electrons (in proportion to the intensity of the radiation) and these are guided, by a simple electron lens system, onto the first dynode. The secondary electrons produced by the impact are directed towards the next dynode, D_2, and so on until the anode is reached. As each primary electron may produce as many as five secondary electrons, the number of electrons reaching the anode may be very large and a considerable current may be measured between the last dynode and the anode even when the incident light intensity is very low. The sensitivity is usually between 10 and 100 μA/lumen, but the maximum current drawn from the tube is limited to about 10 μA to prevent fatigue of the dynode surfaces.

The overall gain of the photomultiplier depends on the number of electrons reaching the anode for each electron leaving the cathode. It is therefore governed by the number of dynodes and the applied voltage. The applied potential between the dynodes controls the kinetic energy of the electrons on impact and thus the number of secondary electrons produced. Unnecessary gain should not be employed as this results in a high 'dark current'; this is the current which is produced by the tube even when no radiation falls on the photocathode and its fluctuations contribute to the signal noise level. Under given tube operating conditions the gain of the photomultiplier varies with the wavelength of the radiation measured. The spectral sensitivity of each tube depends on the nature of the photosensitive coating on the cathode and the material of the entrance

168

window. The cathode coating is frequently an alloy of an alkali metal with bismuth, silver, or antimony.

Most AAS and AFS applications require the use of photomultipliers which have high sensitivity to ultraviolet radiation. These tubes usually employ a dynode chain of 9, 11 or even 13 stages and are fitted with silica entrance windows. Solar blind photomultipliers, which have a photocathode coating and entrance window chosen so that they respond to radiation at wavelengths only below about 310 nm, show promise for use in AFS and with selective modulation or resonance detection in AAS.

(2) Amplification and Measuring System

In principle, the current output from the photomultiplier tube may be measured directly using a galvanometer as with most photomultiplier tubes it will lie between 0.01 and 1 μA. If required it may also be recorded as a signal directly at a conventional strip-chart recorder; in this case the current is applied across a load resistor and the voltage is measured. Almost all work in AAS and AFS has, however, been undertaken with some form of signal amplification prior to measurement of the signal at a meter or recorder. In this way it is possible to use a tuned amplifier to detect a modulated signal. The use of an amplifier also facilitates the provision of 'backing-off' of unwanted signals and the scale expansion of small absorption signals. The degree of scale expansion which can be used depends on the noise level on the signal. As this noise varies with the analysis conditions it is particularly useful if the amplifier unit allows variable degrees of scale expansion, for example in the range 5—50 times. An AAS instrument which has a high quality amplifier and scale expansion facilities may be useful in AFS; the use of cool flames of low background emission results in low noise levels which allow considerable amplification of the signal.

After amplification the signal must be presented to the operator in convenient form. The most widely used read-out device is the microammeter. The most useful meters of this type are calibrated with both a linear 0 to 100 scale and a logarithmic absorbance scale; unless the former is present it is inconvenient to use the instrument for emission or fluorescence work. Several commercially available instruments incorporate meters with linear absorbance scales. This is achieved by fitting an additional circuit to the amplifier so that the output is a logarithmic signal giving direct read-out in absorbance units. The traditional presentation of the absorbance or transmittance values in AAS at a meter is replaced in several commercially available systems by a digital voltmeter read-out. The main advantage of digital display is that the value may be read directly and it is not necessary to read a scale or estimate the value. Some difficulty with this form of display, however, may be encountered when the 100% transmission setting is adjusted or the peak wavelength is set. One instrument currently in use in AAS which employs digital read-out of absorbance incorporates a separate conventional meter to facilitate peaking of the signal from the hollow-cathode lamp. Equipment incorporating a small computer and digital data presentation is currently available. This permits rapid processing of samples by AAS, facilitates zero setting, correction of working curves, direct read-out in absorbance, concentration or emission intensity units and allows the results to be typed out automatically. Such units may also give higher precision than simple meter read-out systems, as it is possible for each reading displayed to be automatically averaged from a given number of individual readings. Similarly fluorescence signals are easily integrated to obtain increased sensitivity and precision.

(3) Modulation of the Source

Modulated sources are widely employed so that when they are used in conjunction with a tuned amplifier the absorption of fluorescence signal may be detected without interference from emission signals originating in the atom cell. Modulation may be achieved either mechanically by placing a rotating or vibrating 'chopper' in front of the source or by supplying the lamp directly with a modulated current. The latter solution is widely adopted with hollow-cathode sources and has also been applied to electrodeless discharge sources [96]. The use of a rotating sector is easily arranged and is applicable to all types of source, even a flame, although it may result in restriction of the usable solid angle of excitation in AFS. Either type of modulation only produces effective results when the modulated signal is accurately synchronised with the narrow band-pass tuned amplifier. With electrical modulation this is conveniently accomplished using an additional output from the power supply to trigger the amplifier. With mechanical modulation an optical or magnetic triggering device fitted to the rotating sector is most frequently employed. The actual modulation frequency which is used varies quite widely in commercially available equipment and does not appear to be critical; low frequencies (less than 50 Hz) are avoided in order to minimise the effect of flame 'flicker' noise.

(4) Double-Beam Operation

Several double-beam AAS instruments have been widely used in routine analysis. The most commonly employed system splits the light from the source into two beams using a set of mirrors and a rotating sector, so that light is passed alternately into each beam. One of these beams is passed through the atom cell and the other, the reference beam, bypasses the atom cell. The two beams are recombined at the detector, which views each beam alternately, and are separated at an amplifier similar to that employed for modulated single beam systems. Both signals could be displayed separately by the read-out device but it is more common to present their ratio so that the indicated absorbance signal is independent of any fluctuations in the intensity of the spectral source. A similar principle is applicable in atomic fluorescence spectroscopy to eliminate the effect of variations in the source intensity. Modern hollow-cathode lamps are very much more stable in their output than the early varieties, however, and the limiting factor which controls analytical precision for some elements has become the flame noise from hot flames rather than the occurrence of instability in the hollow-cathode lamp sources. It is important to bear in mind for both AAS and AFS applications that double-beam operation cannot compensate for any changes in the system other than the total intensity of the source line. For example, a sudden increase in the intensity of the light source may be accompanied by an increase in the spectral line-width. This would result in a decrease in the peak absorbance measured in AAS. Double-beam operation cannot compensate for this effect, however. No compensation is possible for change in the background absorbance from the flame or from molecular absorption of compounds formed in the atom cell. In view of these limitations a second type of double-beam system which uses a deuterium or hydrogen lamp as continuum source for background correction has been introduced [163,164].

Menzies [165] has suggested that background absorption corrections may be made without an additional continuum source by using a non-absorbing line of the same element as that being determined and which is emitted by the source. In this way the advantages of both types of double-beam operation might be secured. The reference line from the source should show similar variations in intensity to the resonance line, and provided that the two lines are of similar wavelength they should be affected similarly by background absorption. Robinson has tested the method with iron [166]. There may be difficulty, however, in finding suitable pairs of lines for some elements. Even then there is no guarantee that intensity changes in the resonance line will be faithfully reproduced at the non-resonance line.

The use of internal standardisation techniques in AAS has received attention. In a true internal standard technique a control element is added to the sample so that the absorption or fluorescence of the standard and the analyte may be monitored simultaneously. Butler and Strasheim [167] improved the precision of copper determinations using gold as internal standard in AAS in this fashion using an instrument designed for multi-element analysis. Feldman [168] has discussed the problems of internal standardisation in both flame emission and AAS in some detail.

(5) Noise

From many of the statements made in the foregoing sections it is apparent that the occurrence of signal noise is of major concern in AAS and AFS. Signal noise determines the detection limit for any determination in both techniques and is an important contributor to the attainable precision in any analytical method. The noise encountered in AAS and AFS may be somewhat different in origin to that of other instrumental techniques. The noise levels obtained from the detector (shot-noise, dark current, instability etc.) and the amplifier generally make only a small contribution to the observed signal noise. A substantial fraction of the noise observed is usually made by the spectral light source and the atom cell. Sophisticated techniques for the detection of low light levels and the reduction of detector noise, such as photon counting, will show useful improvements in signal-to-noise ratios only when stable sources and low noise atom cells are employed. A valuable exercise in the calculation of the relative effects of different types of noise in AAS [169] and AFS [170] has been undertaken by Winefordner and coworkers. A more empirical approach is usually adopted by analytical chemists in practice, however; the spectral source and atom cell are chosen so that the lowest possible noise levels are obtained commensurate with the satisfactory fulfilment of the other requirements of a particular analysis.

4. Flames

Two principal types of flames are employed in atomic absorption and fluorescence spectroscopy. These are (i) premixed flames, in which the fuel and oxidant are mixed and usually constrained to laminar flow before combustion, and (ii) unpremixed flames, in which the fuel and oxidant mix only at the point of combustion. Unpremixed flames may be laminar or turbulent, depending upon the manner in which the fuel and oxidant gases mix. In premixed and unpremixed flames applied to analytical spectroscopy the fuel is usually a gaseous hydrocarbon or hydrogen and the oxidant employed is air, oxygen or nitrous oxide.

(A) PREMIXED FLAMES

The structure and characteristics of premixed flames of hydrocarbons with air may be described by reference to the Bunsen flame burning at an open circular burner port or the Meker flame at a series of smaller circular orifices (see Fig. 26(a)). When a particular hydrocarbon—air mixture is passed at a given velocity along a cylindrical burner tube of length and diameter sufficient to ensure that laminar flow is established in the tube, the flow velocity across the tube diameter assumes an approximately parabolic profile (Poiseuille flow) (see Fig. 26(b)). The flow velocity is therefore a maximum along the central axis and falls virtually to zero along the walls of the cylinder. The initial combustion reaction between fuel and oxidant in the gas mixture emerging from the burner tube with this flow velocity distribution gives rise to the appearance of the primary reaction zone.

(1) Primary Reaction Zone

When the fuel: oxidant concentration ratio of the mixture emerg-

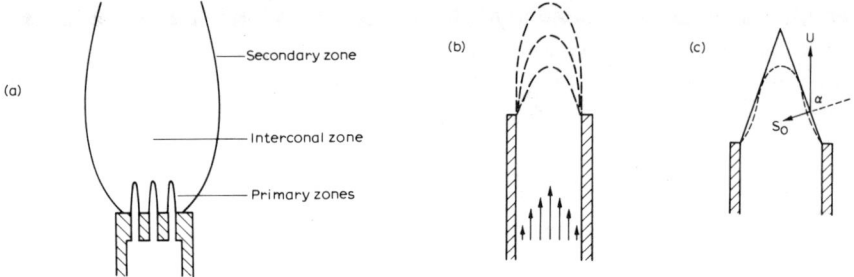

Fig. 26. The structure of premixed flames. (a) Overall appearance; (b) velocity profile of unburnt gas at burner orifice; (c) primary zone shape.

ing from the burner tube lies between certain limits (the *limits of inflammability*) the mixture may be ignited and a continuous reaction sustained. The reaction front which is formed moves normal to its surface through the adjacent unburnt gas with a velocity characteristic of the nature of the fuel and oxidant and their concentration ratio. This velocity is termed the *burning velocity* of the mixture and is nearly constant across most of the flame front. The burning velocity decreases, however, owing to loss of heat and radicals sustaining the chain reaction, near the burner rim itself. This burner rim 'heat sink' effect results in a burning velocity of practically zero at the rim itself. The gas flow velocity is usually equal to the burning velocity at only one distance from the burner rim; at every other distance it exceeds the burning velocity. The flame front adjusts itself so that at any point the component of the gas flow velocity perpendicular to the front is equal to the normal burning velocity at that point. The reaction front therefore assumes a conical profile (see Fig. 26(c)) inclined against the direction of the gas flow at an angle determined by the equation

$$S^0 = U \cos \alpha$$

where S^0 is the standard value of the burning velocity, U is the flow velocity of the gas mixture, and α is the angle between the unburnt gas velocity vector and the burning velocity vector.

The appearance and stability of the flame front thus depends on the balance established between the gas flow velocity and the burning velocity of the mixture. As the fuel: oxidant concentration ratio is varied (i.e. to provide flames containing excess oxidant [fuel-lean]

174

TABLE 4

Burning velocities and calculated and experimental maximum temperatures of some analytically useful flames at atmospheric pressure

Fuel–oxidant mixture	Stoichiometric reaction	Burning velocity (cm/sec)	Maximum calculated temperature (K)	Maximum experimental temperature observed (K)
Town gas–air	Town gas + 0.98 O_2 + 3.9 N_2 → CO_2 + H_2O + 3.9 N_2 108.79 kcal	55	2113	1980
Town gas–oxygen	Town gas + 0.98 O_2 → CO_2 + H_2O 108.79 kcal		3073	3013
Hydrogen–air	H_2 + 0.5 O_2 + 2 N_2 → H_2O + 2 N_2 58 kcal	440	2388	2318
Hydrogen–oxygen	H_2 + 0.5 O_2 → H_2O 58 kcal	3680	2963	2933
Methane–air	CH_4 + 2 O_2 + 8 N_2 → CO_2 + 2 H_2O + 8 N_2 191.8 kcal	70	2228	2148
Methane–oxygen	CH_4 + 2 O_2 → CO_2 + 2 H_2O 191.8 kcal	5502	2993	2950
Propane–air	C_3H_8 + 5 O_2 + 20 N_2 → 3 CO_2 + 4 H_2O + 20 N_2 530.57 kcal	82	2198	2198
Propane–oxygen	C_3H_8 + 5 O_2 → 3 CO_2 + 4 H_2O 530.57 kcal		3123	3123
Butane–air	C_4H_{10} + 6.5 O_2 + 26 N_2 → 4 CO_2 + 5 H_2O + 26 N_2 687.94 kcal	82	2203	2168
Butane–oxygen	C_4H_{10} + 6.5 O_2 → 4 CO_2 + 5 H_2O 687.94 kcal		3183	3173
Acetylene–air	C_2H_2 + 2.5 O_2 + 10 N_2 → 2 CO_2 + H_2O + 10 N_2 300.1 kcal	160	2523	2598
Acetylene–oxygen	C_2H_2 + 2.5 O_2 → 2 CO_2 + H_2O 300.1 kcal	2480	3383	3373
Acetylene–nitrous oxide	C_2H_2 + 5 N_2O → 2 CO_2 + H_2O + 5 N_2 401.5 kcal	180	3203	3025
Acetylene–nitric oxide	C_2H_2 + 5 NO → 2 CO_2 + H_2O + 2.5 N_2	90	3353	
Cyanogen–air	$(CN)_2$ + O_2 + 4 N_2 → 2 CO + 5 N_2 126.68 kcal	90		2603
Cyanogen–oxygen	$(CN)_2$ + O_2 → 2 CO + N_2 126.68 kcal	140	5025	4640

or excess fuel [fuel-rich] the burning velocity of the mixture may vary; if the flow velocity of the gas mixture is maintained constant the appearance of the primary reaction zone (apex angle and stability) will then be determined by the above equation. Similarly, as predicted by this equation, when the gas flow velocity for a given mixture is too great there will be a danger of 'lift-off' of the flame from the burner, and when the flow velocity is too low there will be a danger of 'flash-back' of the flame into the burner tube. Table 4 shows the maximum burning velocities of some fuel: oxidant mixtures used in analytical flame spectroscopy.

The primary reaction zone is only a fraction of a millimetre thick in laminar premixed flames at atmospheric pressure. This flame front, where the initial partial oxidation reaction between the fuel and oxidant occurs, emits quite intense radiation when hydrocarbon fuels are employed. Intense blue or blue-green emission from C_2 and CH radicals is usually observed, and the primary reaction zones of hot flames may also exhibit intense emission from OH, NH, CN, and NO species as well as atomic emission from carbon at 247.8 nm and continuum background from carbon particles and the non-quantised processes of dissociation, association and ionisation. In the primary reaction zone itself thermodynamic equilibrium is not established, so that the concept of a 'temperature' for this zone has little meaning. The suprathermal conditions which prevail are manifested by the presence of above equilibrium concentrations of radicals, degrees of ionisation and exceptionally high emission intensities corresponding to over-excitation of the species present. Owing to intense background emission and high noise levels associated with it, the primary reaction zone is seldom employed in analytical flame spectroscopy.

(2) Interconal Zone

In the primary reaction zone, the initial reaction between fuel and oxidant molecules usually gives rise to only partial oxidation of the fuel. The combustion products for hydrocarbon flames consist mainly of an equilibrium mixture of carbon monoxide, hydrogen, carbon dioxide, water, oxygen, nitrogen and radicals such as H, OH, C_2, CH, and NO. These hot gases, which are subsequently oxidised in the secondary reaction zone, constitute the atmosphere of the *interconal zone* immediately above the primary reaction zone. The composition of the interconal gas mixture varies with the initial fuel: oxidant

176

concentration ratio in the unburnt gas mixture, the amount of carbon monoxide and hydrogen increases and the amount of carbon dioxide and water decreases as the fuel:oxidant ratio increases (i.e. as the flame is made fuel-rich).

Above the primary zone, the flow velocity in the gas mixture increases and the flame becomes broader owing to thermal expansion as the gas passes through the primary zone. The number of moles of product gases may also be greater than the number of moles of reactant gases, e.g. in the air—acetylene flame

$$C_2H_2 + O_2 + 4\,N_2 \rightarrow 2\,CO + H_2 + 4\,N_2$$

and this effect may also give rise to expansion of the volume occupied by the burnt gases.

The interconal zone of premixed flames of hydrocarbons with air is usually in almost complete thermal equilibrium. The maximum flame temperatures determined are usually in quite good agreement with the calculated temperature values. The maximum attainable temperature depends on the amount of energy liberated in the reaction and the amount of heat absorbed by the burnt gas mixture. The expected temperature may be considerably higher than that actually attained if allowance is not made for the dissociation of the combustion products by reactions such as

$$CO_2 \rightleftharpoons CO + \tfrac{1}{2}\,O_2$$

$$H_2O \rightleftharpoons \tfrac{1}{2}\,H_2 + OH$$

$$H_2O \rightleftharpoons H_2 + \tfrac{1}{2}\,O_2$$

$$\tfrac{1}{2}\,H_2 \rightleftharpoons 2\,H$$

$$\tfrac{1}{2}\,O_2 \rightleftharpoons 2\,O$$

The maximum flame temperature is attained, except for some very rich flames, in the interconal zone immediately above the primary zone. The flame temperature in this region varies somewhat with the fuel:oxidant concentration ratio in the initial gas mixture. The highest temperature is usually attained for stoichiometric mixtures, except when the fuel is a highly endothermic or exothermic compound. In these latter cases, the maximum temperature should theoretically occur at slightly fuel-rich or slightly fuel-lean concentration ratios respectively. The temperatures of some common flames used in analytical flame spectroscopy are given in Table 4.

The emissivity of the interconal gases in stoichiometric hydrocarbon—air flames is usually quite low. This may be demonstrated using a simple Smithells flame separator [171] to examine the hot gases between the primary and secondary zone; the emission of the interconal gases may be shown to be almost all thermal in origin. There is some evidence for the persistence of more than equilibrium amounts of radicals such as H and OH in the interconal zone of the acetylene—air flame, but the effect on the flame radiation is small. As the interconal zone may be considered to be in thermodynamic equilibrium in hydrocarbon—air flames, knowledge of its temperature and composition enables prediction of the manner in which dissociation, ionisation and excitation equilibria will control the behaviour of samples introduced into the flame. It is this part of most premixed flames of hydrocarbons which is frequently of greatest utility in AAS and AFS.

(3) Secondary Reaction Zone

When the hot interconal gases come into contact with atmospheric oxygen a secondary combustion reaction occurs in which the excess oxidisable constituents burn. This reaction gives rise to the familiar blue-violet secondary zone (outer sheath) observed with hydrocarbon flames. Air diffuses, or is entrained, into the interconal gases and oxidises carbon monoxide and hydrogen to carbon dioxide and water. Part of the energy liberated in this secondary reaction serves to maintain a high temperature in the interconal zone and prevents a rapid rate of fall of the flame temperature with height above the primary reaction zone. Typical flame temperatures vs. height profiles for a premixed flame are shown in Fig. 27. Although the secondary reaction zone tends to stabilise the flame at the burner over a wider range of fuel:oxidant concentrations than would be attainable in its absence, from an analytical point of view its existence may create problems due to its relatively oxidising nature and high background emission and noise. The radiation emitted by the secondary reaction zone consists largely of emission due to OH in the ultraviolet region and a continuum in the visible region with hydrocarbon flames which arises from the chemiluminescence of the carbon monoxide oxidation reaction ($CO + \frac{1}{2} O_2 \rightarrow CO_2 + h\nu$). The emission intensity natu-

178

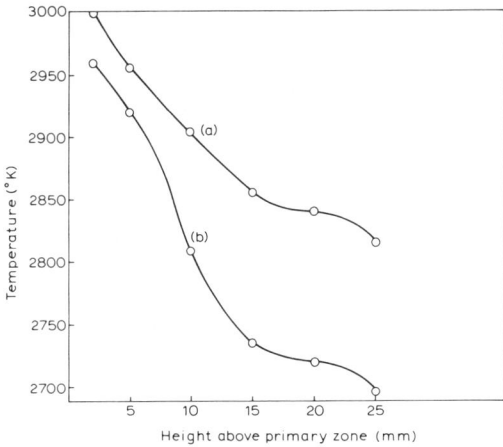

Fig. 27. Variation of temperature with height above primary reaction zone in stoichiometric premixed nitrous oxide—acetylene flame. (a) Conventional flame; (b) argon-separated flame. (Kirkbright and Vetter, unpublished work.)

rally depends on the concentration of excess carbon monoxide and hydrogen in the interconal zone and is thus influenced by the fuel:oxidant mixture strength. As the reaction depends on atmospheric oxygen, some of whose diffusion or entrainment into the flame may be irregular, the flame background may exhibit an appreciable noise component.

(4) Burners for Premixed Flames

The burning of premixed fuel and oxidant places certain restrictions on the burners which may safely be used to support a particular flame. The primary requirement is that the streaming velocity of the gas mixture must be high enough to prevent flash-back of the flame. The streaming velocity must be about three times as high as the mean burning velocity of the fuel:oxidant mixture. This requirement determines the dimensions of the orifice(s) in the burner plate at which the gas mixture is burnt; burner plates designed for use with low-burning velocity mixtures (e.g. air—propane) must not be used for mixtures of high burning velocity (e.g. nitrous oxide—acetylene or oxy-acetylene). In atomic absorption spectroscopy, where the analytical sensitivity is related to the absorption path length, a long slot or one or more rows of circular holes in a massive metal burner plate is commonly employed. In this way premixed flames 5—10 cm long may be produced. Long flames of this type permit increased analytical sensitivity in absorption work compared to a short path length flame provided that suitable optics are employed and the number of absorbing species per unit path length remains the same, i.e. provided that an increase in the amount of sample introduced into the flame per unit time is increased along with the increase in gas flow rates required to support the longer flame. Otherwise the sample will simply be diluted in the larger volume of gas burnt per unit time and even with the greater path length the analytical sensitivity will not increase. The production of long narrow flames may also result in some difficulty in ensuring that all the radiation received from the source passes through absorbing atoms in the flame; this is necessary to avoid low sensitivity and deviation from linearity of the analytical working curves. In order to avoid this problem burners which have several slots parallel [172] or perpendicular [173] to the optical axis have been recommended.

Although premixed flames have several advantages compared with unpremixed flames with regard to stability, efficiency, low noise levels etc., one of their principal disadvantages is the difficulty encountered with design of burners which can support hot premixed flames of high burning velocity, e.g. oxy-hydrogen and oxy-acetylene. In order to ensure a high streaming velocity and safe operation with these mixtures very narrow slots and good heat dissipation are required. This may introduce difficulties with clogging of the burner slot by deposition of material from samples containing large amounts of dissolved solids. Such high burning velocity mixtures are normally supported as unpremixed flames, although burners to produce premixed flames of oxy-acetylene for both atomic emission and absorption spectroscopy have been designed [174—176].

In most cases the burners used for AFS to date have been similar to those commonly employed to produce premixed flames for atomic emission spectroscopy. Where fluorescence emission from the flame is to be collected over an appreciable solid angle, cylindrical premixed flames supported on conventional Meker-type burners have most frequently been employed. Figure 28 shows some common burner types for premixed flames used in AAS and AFS.

In addition to the primary requirement of a sufficiently high streaming velocity in the gas as it emerges from the burner plate to

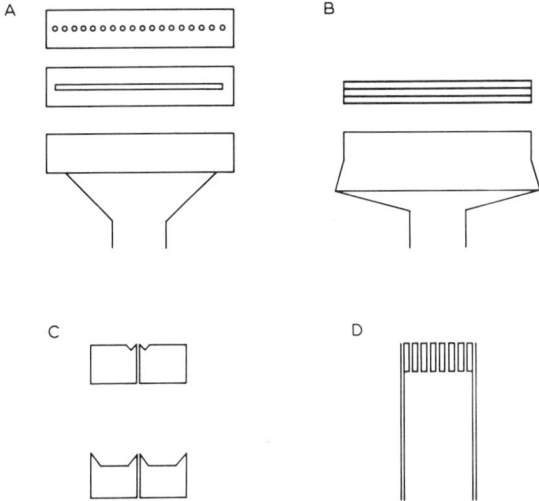

Fig. 28. Some commonly employed burner heads. A. Long path burner head with slot or holes. B. Triple slot burner. C. End-on view of grooved burner heads for long-path, nitrous oxide—acetylene flame. D. Meker-type burner head to produce a cylindrical flame for AFS.

ensure safe handling, several other requirements for the design of efficient burners exist. These are listed below.

(i) The burner plate should have sufficient thickness to ensure as near laminar flow as possible in the emergent gas stream and to act as an efficient 'flame trap' to minimise risk of flash-back.

(ii) Efficient dissipation of heat from the burner should be attained by the use of a material of good thermal conductivity.

(iii) The burner should be free from clogging effects at the slot or orifices caused by deposits produced on nebulisation of samples of high dissolved-solid content. This requirement can be satisfied by ensuring that the burner slot (or individual orifices) is not too narrow, but a compromise must be reached for the width which enables safe operation without tendency to excessive clogging.

(iv) The edge of the slot should not be prone to carbon deposition when hot flames are used. Carbon may be formed at the slot by partial degradation of the fuel, particularly with hot, fuel-rich hydrocarbon flames such as nitrous oxide—acetylene. This may cause flame noise and instability. This problem is overcome in several commercially available burners by the use of a raised edge at the burner slot (see Fig. 28) [177].

(v) The burner should be free from tendency to corrosion by acid solutions nebulised into the flame. Apart from limiting the useful life of the burner plate, corrosion may lead to errors and low precision in analysis due to volatilisation of the corrosion products into the flame [178]. Burner bodies are usually constructed of stainless steel, but glass [179] and plastic [180] have also been used. Stainless steel has most frequently also been used for the burner top plate, although gilded brass [181] and tantalum have been employed; the use of titanium for the burner top plate is becoming more popular [182]. Water cooled burner heads have been employed with various instruments, and some workers claim that this results in elimination of slow drift in the analytical signals in AAS [183]; excessive cooling, however, may result in deposition in the slot of solid materials from the samples nebulised.

These requirements for safe-handling, long operating lifetime and efficient nebulisation of liquid samples into the flame impose burner design restrictions which result in the very similar appearance of many of the premixed flame burners used with commercial analytical flame spectrometers.

(B) INTRODUCTION OF SAMPLES INTO PREMIXED FLAMES

In order to permit the quantitative analytical application of AAS or AFS when a flame is employed as the 'sample cell', it is necessary to ensure the uniform and efficient transfer of the samples into the flame. Flame sample cells provide best reproducibility when supplied with liquid samples; the most widely used form of transfer for such samples is as a fine aerosol mist produced by a nebuliser*. In the most commonly encountered form of this device the liquid sample is introduced into a high-velocity jet of the gas employed as oxidant in the gas mixture. The mist produced is then mixed with the fuel gas (and frequently additional oxidant gas) and passed via a mixing/expansion chamber to the burner base. The principle and application of this

---

* The word *nebuliser* has virtually replaced the older misleading term *atomiser* to describe the device employed to produce the sample mist; the word atomiser is now used to describe the system, in this case a flame, in which free atoms are generated from the analytical sample.

182

type of pneumatic nebulisation for use with premixed flames is well-established and has been widely exploited in analytical flame emission spectroscopy.

(1) Pneumatic Nebulisers and Spray Chambers

Many different arrangements for pneumatic nebulisation have been described. The two most commonly used nebuliser geometries are shown in Fig. 29. In the simplest *right-angle nebuliser*, a duct for the gaseous stream is placed at 90° to a duct for the sample solution.

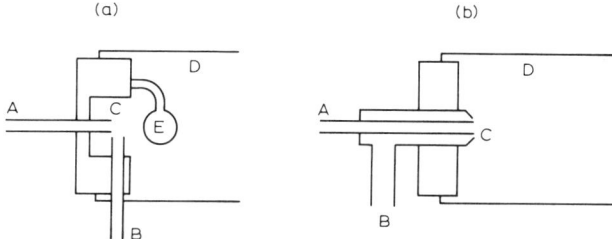

Fig. 29. Diagramatic representation of typical indirect nebuliser heads. (a) Right-angle type with impact ball. (b) Concentric type. A, Sample uptake capillary; B, nebuliser gas inlet; C, nebuliser jet; D, expansion chamber; E, impact ball.

The output of the nebuliser can be increased by surrounding the gas outlet with a circular orifice for the liquid to form a *concentric nebuliser*. In both types the stream of gas emerges from its nozzle with high velocity, entrains some liquid from the outlet of the liquid duct, and tears from the liquid surface fragments which then disperse into a cloud of fine droplets.

The most important general requirements for the nebuliser used to supply a premixed flame with sample are

(i) The nebuliser should produce a stable and reproducible rate of sample aspiration. The spray-rate should not depend on the hydraulic head, i.e. the sample uptake rate should not change as the solution level changes in the sample container. The density, surface tension, and viscosity of the sample should exert minimal effect on the spray rate.

(ii) The mean drop size in the mist produced should be reproducible and as small as possible. The sensitivity attainable and the severity of chemical interferences depend on the mean drop size; several workers have shown that less serious interferences in flame emission and absorption spectroscopy are encountered when the drop size is as small as possible [184,185].

(iii) The highest possible sample concentration per unit volume of nebulising gas should reach the base of the burner after the larger drops have fallen out of the stream in the spray-chamber. The supply of solution should be provided automatically, through a Venturi effect, by the gaseous stream used to operate the nebuliser. The flow rate of oxidant gas producing the optimum spray rate should be as close as possible to that needed to support the required flame.

(iv) A constant concentration of analyte should be obtained in the flame rapidly after the commencement of nebulisation. The speed with which a constant signal is obtained depends on the homogeneity of the aerosol mist as a function of time from the moment the sample is introduced. This factor also depends on the shape and size of the spray chamber and the distance between the nebuliser and burner. The nebuliser and spray chamber should be easily and rapidly washed free of the residue of one sample before the introduction of the next.

(v) The nebuliser and spray chamber should not tend to become clogged after long periods of use; it should be constructed of a material which resists corrosion, and it should be easily dismantled for cleaning and maintainance.

Glass, quartz, plastic, stainless steel, platinum, and silver nebuliser units have all been recommended for analytical work, but the most common nebulisers used on modern flame spectrometers are made of platinum—iridium alloy (90%:10%). Glass or plastic spray chambers of several different designs are usually employed with these nebulisers. The spray chamber is generally fitted with a drain tube which has a U-tube or other means of maintaining a hydraulic closure, and in some instruments an additional side tube is fitted to the chamber to permit the supply of an auxiliary flow of oxidant gas. The fuel gas is often also introduced into the oxidant—analyte mist in the spray chamber to ensure thorough mixing before the gas passes into the burner. For reasons of safety, however, particularly when a glass spray chamber is employed, the fuel gas is introduced into the oxidant—analyte mist only after it has passed to the base of the burner head in most instruments. The dimensions of the spray chamber affect the fraction of the sample nebulised which reaches the flame, the analytical signal reproducibility and the time required after commencement of nebulisation before maximum analytical signal is obtained. The use of a large-volume spray chamber decreases the fraction of the sample nebulised which reaches the flame and lengthens the time required to establish the maximum signal. It may, however, produce a damping effect on fluctuations in aerosol density which occur due to poor nebuliser design or pressure variations in the nebuliser gas supply; this damping effect leads to improvement in flame stability and signal noise levels. The use of a small-volume spray chamber, on the other hand, permits rapid attainment of the maximum signal and speeds up the analysis as the 'wash' time is also decreased, while a slight improvement in analytical sensitivity may also be obtained due to its somewhat higher efficiency.

The sample uptake rate at a pneumatic nebuliser depends on the capillary diameter and oxidant gas pressure employed. In many units the gas attains supersonic velocities as it emerges from the jet. This high velocity produces the small size droplets which are required for efficient nebulisation. The sample uptake rate and the volume of gas produced after the jet both increase almost proportionally when the gas operating pressure before the jet is increased. The concentration of aerosol in the gas, and therefore the concentration of analyte per unit volume of flame gases, may therefore not increase substantially. Little practical advantage in analytical detection limits may therefore be obtained.

The performance of an indirect nebuliser—spray chamber unit is partly characterised by the fraction of the original solid content of analytical samples nebulised which ultimately reaches the flame. This depends on the drop size distribution which is produced during nebulisation. The presence of a high proportion of large droplets in the aerosol mist is undesirable. Many of the large droplets collect on the walls of the spray chamber and burner connections and do not reach the flame. Some nebulisers are fitted with an impact wall near the jet (see Fig. 29); this serves to further break down the larger droplets in the mist near the jet. Several nebuliser spray chambers produced commercially are designed to separate the larger droplets efficiently by centrifugal force or by use of suitably shaped baffles in the spray chamber. Nebuliser and spray chamber assemblies are therefore designed with the primary objective of transferring the highest possible concentration of the analyte to the flame in the form of small droplets of the nebulised solution. All droplets in the mist lose some solvent by evaporation during their passage through the spray chamber and burner connections. With many indirect nebuliser and spray chamber assemblies which produce a small mean drop size it may be expected that a substantial fraction of the droplets will lose all their solvent by evaporation during their residence time between nebuliser and burner head. Under these conditions a 'solid aerosol' of solute crystals is formed which is then directly vaporised in the flame.

The efficiency of indirect nebuliser and spray chamber assemblies, defined as the fraction of the nebulised analyte which passes into the flame, is usually low. With aqueous solutions an efficiency in the range 1—10% is commonly obtained. When organic solvents are employed, however, the efficiency may be as high as 20%. This organic solvent effect results from the effect of the surface tension, viscosity and vapour pressure of the organic solvent in decreasing the mean drop size and increasing the rate of evaporation of solvent from the droplets. This effects an increase in the yield of sample reaching the flame. A rough assessment of the efficiency of a nebuliser and spray chamber assembly may be obtained by measurement of the rate of collection of unused nebulised sample which collects in the drain tube from the nebuliser and of the sample uptake rate at the nebuliser capillary. The difference between these values represents roughly the rate of delivery of aerosol mist to the burner and this may be expressed as a percentage of the sample uptake rate to obtain the percentage efficiency. This procedure does not take into account evaporation of solvent from the droplets or loss of solute by deposition of solid particles in the burner connections or the burner itself. The value obtained, therefore, which is commonly termed the nebuliser efficiency, does not accurately express the rate of introduction of the analyte element into the flame.

(2) Non-pneumatic Nebulisers

The simple indirect pneumatic nebuliser has found widespread application in analytical flame spectroscopy. This method of nebulisation has several attractive features when due attention is paid to the design of the assembly. These include simplicity, ease of operation, reproducibility, small sample volume requirements and little tendency to memory effects. Several disadvantages, however, exist with pneumatic nebulisation. As described above, the efficiency is usually low. The nebuliser performance varies with the viscosity and surface tension of the solution, temperature of the nebulising gas and solution and the vapour pressure of the solvent. Sample solutions of high salt content may cause problems by blocking the nebuliser jet outlet. There is much interest in the development of other forms of non-pneumatic nebuliser to overcome some of these disadvantages. Centrifugal nebulisers, in which the sample solution drops onto a rapidly spinning disc and is divided into fine droplets by the centrifugal force, have been investigated [186,187]. Walton and Prewett [188] reported the formation of a uniform spray of droplets from 3—15 μm in diameter with this type of nebuliser, and several other workers have described the performance of nebulisers of this type. Ultrasonic nebulisation methods have also been employed [189,190]. The liquid sample is placed in contact with a rapidly vibrating plate of suitable material (quartz, barium titanate) and is dispersed into very fine droplets. The mean droplet size achieved may frequently be smaller than that obtained with pneumatic nebulisers and these methods show considerable promise. Several forms of electric and electrostatic nebulisers have been described. In these devices the sample is nebulised by the direct action of a spark at the solution surface [191,192] or by the production of a very high electrostatic field at the surface which gives rise to forces which effect the nebulisation [193,194]. Although these and other types of non-pneumatic devices which have been studied, each possess particular advantages for some applications, these benefits are frequently obtained at the expense of simplicity, ease and speed of operation and freedom from memory effects. No one of these methods, with the possible exception of ultrasonic nebulisation, shows promise of becoming competitive with pneumatic nebulisation for widespread adoption to routine analytical applications.

(C) UNPREMIXED FLAMES

In unpremixed, or turbulent, flames the fuel and oxidant mix only at the point of ignition. No state of laminar flow is therefore established in the gas mixture before the combustion reaction, although in some diffusion flames both the fuel and oxidant streams may be constrained to laminar flow before the flame front. A primary reaction zone may be observed, but it is frequently not well-defined, and with total consumption nebuliser—burners the inhomogeneity and turbulence causes the burning velocity and direction of propagation of the flame front to show irregular local variations. This irregular and diffuse nature of the primary reaction zone is responsible for the high level of acoustic noise produced and the extensive volume of the flame in which high and noisy background radiation is observed. Air is entrained into the flame to produce a secondary reaction zone similar to, but less well-defined than, that of a premixed flame. In analytical flame spectroscopy only fuel—oxidant mixtures of very high burning velocity are usually burnt as unpremixed flames; oxy-hydrogen and oxy-acetylene flames may be supported with safety as unpremixed flames in this way. Recently, however, unpremixed flames of lower temperature and burning velocity, such as hydrogen-air, hydrogen—entrained air and hydrogen—argon—entrained air flames, have been used at total consumption nebuliser—burners for atomic fluorescence spectroscopy with good results.

(1) Nebuliser-burners for Unpremixed Flames

In virtually all of the applications of turbulent oxy-hydrogen and oxy-acetylene flames to AAS and AFS nebuliser-burner units of the total consumption type have been employed. In these devices the sample aerosol is introduced directly into the flame at the point where the fuel and oxidant mix and combustion occurs. The device in most widespread use, which was introduced by Gilbert [195] and is known as the total consumption nebuliser—burner, is shown diagramatically in Fig. 30. A central capillary tube, which carries the

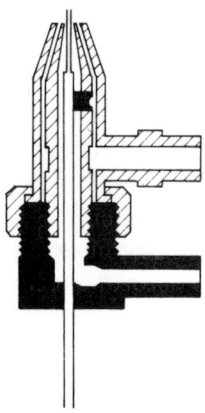

Fig. 30. Nebuliser—burner for unpremixed flames. (Beckman Instruments, Fullerton, California.)

188

solution to be nebulised, is surrounded by a nozzle for oxidant gas. This functions in a similar manner to a concentric nebuliser unit (see Section 4(B)(1)) and the oxygen or other oxidant gas produces an aerosol from the sample solution. The fuel gas (hydrogen or acetylene) is introduced through a second nozzle which surrounds the first. A turbulent diffusion flame is produced above the nozzle outlets where the gases and sample mix. The hot flame thus does not touch the burner itself, although before the fuel has mixed with the oxygen to give rise to the main part of the flame some fuel is consumed by combustion with atmospheric oxygen to produce a small diffusion flame which rests on the burner rim. The turbulence of the flame produces acoustic noise which is particularly pronounced when solutions are nebulised into the flame.

Total consumption nebuliser—burner units show several advantages over laminar premixed flames used with indirect nebulisers for analytical work. Their principal advantages are

(i) The nebulisation efficiency is 100%, i.e. all the sample solution reaches the flame and there are no losses as by condensation in the spray chamber of an indirect nebuliser.

(ii) The nebuliser—burner unit is relatively simple to manufacture and operate and flash-back cannot occur with flames of high burning velocity. Organic solvents may be safely aspirated.

Several relatively serious disadvantages exist for these burner units, however, compared with premixed flames with indirect nebulisers.

(i) A relatively large fraction of the nebulised sample may enter the flame as large droplets. These are incompletely evaporated by the flame [196,197] and pronounced chemical interferences may be encountered for the analyte element. The flame may also be cooled appreciably by the high sample flow rate [198].

(ii) The flames produce acoustic noise and frequently also give rise to relatively high noise levels at the detector of the flame spectrometer employed.

(iii) The units may suffer from incrustation of the tip of the burner or blockage of the sample capillary on prolonged usage.

Several workers have modified total consumption nebuliser—burner units to improve their performance in analytical work. Thus Gilbert [199] has described the use of the nebuliser—burner with oxygen or air sheathing to improve the flame stability and increase emission intensities. Kniseley et al. [174,175] have reported modification of the Beckman burner to produce a premixed oxy-acetylene flame. Several workers have recommended modifications of this type of burner which permit the supply of the liquid sample to the flame at a reproducible rate [198,200].

Total consumption nebuliser—burner units have been used extensively in analytical flame emission spectroscopy. Their application in AAS has been limited owing to the preference for flames which provide long optical path lengths. The effective path length may be increased, however, by passing the light beam through the flame several times [201]. Slavin and Manning [202] and Dowling et al. [203] have used the turbulent oxy-acetylene flame for AAS with organic solvents. At the time of writing, total consumption nebuliser—burner units are also finding application in AFS to support cool flames in addition to their more conventional use with hot oxygen supported flames.

(D) FLAMES USED IN ANALYTICAL FLAME SPECTROSCOPY

The preceding sections have described the two main flame types useful in analytical work. In the section on premixed flames (Section 4(A)) reference has been made throughout to premixed hydrocarbon—air flames. These have been the most commonly used flames to date. The flames employed in atomic absorption or fluorescence spectroscopy, however, must provide for the formation and maintenance of free atoms from elements introduced as solutions of their salts. While the turbulent oxy-acetylene flame has this capability for a wide range of elements, the disadvantages of the use of turbulent flames, particularly in AAS where a long-path length flame is preferred, have led to more interest in the development of premixed flames which similarly permit efficient atomisation. The most commonly employed flames used until recently in AAS were the premixed flames of air with town gas (ca. 2100°C), propane (ca. 2200°C) or acetylene (ca. 2500°C). The efficiency of free atom formation, which depends on the composition and temperature of the flame, is low for some elements in these flames and chemical interferences are also prevalent. Sensitive determination by AAS is only possible for about 35 elements in these flames, and elements which form thermally stable oxide species are not efficiently atomised. The sensitive determination of elements such as aluminium, beryllium, boron, germanium, hafnium, molybdenum, niobium, silicon, tantalum, titanium, vanadium, tungsten, zirconium, uranium and the rare earths is

190

therefore not possible. The considerable attention paid to the development of premixed flames to provide more efficient atomisation of these elements has had ccnsiderable success. In atomic fluorescence spectroscopy, in addition to the criterion of efficient atomisation and minimisation of interferences, the flame medium employed should also exhibit low background absorption and emission and minimal quenching properties. Although these two sets of requirements are difficult to realise with the same flame, at the time of writing the development of flames for AFS work which strike an effective compromise between these requirements is proceeding rapidly.

The following section presents a brief outline of the properties of some of the premixed flames other than those of the hydrocarbons with air which have found analytical application.

(1) Premixed Nitrous Oxide-Acetylene Flame

Undoubtedly the most useful flame which has been introduced since 1965 has been the premixed flame of nitrous oxide and acetylene. Willis [204] pointed out its high temperature (ca. 3000°C) and relatively low burning velocity and described its use for the sensitive determination by AAS of several refractory oxide-forming elements. The flame can be supported safely at a slot burner of length as great as 8—10 cm, although most current commercial burners are fabricated from stainless steel or titanium and employ a slot 5 cm in length and about 0.5 mm wide.

The flame shows an intense white-blue primary reaction zone and blue secondary reaction zone typical of most premixed hydrocarbon flames. The interconal zone of the fuel-rich flame* exhibits a red colour, commonly known as the 'red-feather'. The height of this zone may become as great as 3—4 cm as the fuel:oxidant ratio is increased before it becomes obscured by the yellow luminosity from incandescent carbon particles which extends throughout the flame when the mixture is very fuel-rich. It is apparent from both emission [205] and absorption [206] measurements that free atoms of many elements exist only in this interconal zone of the fuel-rich flame and

---

* A stoichiometric mixture is defined here as 1 mole acetylene and 2 moles of nitrous oxide for the reaction

$$C_2H_2 + 2\,N_2O \rightarrow 2\,CO + H_2 + 2\,N_2$$

not in the secondary reaction zone or in lean flames. Thus although little temperature difference has been observed above the primary reaction zone for lean, stoichiometric and fuel-rich flames [207,208], a fuel-rich flame is required for atomisation of many elements. The interconal zone of the fuel-rich flame contains substantial concentrations of carbon containing species [207] and very low concentrations of oxidising species. It seems probable that this reducing environment assists the dissociation of metal oxide species to provide free atoms of the metal.

The use of this flame in AAS has now become widespread for the determination of about 35 elements which prove difficult or impossible to atomise in cooler flames. Several chemical interferences which are troublesome in cooler flames, such as that of phosphate on the determination of calcium, are also completely eliminated in this flame [209]. The high flame temperature results in appreciable ionisation for elements such as calcium, barium, aluminium, and some rare earth elements. This is normally suppressed in practical analysis by the addition of a large excess of an element (such as potassium) with a lower ionisation potential.

Suitable burners for the use of this flame in AAS are provided by all the manufacturers of commercial atomic absorption spectrophotometers and detailed accounts of their performance have been published. The use of this flame in atomic emission spectroscopy with a long-slot [210] or circular-slot burner has been described [211]. A premixed nitrous oxide—acetylene flame burning at a total consumption nebuliser—burner unit has also been recommended for atomic emission spectroscopy [212]. Several reports of the use of this flame in atomic fluorescence spectroscopy have been published [213—216].

(2) Premixed Oxy-Acetylene Flames

The use in AAS of premixed flames of acetylene with oxygen diluted with nitrogen has been described by Amos and coworkers [206,217]. Amos and Thomas [217] studied the absorption of aluminium with such flames, and Amos and Willis [206] extended the investigation to the detection of other elements forming refractory oxides. Mixtures containing as much as 50% oxygen were burnt at a stainless steel burner head 1.25 cm thick containing a slot 3 cm in length and 0.45 mm wide.

192

Fassel and coworkers [176] have described a long-path burner suitable for use with fuel-rich premixed oxygen—acetylene flames containing no nitrogen diluent. These flames may attain a temperature of 3300 K and exhibit an interconal zone in which carbon-containing species are abundant. The high temperature and the reducing environment provide for the formation of free atoms from refractory oxide-forming elements. The limits of detection obtained in AAS with this flame for twelve such elements were mostly similar to those obtained with a nitrous oxide—acetylene flame supported at the same burner; detection limits for lanthanum and uranium were appreciably better in the oxy-acetylene flame, however. In order to avoid flash-back, a high flow rate of gas mixture must be used at the narrow slot (8 cm long, 0.25 mm wide) and the mixture must always be maintained fuel-rich.

(3) Premixed Nitrous Oxide-Hydrogen Flame

The use of the nitrous oxide—hydrogen flame (ca. 2900°K) in emission spectroscopy has been reported to yield high sensitivity in the detection of the elements cadmium, calcium, and zinc [218]. The flame exhibits low background radiation in the visible region and should prove useful in emission and fluorescence analysis. This flame has been employed for the AAS detection of the alkaline earth elements, but with lower sensitivity than that obtained using the nitrous oxide—acetylene flame [219]. Elements which form refractory oxides do not appear to be appreciably atomised in this flame.

(4) Hydrogen Flames

Premixed air—hydrogen flames are useful in the AAS determination of elements which are easily atomised. An exception to this is tin, which although relatively difficult to atomise, gives better sensitivity in the air—hydrogen flame than in the air—acetylene flame. Tin atom formation in this case may proceed via an intermediate tin hydride species.

Unpremixed air—hydrogen and hydrogen—oxygen flames have found extensive application in atomic fluorescence spectroscopy [14,92,107,220,221]. Diffusion flames in which combustion proceeds with the entrainment of air into hydrogen [221], hydrogen—nitrogen [94], or hydrogen—argon [57,222] mixtures have also been employed in AAS and AFS. The principal advantage of this type of flame in AAS is the low flame background absorbance in the ultraviolet near and below 200 nm for the elements arsenic, selenium and tellurium. Rains [223], for example, shows flame background absorbance values at the arsenic resonance line at 193.7 nm for premixed air—acetylene, air—hydrogen, and argon—hydrogen entrained air flames of 1.35, 0.96 and 0.27 respectively. As the line intensity output of arsenic and selenium hollow cathode lamps is frequently poorer than that of other hollow cathode sources the gain in energy received at the detector resulting from the use of a flame of low background absorption is a significant advantage. In addition to this advantage in AAS, these flames show the desirable property in AFS of low flame background emission and low atomic emission from analyte elements. This may result in improved AFS detection limits because of the lower noise levels experienced. The use of argon diluent rather than nitrogen also results in a somewhat higher flame temperature and less severe quenching of the fluorescence by the flame species. These flames are relatively cool, however, and interference from foreign ions may be encountered in AAS and AFS due to inefficient vaporisation, and to scattering of incident radiation in AFS.

(5) Absorption Tube Devices

In order to provide for increased sensitivity in AAS via the use of a long path length when a total consumption nebuliser—burner is employed, several authors have described burner systems in which the flame gases are directed along the optical axis. Thus Robinson [147] and Feldman and Dhumwad [224] passed the flame into a T-piece adaptor which directed the flame horizontally to produce a long absorption cell. Fuwa and Vallee [225] used a long horizontal ceramic tube 90 cm X 1 cm and pointed an air—hydrogen flame from an inclined total consumption nebuliser—burner into the tube. They have reported very high sensitivity for cadmium, zinc, magnesium, copper, cobalt and nickel. Several other groups of workers have utilised similar techniques and extended the range of application of these systems to other elements [226—231]. The flames most commonly used have been unpremixed hydrogen—air and hydrogen—oxygen. In general the absorption tube technique improves analytical detection limits in AAS for elements whose oxides are readily dissociated. For elements which form more stable oxides the absorbance falls rapidly along the length of the tube and little or no overall increase in sensitivity may be observed compared with measurements made in a conventional flame. A disadvantage of these absorption tube systems is the susceptibility to 'memory effects' due to deposition of nebulised material onto the walls of the tube. High flame background absorption may also be encountered from the flame gases and from gaseous molecular species of concomitant elements in the flame.

(6) Separated Flames

The application to analytical flame spectroscopy of separated flames, in which the secondary reaction zone is prevented from surrounding the interconal zone as in conventional premixed hydrocarbon—air flames, has been described [232]. A Smithells-type glass or silica separator may be employed to prevent the access of the atmospheric oxygen necessary for secondary combustion above the primary zone, but the use of an inert nitrogen or argon shielding gas is preferred. The atomic population produced in the hot interconal zone from samples nebulised into these flames may be utilised in flame atomic emission [233—236], atomic absorption [237,238], and atomic fluorescence spectroscopy [110,239,214,216,213] without interference from much of the background emission or absorption and noise contributed by the secondary reaction zone in conventional flames. The principal advantages of separated flames are exploited in their application to flame emission and atomic fluorescence spectroscopy. The nitrogen-separated air—acetylene flame has been found useful in the determination of arsenic and selenium by AAS, however, where a substantial reduction in flame background and noise has been reported on separation [237]. Improved detection limits and somewhat higher sensitivities compared to the conventional flame have also been reported with the use of a nitrogen or argon shielded premixed nitrous oxide—acetylene flame for determination by AAS of elements which form refractory oxides [238]. Separation of the fuel-rich flame leads to protection of the reducing environment in the interconal zone. A greater total number of free atoms is available for use in the extended interconal zone. Superior detection limits in AFS compared to the conventional flame have been reported for the premixed nitrous oxide—acetylene [213,214,216] and air—acetylene flames [110,239].

(E) FORMATION OF ATOMS IN FLAMES

In each of the three techniques of analytical flame spectroscopy the most basic function of the flame is to provide free atoms of the element to be determined. The manner in which the atoms are formed in the gaseous phase from samples nebulised into the flame as droplets of sample solution is quite complex. It is first of all necessary for the sample droplets in the aerosol mist to lose their solvent and form dry particles of the solute. The particles are then vaporised at the flame temperature and may give rise to molecules in the gaseous state. Any molecular species formed must then undergo dissociation to form the free atomic species which are to be used for analysis. This chain of processes which are involved is illustrated diagrammatically in Fig. 31. The rate at which the relevant processes occur, and the extent to which they occur, is strongly dependent on the nature of the sample, solvent, and flame type. The following sections outline briefly the manner in which desolvation, vaporisation, dissociation, and ionisation affect the atomic population finally produced in the flame from nebulised sample solutions.

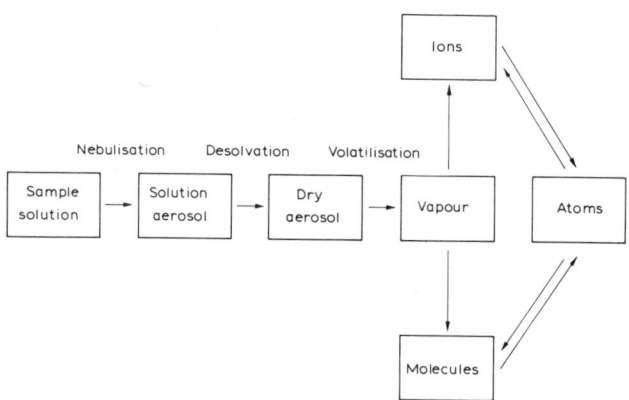

Fig. 31. Processes involved in flame spectroscopy between sample introduction and measurement.

196

(1) Desolvation

When indirect nebuliser units are employed with premixed flames the sample droplets begin to lose solvent by evaporation as soon as they are formed in the spray chamber. In this way with an efficient nebuliser the smallest droplets which are produced may lose virtually all of their solvent during their transit time from nebuliser spray head to the burner slot. They then enter the flame as dry particles of the solid material contained in the sample solution. The larger droplets, however, may only lose their solvent completely when they enter the flame. Some desolvation occurs as the aerosol droplets pass through the preheating zone immediately before they enter the primary reaction zone, but many of the droplets are only completely dried out as they pass through the primary reaction zone and enter the flame proper. The rate of desolvation in the flame is determined principally by the rate of heat transfer from the flame to the droplet [251]. Heat is transferred by conduction through the gas to evaporate the solvent and heat the solute for vaporisation. It has been demonstrated [241,242] that assuming conduction to be the principal method of heat transfer and the approximate independence of the thermal conductivity of the gas on the temperature, the rate of loss of mass during desolvation is

$$-\frac{dm}{dt} = (4\pi r\lambda/C_p)\ln[1 + C_p(T-T_b)L]$$

where λ is the thermal conductivity of the gas, r is the radius of the droplet, L is the specific heat of vaporisation of the solvent, C_p is the average specific heat of the vapour at constant pressure, and T and T_b are the flame temperature and boiling point of the solvent, respectively.

As the rate of loss of mass depends on the droplet radius, the rate of loss of surface area, $-dA/dt$, per unit time can be shown to be constant. The total time t taken for a droplet of liquid of density ρ and initial surface area A_0 to be completely desolvated becomes

$$t = \frac{\rho A_0 C_p/8\pi\lambda}{\ln[1 + C_p(T-T_b)L]}$$

Alkemade [242] demonstrates in this way that water droplets of initial diameter 10 μm will be evaporated completely within 0.3 msec in the air—acetylene flame. For a flame of vertical rise velocity of 10 msec^{-1} this time corresponds to a height interval of 3 mm. Well-designed indirect nebuliser units frequently provide average initial droplet diameters smaller than 10 μm so that problems of incomplete desolvation are seldom encountered. The abundance of large droplets introduced into unpremixed flames of high rise velocity as total consumption nebuliser-burner units, however, may result in the incomplete desolvation of sample solutions during their residence time in these flames.

The rate of desolvation of droplets in the flame is affected if the solvent is combustible. In addition to the enhanced desolvation which occurs in the spray—chamber when organic solvents are nebulised due to their low surface tension and high vapour pressure at room temperature, the desolvation rate in the flame itself is increased by the heat released when the solvent vapour is combusted.

Incomplete desolvation results in a fraction of the analyte element in the sample remaining unconverted finally into free atoms required for AAS or AFS. The percentage loss of atoms of the analyte element in this way, however, does not depend on the concentration in the sample solution. Incomplete desolvation should not therefore lead to non-linearity of analytical working curves. The depressive effect of incomplete desolvation on the analytical sensitivity in this way may be minimised by observation of the analytical signal higher in the flame (i.e. after increased residence time), or by the use of organic solvents or a hotter flame.

The energy which must be supplied by the flame in order to evaporate the solvent may be a significant fraction of the energy available in the flame. This effect, which is pronounced with total consumption nebuliser—burner units, results in a lowering of the flame temperature when sample solutions are nebulised. The temperature decrease is well-established experimentally and a considerable amount of data is available concerning the effect of sample aspiration on flame temperature for unpremixed flames [198,243—246]. In a premixed flame with indirect nebuliser, where partial desolvation has occurred before the aerosol enters the flame, however, the cooling effect of solvent evaporation is usually negligible.

198

(2) Vaporisation

The processes which occur during the final stages of desolvation and the vaporisation of the dry aerosol particles are complex and may vary considerably depending on the nature of the analyte and concomitant elements in the sample nebulised. The chemical composition of the dry particle produced after the final solvent has evaporated may also depend on the solvent used, the flame gas composition and the flame temperature.

The rate of vaporisation of the analyte element from the dry particle is also usually influenced by the same factors. The rate of formation of molecules and atoms in the vapour phase may be maintained constant for the purposes of analytical flame spectroscopy when the above factors are carefully controlled. The nature and concentration of the concomitant elements present in the sample may not be held constant in practical samples, however, and their effect on the analyte vaporisation rate may then give rise to *condensed phase interference*. Obviously, the optimum working conditions in analysis will result when all of the material present in the dry particle evaporates rapidly, and consequently the use of the highest temperature flames usually results in less prevalent condensed phase interferences.

It is not essential for the melting and boiling points of the material of the dry particle to be lower than the flame temperature for complete evaporation to be obtained. The final partial pressure of the analyte element in the flame is usually very low and seldom exceeds ca. 10^{-3} atm. [242]. This low pressure is usually much less than the saturation pressure of the analyte in the solid or molten state in the flame and vaporisation must therefore ultimately become complete. When the melting and boiling points of the compound in the dry aerosol lie well below the flame temperature a high rate of evaporation is normally obtained. This type of behaviour is observed with elements such as the alkali metals, copper, zinc, cadmium and silver. Complete evaporation into the gaseous state of the compounds formed from these elements in the dry aerosol is thus obtained well within their residence time in most premixed flames of hydrocarbons with air. When the melting and boiling points of the compound in the dry aerosol are comparable to or greater than the flame temperature vaporisation will tend to proceed slowly and the residence time of the particle in the flame may be insufficient to allow complete volatilisation. Under these conditions the use of a hotter flame, observation of the signal at a greater height above the burner, or reduction in the mean drop diameter in the nebuliser is desirable. The use of high solution concentrations, which will produce a larger mean particle size, should also be avoided if possible as this effect also increases the time required for complete vaporisation in the flame.

The manner in which the rate of vaporisation depends on the analyte and concomitant elements present in the sample is complicated by the fact that the salts of the elements introduced into the flame may decompose into other species depending on the flame temperature and composition. For example, the rate of attainment of complete vaporisation for magnesium as analyte element is markedly dependent upon the nature of the magnesium salt introduced into the flame. Magnesium sulphate decomposes into magnesium oxide with the release of oxides of sulphur at ca. 1150 K: magnesium oxide has a melting point of ca. 3070 K and a boiling point of 3850 K. Solid magnesium oxide, however, has a sublimation temperature of 3040 K. In flames of hydrocarbons with air, therefore, the rate of volatilisation of magnesium oxide would be slow, whereas in hot flames (e.g. oxy-acetylene) solid magnesium oxide might sublime before melting [247]. Halls and Townshend [248] have suggested that in a hydrocarbon—air flame a reaction of H radicals with solid magnesium oxide to produce gaseous OH radicals and magnesium atoms might occur at the surface of the solid. Magnesium nitrate melts at 500 K and decomposes at ca. 600 K to magnesium oxide. If the molten state of the nitrate can be produced rapidly by heating, however, before decomposition to the oxide, vaporisation may proceed directly from this liquid state. Magnesium chloride in solution in nebulised droplets will form $MgCl_2.H_2O$ in the dry aerosol particle on heating to 430 K. This compound may then decompose in two ways: it may form solid magnesium chloride and water vapour; or it may form solid magnesium oxide and gaseous HCl. Magnesium chloride boils at about 1650 K and volatilisation of this compound via the former mechanism is an effective method of populating the flame with gaseous magnesium species. If the latter route is followed, however, again volatilisation of magnesium oxide would be expected to be a slow process in hydrocarbon—air flames.

Similar phenomena are encountered for other analyte elements whose salts decompose readily at low temperatures to form involatile oxide species. Thus aluminium nitrate decomposes at quite a low temperature (ca. 400 K) to form aluminium oxide which is difficult to vaporise. Aluminium fluoride, however, has a melting point of 1300 K and sublimes at 1475 K. It is thus more stable than the nitrate and vaporisation may proceed directly from the molten state. Hydrofluoric acid is frequently added to aluminium sample solutions to effect enhancement of atomic absorption sensitivity for this element in this way. An illustration of the dependence on the volatility of the oxide particles for the volatilisation process rather than on the volatility of the metallic states is given by consideration of molybdenum and aluminium. In hydrocarbon—air flames only low concentrations of free aluminium atoms are obtained compared to the concentrations of molybdenum atoms produced using solutions of equal concentration. Metallic aluminium, however, is much more volatile than molybdenum, but molybdenum oxide is more volatile than aluminium oxide [249].

Incomplete vaporisation of the aerosol particles containing both the analyte and concomitant elements can give rise to loss of analytical sensitivity and selectivity in AAS and AFS. The highest possible concentration of free atoms of analyte, and therefore the highest sensitivity, can only be produced if vaporisation is complete. The concomitant elements present may form compounds in the condensed phase with the analyte element which are relatively less volatile than those formed in the absence of the concomitant element. Effects of this type are quite commonly encountered in air—hydrocarbon flames, e.g. the effect of phosphate or aluminium on calcium emission or absorption at 422.7 nm. due to formation of calcium-oxygen-(P) or (Al) compounds, and are observed as 'interference' by a systematic change in the analytical signal. Owing to the high rise velocity and larger mean size of the dry aerosol particles in unpremixed flames supported at total-consumption nebuliser—burners, this type of effect due to incomplete vaporisation is usually more pronounced in these flames than in premixed flames with indirect nebulisers.

Further consideration of reactions in the condensed phase which affect the rate of vaporisation of the dry aerosol is given in Section 6(B)(1), concerned with chemical interferences.

(3) Dissociation

In many premixed hydrocarbon—air flames used in analysis, thermal equilibrium may be considered to prevail above the primary reaction zone. A characteristic temperature may thus be assigned to the flame gases in this region. Under these conditions equilibrium concentrations at the flame temperature of the flame molecules and atoms are present (e.g. OH, O, H_2, H, CO, CO_2, O_2 and N_2). When the dry aerosol produced from the analytical sample vaporises, the equilibrium concentrations of the molecular and atomic analyte species which are formed in the gaseous phase are governed by the thermal equilibrium in the burnt gases.

The dissociation equilibrium of a molecule MA containing the analyte element M

$$MA \rightleftharpoons M + A \tag{61}$$

is characterised by a temperature-dependent equilibrium constant K_D, where

$$K_D = \frac{[M][A]}{[MA]} \tag{62}$$

The value of the constant depends on the dissociation energy of the molecule as well as the temperature. In order to promote a high degree of dissociation, a_D, where

$$a_D = \frac{[M]}{[M] + [MA]} \tag{63}$$

a high flame temperature and low value for the dissociation energy are necessary.

The equilibrium constant is related to the temperature T (K) and dissociation energy D_0 (eV) of the molecule by the relationship,

$$\log K_D = 5040(D_0/T) + \tfrac{5}{2} \log T - 1.585 + 1.5 \log(g_M g_A / g_{MA})$$
$$+ \log U_M U_A / U_{MA}$$

where g is the statistical weight of the species indicated by the suffix, and U is the partition function for each species marked as a suffix. Although the value of the equilibrium constant thus depends on the vibrational and rotational energies of the molecule, it can be seen

202

that it is the values of D_0 and T which principally control the value of K_D.

The atoms of the analyte element, M, may be bound as molecules with one of the bulk constituents of the flame gases (e.g. O or OH) or with one of the atoms introduced into the flame in the sample nebulised (e.g. Cl or F). The fraction of analyte atoms bound as molecules MA depends also on the concentration of A in the flame. When A is introduced into the flame in the solution aspirated it must usually be present in very high concentration in the solution in order to give rise to an appreciable equilibrium concentration of molecules with the analyte element. When A is a bulk component of the flame gases the concentration present is usually much higher than the concentration of MA. The concentration of A then remains substantially unaffected by the binding of some free A species to form MA. Under these conditions, therefore, for stable MA molecules at flame temperatures, the fraction of M atoms (and the equilibrium concentration) depends on the equilibrium constant for the dissociation and may be large or small. When the component A is present in only very low equilibrium concentration in the flame gases, however, its concentration may be lower than or of the same order as that of M. This then results in a larger free atom fraction of M in the flame.

In hydrocarbon flames with air, oxygen, or nitrous oxide, several elements (e.g. silver, copper, and zinc) are completely atomised and do not form stable molecules with constituents of the flame gases. The sensitivity of the determination of these elements by AAS is therefore not greatly dependent on the temperature of the burnt gases above the primary reaction zone (and therefore on the fuel:-oxidant concentration ratio supplied to the flame). In flame atomic emission or fluorescence spectroscopy, however, the analytical sensitivity may still depend on the burnt gas composition even for these elements owing to change in flame temperature or quenching factors when the fuel:oxidant concentration ratio is varied.

In hydrogen or hydrocarbon flames with air, molecular hydroxide species are formed between OH in the flame gases and some alkali metal and alkaline earth elements. Potassium, rubidium, caesium, and lithium may each form monohydroxide species, MOH; the effect is most marked with lithium. The alkaline earth elements also form monohydroxide species from which molecular band emission in the visible region of the spectrum is quite intense in most flames. Dihydroxide molecules of the alkaline earth elements, $M(OH)_2$, may also be formed in hydrogen flames.

The most common reason for incomplete atomisation of analyte elements introduced into the flames presently used in analytical work is the formation of metal monoxides. For elements other than those which are virtually completely atomised (e.g. silver, copper, sodium, and zinc), even though sufficient free atoms may exist in flames of moderate temperature to permit analytical flame spectroscopy, a substantial fraction of the total metal supplied to the flame is usually present as metal oxide. This is the case even for the alkaline earth elements, where even though most of the observed molecular radiation is attributable to the formation of metal hydroxide [46] it has been shown that considerable amounts of the metal oxide are present [250]. Elements such as iron, cobalt, nickel and chromium are present partly as oxides in hydrocarbon—air flames, while elements such as aluminium, boron, beryllium, molybdenum, niobium, tantalum, titanium, vanadium, tungsten, zirconium, lanthanum, uranium, silicon, and germanium which form oxides which are either difficult to vaporise or dissociate (or both) are so little atomised that they are difficult, or impossible, to determine by flame spectroscopy in these flames. When moderate temperature hydrocarbon—air flames ($\gtrdot$ ca. 2500 K) are made fuel-rich, however, so that the equilibrium concentrations of the oxidising species in the flame gases are lower, dissociation of gaseous oxides of the above elements may be promoted slightly. In the equilibrium

$$MO \rightleftharpoons M + (O) \tag{64}$$

the partial pressure of atomic oxygen, p_o, is related to the partial pressures of other oxidising species, e.g. oxygen and OH in thermal equilibrium. The partial pressure of the atoms of the analyte element, p_M, is given by

$$p_M = K_{MO}(p_{MO}/p_O) \tag{65}$$

where K_{MO} and p_{MO} are the temperature-dependent equilibrium constant and partial pressure of the gaseous metal monoxide, respectively.

When the particular metal oxide is easily vaporised at the flame temperature, so that its partial pressure may be relatively high, p_{MO}

decreases when a fuel-rich flame is employed. The partial pressure of the atoms of the analyte element, p_M, therefore is not exactly in inverse proportion to p_O. The oxides of some elements, however, have vapour pressures which are very low at the flame temperature. Under these conditions p_{MO} is then equal to the saturated vapour pressure of the metal oxide at the flame temperature and unevaporated oxide particles are also present in the flame gases. The partial pressure of the gaseous metal oxide, p_{MO}, is then constant, and when the flame is made fuel-rich the value of p_M may increase in inverse proportion to that of p_O. The increase in p_M which may be obtained when fuel-rich flames are employed by the effect described outweighs the tendency for p_M to decrease owing to the lower value of the equilibrium constant for the dissociation reaction in the fuel-rich flame (which may be somewhat cooler than a stoichiometric flame). The improvement in analytical sensitivity to be expected for many elements in fuel-rich flames was predicted by several workers and has been widely confirmed experimentally.

For elements whose oxides have high dissociation energies and low vapour pressures at about 2500 K fuel-rich hydrocarbon-air flames may not provide for appreciable atomisation. High temperature fuel-rich flames, such as oxy-acetylene or nitrous oxide—acetylene, are then necessary to give rise to adequate atomisation for analysis. These flames are widely employed for the determination of the elements aluminium, boron, beryllium, germanium, molybdenum, niobium, silicon, tantalum, titanium, tungsten, vanadium, and zirconium, but do not permit high sensitivity for elements such as uranium, thorium and lanthanum whose oxides are extremely difficult to decompose.

Incomplete dissociation of molecules containing atoms of the analyte element obviously reduces the available sensitivity of the determination of the element by AAS or AFS. The shape of the analytical calibration curve should not be affected, however, as the fraction of the analyte atoms bound in molecular form does not depend on its concentration in the sample solution (see eqn. (63)). Incomplete dissociation, however, does give rise to more pronounced temperature dependence of the analytical atomic absorption or fluorescence signal. Both the equilibrium constant for the dissociation reaction and the concentration of the flame species which forms the molecule with the analyte atom are temperature dependent.

Interference effects from concomitant elements which result from their participation in the dissociation equilibria in the flame which govern the concentration of the atoms of the analyte element are discussed in Section 6(C).

(4) Ionisation

In most flame gases in thermal equilibrium it would not be expected that appreciable ionisation would occur, owing to the high ionisation energies of the flame gas constituents ($\lessdot$ 8—9 eV). Arrhenius reported the occurrence of ionisation in flames as early as 1891, however, and its enhancement by the addition of metal salts to the flame. In the primary reaction zone of hydrocarbon—air flames high concentrations of ions are formed during the initial combustion reaction. These flame ions and electrons may persist above the primary reaction zone at a concentration which greatly exceeds that expected from thermal ionisation at the temperature of the interconal zone. The positive ion and electron concentrations in the flame gases decrease only slowly to approach the equilibrium concentration with increasing height in the flame owing to the slowness of their recombination reactions. The principal flame ion involved is H_3O^+ [251,252]. Lewis and Von Elbe [253] have suggested that the free flame electrons may be accounted for in some hydrocarbon flames by the presence of carbon particles. The work function of solid carbon is only 4.35 eV, and small aggregates of carbon in incandescent hydrocarbon flames might exhibit behaviour intermediate between that of solid carbon and the C_2 radical (which has a very high ionisation potential, > 11 eV) and give rise to an appreciable free flame electron concentration.

At the temperatures of many flames used in analytical work the free metal atoms produced from nebulised samples may be ionised to form positive ions and free electrons

$$M \rightleftharpoons M^+ + e^- \tag{66}$$

Assuming that thermal equilibrium prevails, the rates of ionisation and recombination are balanced; the concentrations of the metal ions, M^+, electrons, e^- and metal atoms, M, are then related to each other via the Saha equation [254]

$$K_I = \frac{[M^+][e^-]}{[M]} \tag{67}$$

206

where K_I is the temperature-dependent ionisation constant. The temperature dependence of K_I is given by

$$\log K_I = -\frac{5040\,E_I}{T} + \tfrac{5}{2} \log T - 6.5 + \log \frac{g_{M^+} \cdot g_e}{g_M} \tag{68}$$

where E_I is the ionisation potential of the metal in eV, T is the temperature in K, and g_{M^+}, g_e and g_M are the statistical weights of the ionised atom, the electron and the neutral atom. For the electron, owing to its spin, $g_e = 2$.

The degree of ionisation, a_I, may be defined as

$$a_I = \frac{[M^+]}{[M^+] + [M]} \tag{69}$$

for the metal species considered. When the metal atom M is the only ionisable species introduced into the flame, $[M^+]$ is equal to $[e^-]$ for electrical neutrality; thus

$$K_I = \frac{[M^+]^2}{[M]} \tag{70}$$

The Saha equation then gives

$$\frac{a_I^2}{1 - a_I} = \frac{K_I}{[M^+] + [M]} \tag{71}$$

Thus an increase in the total concentration of analyte element, $[M^+] + [M]$, supplied to the flame results in a lower value of a_I, whereas for low total analyte element concentrations the value of a_I approaches unity. The degree of ionisation thus varies with the concentration of analyte supplied to the flame. The fractional loss of free atoms increases with decreasing analyte element concentration, and this leads to curvature of analytical growth curves *away* from the concentration axis in AAS. When $[M^+] = [e^-]$, eqn. (70) shows that $[M^+]$ is proportional to $[M]^{1/2}$. When $[M]$ is approximately equal to the concentration of M in the sample solution, i.e. when the degree of ionisation is low, the population of metal ions $[M^+]$ in the flame should show a square root dependence on the sample concentration.

References pp. 253—262   207

When two ionisable elements are present

$$M_1 = M_1^+ + e^- \tag{72}$$

$$M_2 = M_2^+ + e^- \tag{73}$$

the requirements of electrical neutrality give

$$[M_1^+] + [M_2^+] = [e^-] \tag{74}$$

In effect, the presence of the second ionisable element, M_2, which produces additional electrons, tends to suppress the degree of ionisation of the first element, M_1, via the Saha equation. This effect is widely employed in AAS to suppress ionisation of atoms of analyte elements in flames. A relatively high concentration of an easily ionised element such as potassium is frequently added to the sample solution in the determination of elements which give low sensitivity due to ionisation.

The presence of free flame ions and electrons in most unsalted air—hydrocarbon flames leads to shifts in the metal ionisation equilibria described here. When the concentration of free flame electrons is high relative to the concentration of the analyte element, it may be regarded as constant for varying concentrations of M. Under these conditions the degree of ionisation is also unaffected by the concentration of atoms of the analyte element. This effect of the natural flame electrons results in a buffering action at low metal atom concentrations similar to that of adding an easily ionised element at higher concentrations. Thus for low metal atom concentrations, when an abundance of flame electrons are present, the value of a_I does not become zero as predicted from eqn. (71).

Partial ionisation of the analyte element introduced into the flame gives rise to loss of sensitivity in AAS and AFS. In the hot flames which are required to obtain freedom from chemical interferences in the condensed or vapour phases for many elements, the degree of ionisation for the element may be extremely high. In the nitrous oxide—acetylene flame, for example, barium (E_i = 5.2 eV) at a partial pressure of 10^{-6} atm in the flame at 3223 K, should exhibit a degree of ionisation of 0.92; a value of a_I of 0.92 has actually been determined experimentally under similar conditions [255]. With such a high degree of ionisation a higher sensitivity may actually be

TABLE 5

Calculated and measured degrees of ionisation of some metals in the nitrous oxide—acetylene flame[*]

Metal	Ionisation potential (eV)	Degree of ionisation, a_I	
		Calculated for 3223 K ($p_M = 10^{-6}$ atm)	Measured in $N_2O—C_2H_2$ flame
Na	5.14	0.82	
K	4.34	0.98	
Be	9.32	0.00	0.00
Mg	7.64	0.03	0.06
Ca	6.11	0.43	0.43
Sr	5.69	0.71	0.84
Ba	5.21	0.92	0.88
Al	5.98	0.17	ca. 0.14
Eu	5.67	0.61	ca. 0.60
Yb	6.2	0.36	0.20

[*] From ref. 256.

obtained in AAS when the ionic Ba^+ line at 455.4 nm is employed rather than the atomic line at 553.6 nm [255]. Table 5 shows the calculated and measured degrees of ionisation of some metals at a partial pressure, p_M, of 10^{-6} atm in the nitrous oxide—acetylene flame [256].

Mutual interference may result via their ionisation equilibria when both the analyte element and a concomitant element in the sample for AAS are ionised in the flame. This type of interference effect may again be minimised by suppression of the ionisation of the elements by the addition of excess potassium to the sample. Alternatively, the concentration of the concomitant element may be raised substantially to suppress the ionisation of the analyte element; above a certain concentration of concomitant element the analyte ionisation is effectively suppressed, and variation in the concentration ratio of concomitant: analyte element has little effect on the signal in AAS.

(5) Distribution of Atoms in Flames

Each of the processes of desolvation, volatilisation, dissociation, and ionisation affect the spatial distribution of the concentration of the atoms of the analyte element in the flame. These processes are also governed by equilibria which are temperature dependent and may be affected by the flame gas composition. The variation of temperature and gas composition within the interconal and secondary zones of flames, therefore, has an important bearing on the spatial distribution of the concentration of atoms of the analyte ele-

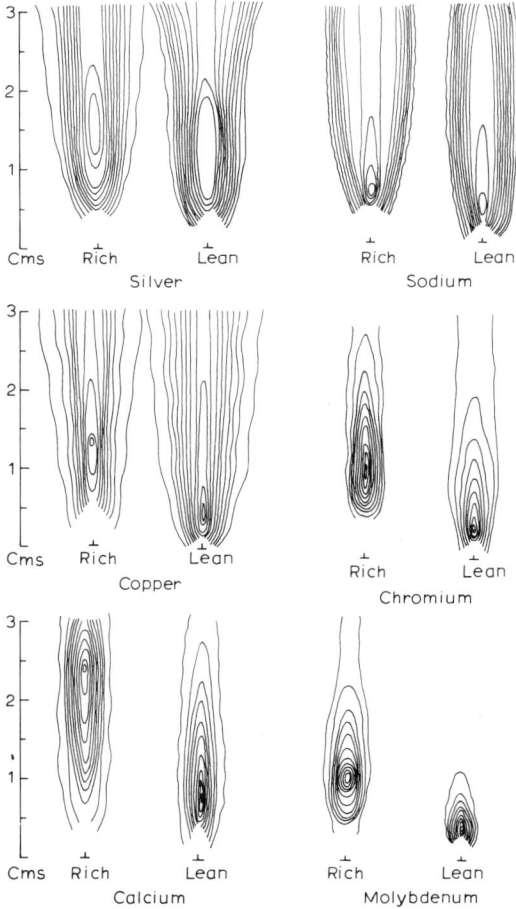

Fig. 32. Distribution of atoms of six elements in a premixed air—acetylene flame burning at 10 cm slot burner. (C.S. Rann and A.N. Hambly Anal. Chem., 37 (1965) 879.)

210

ment. Obviously the region of the flame above the primary zone which provides the highest sensitivity in AAS and AFS is that which contains the highest concentration of free atoms; the optimum region varies from element to element depending on the particular characteristics for that element of ease of volatilisation, dissociation and ionisation.

Rann and Hambly [257] have made a careful study of the distribution of the atoms of several elements in a premixed air—acetylene flame burning at a 10 cm slot burner. The contours they obtained for six elements in lean and rich flames are shown in Fig. 32. In the case of elements which are relatively easily atomised in this flame, for example silver and sodium, it can be seen that a high atomic population is present over quite a large volume of the flame gases even though its temperature and composition varies appreciably within both the rich and lean flames horizontally and vertically and from one flame to the other. For those elements which are difficult to atomise in this flame, however, such as molybdenum, the maximum atomic concentration persists over only a limited flame volume and is generally greatest in the hotter and/or most reducing regions of the flame.

The distribution of atoms of analyte elements in the nitrous oxide—acetylene and oxy—acetylene flames shows similar variation from element to element and depends on the temperature and burnt gas composition. For elements which are easily atomised in these higher temperature flames the highest atomic populations (after the suppression of ionisation) are found in the hottest part of the flame and no very substantial dependence on the flame gas composition is observed. For those elements such as aluminium, titanium, zirconium, etc., which form very refractory oxides, however, a highly reducing flame gas composition is more important than a high temperature. Fuel-rich flames, whose maximum temperature is somewhat lower than that of a stoichiometric flame, are required. Observations in AAS and AFS for these elements are normally confined to the interconal region in these flames as this part of the flame is strongly reducing in character.

5. Experimental techniques

The following section describes some of the considerations necessary in the practice of AAS and AFS. These include the selection of spectral lines used for analysis, the general technique of instrument operation, the attainable sensitivity, the preparation and measurement of samples and standards and the precision and accuracy of the techniques.

(A) WAVELENGTH OF MEASUREMENT

(1) Atomic Absorption Spectroscopy

The atomic line of an element which is most frequently employed in AAS will correspond to the ground state line which has the highest absorption oscillator strength, f_{ji} (See eqn. (45)). As described in Section 2(G), however, when the ground state and the next level above it are very close, the latter may have a higher population even in relatively low temperature flames. For example, in the case of aluminium where the 309.3 nm line ($g = 4$), which is 0.014 eV above the ground state, provides greater sensitivity than the 308.2 nm line ($g = 2$) although the lines have similar oscillator strengths (ca. 0.22). Several practical reasons may also lead to the choice of a line other than the ground state line of highest oscillator strength. Thus when a line from the filler gas of the hollow-cathode lies close to the most sensitive atomic line of the element concerned, this may be transmitted by the monochromator and lead to low sensitivity and non-linear calibration curves (see Section 6(C)). An alternative, if less sensitive, line of the element might then be preferred for analytical use. Similarly, the presence of high flame background absorbance and noise at the most sensitive absorption line might lead to the choice of an alternative line for AAS of the element concerned. Thus flame and atmospheric absorption prevent the use of the mercury line of greatest oscillator strength at 184.9 nm ($f_{ji} = 1.2$) and the line at 253.7 nm (f_{ji} = ca. 0.03) is employed for analysis in most flames. In the same way the high flame background absorbance below 200 nm combined with the relatively low intensity output from arsenic hollow cathode lamps (which necessitates the use of high gain at the detector) usually results in superior detection limits at the arsenic 193.7 nm line over those obtained at the arsenic 189.0 nm line. The application of inert gas separated nitrous oxide—acetylene flames, which exhibit low background absorption below 200 nm, however, may lead to the use of the mercury 184.9 and arsenic 189.0 nm lines for AAS.

212

(2) Atomic Fluorescence Spectroscopy

In AFS the nature of the fluorescence process stimulated for the element concerned governs the selection of the atomic line used for quantitative intensity measurements. The resonance fluorescence emission has been employed for the analytical AFS of most elements as it may provide higher absolute signal intensities than direct line or stepwise line emission. In some cases, however, direct-line emission may be more intense than the resonance emission. Thus, for example, the lead 405.8 nm direct-line fluorescence signal strengths in a number of flames have been observed to be more intense than the resonance fluorescence emission at 283.3 nm [258]. The relative intensities of different lines of an element emitted by the source and the variation in photomultiplier response with wavelength are also important factors which may influence the atomic line used for analysis by AFS. This latter effect of detector sensitivity, particularly, may enhance the apparent signal intensity obtained at non-resonance lines compared to that observed for the resonance line. It may be desirable in some instances to avoid the measurement of resonance fluorescence emission. As in AAS, when high flame background absorption or flame emission noise is registered at the wavelength of the resonance line an alternative line of lower absolute emission intensity may yield higher signal: noise ratios. When the existence of particulate material which would give rise to scatter of radiation is anticipated in the flame during sample introduction this is difficult to differentiate from fluorescence when resonance fluorescence is employed. Under these conditions the use of a simple filter in the incident light path (to select the resonance excitation line and reject any emission from the source at the wavelength to be measured) and measurement of direct-line or stepwise line emission at longer wavelengths avoids the detection of scattered radiation. Similarly the range of concentrations over which linear calibration graphs are obtained may be inadequate for a particular purpose when resonance fluorescence is measured. This effect, which results from the re-absorption of the fluorescence emission before it leaves the flame, frequently may be eliminated by measurement of direct-line fluorescence; unless a very hot flame is employed the population of atoms in the lower state of the direct-line emission will usually be too low to result in appreciable re-absorption of the emitted radiation.

The fluorescence emission lines reported by various workers for use in the determination of elements by AFS are shown in Table 7 in Section 5(D) (p. 220).

Section 5(D) (p. 220).

(B) INSTRUMENT OPERATION

General instructions for the setting up of AAS instruments are invariably provided by the instrument manufacturers. These instructions include advice on the positioning of the instrument in the laboratory to avoid draughts and bright sunlight, and the provision of an exhaust chimney for the removal of the burnt gases which frequently contain toxic materials from the samples nebulised into the flames employed. Similar considerations naturally apply to the setting-up and operation of instruments for use in AFS studies.

In an experimental procedure by AAS or AFS the following operations are normally performed.

(i) The source and detector are switched on and the source power is set to the recommended value. With modern hollow cathode lamps and photomultiplier—amplifier systems very little warm-up time is required. When electrodeless discharge tubes are used as sources in AFS some additional attention may be required to the tuning of the assembly to minimum reflected power and the time taken to achieve stable output intensity may vary from element to element.

(ii) The analytical line and spectral slit width to be employed are selected at the monochromator.

(iii) The flame is ignited and the fuel—oxidant flow rate ratio adjusted to the recommended settings. The position of the incident light beam in the flame is optimised by variation of the height of the burner in the light beam. Information concerning the correct procedures for igniting and extinguishing the flame is invariably provided in the instruction manual with commercial instrumentation.

(iv) The detector sensitivity is set by adjustment of the amplifier gain or photomultiplier EHT. In AAS the sensitivity is adjusted to obtain a reading of 100% transmission (0 absorbance) at the meter or recorder. The zero transmission setting is checked by obstructing the light path between the source and detector with a shutter. In AFS the sensitivity is set to an established optimum setting for the sample concentration range anticipated, and any stray or scattered light is 'backed-off' electrically so that the meter or recorder reads zero deflection.

(v) Pure solvent is nebulised into the flame and the 100 and 0%

transmission settings (in AAS) and zero deflection setting (in AFS) are re-checked. The instrument is then ready for measurement.

(vi) The sample solution is nebulised into the flame and the absorbance or fluorescence emission reading is noted. The solution should be aspirated for sufficient time to enable the analytical signal to reach a steady value.

(vii) The nebuliser and burner assembly are washed by aspiration of pure solvent between each sample measured. This is particularly important when dealing with liquid samples of high dissolved solid content.

The concentration of the element determined in the sample solution is found from relating its absorbance or fluorescence to the absorbance or fluorescence signals produced under the same conditions by solutions of known concentration. These latter values are plotted as a calibration curve of analytical signal vs. concentration of element in the standard solution (see Section 5(E)); the sample solution concentration is then established from this graph.

(C) PREPARATION OF SAMPLE AND STANDARD SOLUTIONS

As mentioned above, for a particular instrumental assembly and flame type the optimum concentration range for the determination of an element by AAS is decided by the sensitivity attained under these conditions. It is thus necessary to prepare sample solutions so that the concentration of the element to be determined lies in this range. For solid samples the weight of sample which contains the analyte element which is taken, and the final solution volume, must be adjusted accordingly. For liquid samples dilution or concentration (for example by solvent extraction or ion exchange) of the analyte element may be necessary. Standard solutions are prepared which contain the analyte element in the optimum concentration range. When a linear calibration graph is obtained for the analyte element the use of a single standard solution may be permissible; for high precision, or when the graph is non-linear, several calibration standards should be chosen to produce different absorbance values (usually between 0.02 and 0.6 absorbance).

The standard solutions prepared may contain only a salt of the analyte element when the sample solutions to be analysed contain only low dissolved solid content (less than ca. 0.1—0.2 wt.%). In order to compensate for the different nebulisation and vaporisation characteristics encountered when the sample solutions contain large amounts of dissolved solid, however, it is necessary to add similar amounts of a salt of the matrix element (and possibly any releasing or protective agent employed) to the standard solutions. Usually only an approximate 'matching' of the matrix element concentration in the sample and standards is necessary in AAS. Similarly, some acids used to dissolve samples, e.g. hydrofluoric acid, may cause an enhancement or suppression of the absorbance for the analyte element in the sample relative to that obtained for the same element in the standards; a similar amount of acid should then be used in both sample and standard solutions. In routine AAS work, where several elements are frequently determined in a particular type of sample, it may be convenient to prepare one set of solutions which contain calibration standards for all the elements determined. The blank solutions nebulised, in order to establish the bottom of the calibration graph or set the 100% transmission value, contain no added analyte element but should contain the matrix element salt and releasing or protective agent if these are present in the sample and standard solutions.

Considerations similar to those above with regard to matching of the standard solution and sample composition to compensate for changes in nebulisation and vaporisation characteristics apply also in AFS when flame cells are used. A more precise match of sample and standard solution compositions would only be necessary if samples of high dissolved solid content cause scattered light signals when resonance fluorescence is to be measured and the scattered light backed-off. This procedure normally leads to poor precision, however, and is not generally applicable. In AFS the variable gain of the detector system may be employed to influence the sensitivity of a particular determination. In contrast to AAS, therefore, where sample dilution or preconcentration may be required to adjust the analyte concentration to within the optimum range, in AFS the detector sensitivity may be increased or decreased to achieve a similar result. This may constitute a major advantage of AFS over AAS in simultaneous multi-element analysis; under these conditions it is difficult to ensure that each analyte element lies within its optimum concentration range for determination by AAS by a single dilution or concentration step.

(D) ANALYTICAL SENSITIVITY

The *sensitivity* attainable for the determination of an element by AAS is defined as the concentration of the element (usually in aqueous solution) which produces a 1% absorption signal (0.0044 absorbance) under optimum experimental conditions. This sensitivity value depends primarily upon the atomic absorption coefficient for the atomic line employed, the absorption path length and the nebulisation and atomisation efficiency. As the sensitivity defines the slope of a linear calibration graph it is frequently useful when a knowledge of the required concentrations of standards and samples is needed. The attainable sensitivity values for particular elements under optimum experimental operating conditions remain fairly constant between instruments of the same type. The values of the sensitivities obtainable for different elements by AAS with a typical commercial instrument are shown in Table 6.

Adequate precision may be obtained for many elements with modern AAS instruments at concentrations which produce less than 1% absorption. The term *detection limit* is used in AAS to define the lowest detectable concentration of the element in the sample solution. This value is useful for the comparison of instrument performance and analytical procedures for different elements. It may be defined for a particular element as that concentration of the element which produces an absorption signal equivalent to twice the standard deviation in the background noise under the experimental conditions employed. The detection limit for an element may vary considerably from instrument to instrument and from one type of sample matrix and solvent to another. Values of detection limits for different elements obtained with a commercial AAS instrument are shown in Table 6. The absolute detection limit, defined as the minimum detectable *amount* of the element, depends on the relative detection limit and also on the minimum volume of sample solution required for the measurement. The time required for the instrument to achieve a constant and reproducible absorbance reading for the element depends on the nebuliser characteristics and the response time of the detector and read-out system.

Sometimes the 1% sensitivity attainable in the determination of two elements A and B might be similar, but owing to background noise from the flame or detector the available detection limit for each element may be quite different. This is illustrated in Fig. 33.

TABLE 6

Analytical sensitivities and detection limits in flames by AAS*

Element	Wavelength (nm)	Sensitivity (ppm/1% abs.)	Detection limit (ppm)
Ag	328.1	0.08	0.005
Al	309.3	1.1	0.1
As	193.7	1.0	0.2
Au	242.8	0.5	0.02
B	249.7	35	6
Ba	553.6	0.4	0.05
Be	234.9	0.03	0.002
Bi	223.1	0.7	0.05
Ca	422.7	0.03	0.002
Cd	228.8	0.03	0.005
Co	240.7	0.1	0.005
Cr	357.9	0.15	0.005
Cs	852.1	0.5	0.05
Cu	324.7	0.1	0.005
Dy	421.2	0.7	0.4
Er	400.8	0.85	0.1
Eu	459.4	0.75	0.2
Fe	248.3	0.15	0.005
Ga	287.4	2.3	0.1
Gd	368.4	17	4
Ge	265.2	2.5	1
Hf	307.3	15	15
Hg	253.7	15	0.5
Ho	410.4	1.4	0.3
In	303.9	0.9	0.05
Ir	264.0	13	2
K	766.5	0.1	0.005
La	550.1	30	2
Li	670.8	0.07	0.005
Lu	331.2	15	3
Mg	285.2	0.008	0.0005
Mn	279.5	0.08	0.003
Mo	313.2	1	0.1
Na	589.0	0.04	0.005
Nb	334.4	21	5
Nd	463.4	10	2
Ni	232.0	0.1	0.005
Os	290.9	1	1
Pb	283.3	0.5	0.01
Pd	247.6	0.3	0.02
Pr	495.1	13	10

TABLE 6 (continued)

Element	Wavelength (nm)	Sensitivity (ppm/1% abs.)	Detection limit (ppm)
Pt	265.9	2	0.1
Rb	780.0	0.2	0.005
Re	346.0	20	1.5
Rh	343.5	0.35	0.03
Ru	349.9	2	0.3
Sb	217.5	1	0.2
Sc	391.2	0.5	0.2
Se	196.0	2	0.5
Si	251.6	1.2	0.1
Sm	429.7	8.5	5
Sn	224.6	1.2	0.06
Sr	460.7	0.2	0.01
Ta	271.5	30	6
Tb	432.6	7.5	2
Te	214.3	2	0.3
Ti	364.3	1.4	0.2
Tl	276.8	0.2	0.8
Tm	371.8	1	0.15
U	351.4	200	30
V	318.4	1.3	0.04
W	400.9	35	3
Y	407.7	1.1	0.3
Yb	398.8	0.17	0.04
Zn	213.8	0.04	0.002
Zr	360.1	20	5

* This table shows typical AAS data obtained in conventional flames and does not include those obtained with flame adaptors, non-flame cells etc.

The detection limit for element A is superior to that for element B although the 1% sensitivity figure for element B is slightly greater than that for element A.

The sensitivity of a procedure for the determination of an element by AFS is usually defined only by the *relative* detection limit, i.e. the concentration of the element in solution which produces a fluorescence signal equivalent to twice (or three times) the standard deviation in the background noise. Whether twice or three times the standard deviation is used depends on the confidence level required (95% or 99%). Table 7 shows some AFS detection limits taken from the literature. These values represent the best detection limits for each

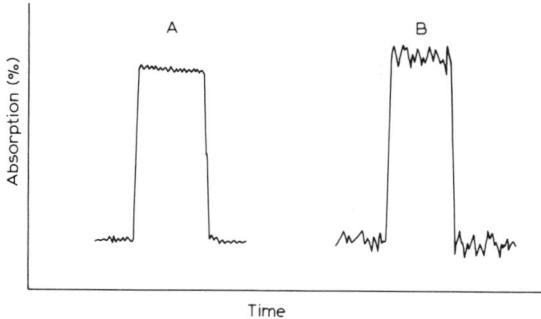

Fig. 33. Absorbance readings for two elements A and B at the same concentrations in aqueous solution nebulised for equal periods of time. The detection limit for element A is superior to that for element B although the 1% sensitivity is similar.

TABLE 7

AFS detection limits for elements in flames

Element	Wavelength (nm)	Flame	Source	Detection limit (ppm)	Ref.
Ag	328.1	Air/H_2	EDT	0.0001	101
Al	396.1	Argon. sep. $N_2 O/C_2 H_2$	EDT	0.1	216
As		Ar/H_2/air	EDT	0.2	
Au	267.6	Air/H_2	EDT	0.2	101
	242.8	O_2-H_2-Ar	H–HCL	0.005	282
Be	234.8	Sep. $N_2 O/C_2 H_2$	EDT	0.01	213
Bi	302.6	Ar–H_2–air	Iodine EDT	0.05	268
	306.8	N_2 sep. air/$C_2 H_2$	EDT	0.09	269
Ca	422.7	Air/H_2	EDT	0.02	101
		H_2–air	450W Xe	0.02	270
Cd	228.8	Air/H_2	EDT	0.000001	101
Co	240.7	Air/$C_3 H_8$	EDT	0.005	271
Cr	357.9	Air/H_2	Demountable cooled HCL	1.0	69
Cu	324.7	Air/H_2	H–HCL	0.001	272
	327.4	Air/H_2	Hot HCL	0.001	273
Fe	248.3	Air/$C_2 H_2$	EDT	0.009	274
		Ar/O_2/H_2	H–HCL	0.02	275
Ga	403.3	Air/H_2	H–HCL	0.5	276
	417.2				
Ge	265.1	Ar sep. $N_2 O/C_2 H_2$	EDT	0.1	215
Hg	253.7	O_2/H_2 org. solv	Hg pen light	0.002	277

TABLE 7 (continued)

Ele-ment	Wavelength (nm)	Flame	Source	Detection limit (ppm)	Ref.
In	410.5	Air/H_2	H—HCL	0.2	276
	451.1				
Mg	285.2	Air/propane	H—HCL	0.001	278
Mn	279.5	Air/C_2H_2	EDT	0.001	239
Mo	313.3	Ar sep. N_2O/C_2H_2	EDT	0.46	216
Ni	232.0	Ar/O_2/H_2	H—HCL	0.003	275
		Air/H_2	H—HCL	0.003	272
Pb	405.8	Ar/O_2/H_2	EDT	0.01	258
		Air/H_2	H—HCL	0.02	276
Pd	340.4	Air/H_2	H—HCL	2.0	276
Pt	265.9	Air/H_2	H—HCL	50	276
Sb	231.1	Air/C_3H_8	EDT	0.05	279
Se	196.0	Air/C_3H_8	EDT	0.25	280
Si	251.6	Ar sep. N_2O/C_2H_2	EDT	0.55	214
Sn	303.4	N_2 sep. Ar/O_2/H_2	EDT	0.1	281
Sr	460.7	Ar/H_2—air	EDT	0.03	101
Te	238.3	Air/C_3H_8	EDT	0.12	280
	238.6				
Tl	377.6	Ar/H_2—air	EDT	0.008	101
V	318.4	Ar sep. N_2O/C_2H_2	EDT	0.07	216
Zn	213.9	Air/H_2	EDT	0.00004	101

H—HCL = high intensity hollow-cathode lamp.
EDT = electrodeless discharge tube.
450W Xe = 450 W xenon arc source.
Ar sep. N_2O/C_2H_2 = Argon separated premixed nitrous oxide—acetylene flame.

element available in the literature at the time of writing; it should be emphasised that while these values all apply to flame cells, a variety of different instrumentation was employed by different workers to obtain these values and various authors do not necessarily present their values according to the definition of detection limits given above. Table 7 also shows brief details of the source and flame type used to obtain each detection limit.

(E) CALIBRATION GRAPHS

The possibilities of absolute analysis by AAS are discussed in Section 5(F)(4) but at the present time absolute analysis by AAS and

AFS is possible only with experimental difficulty and under limited analytical circumstances (i.e. low sample solution concentrations). In practice, when either technique is used for routine analysis a calibration of the signal strength vs. solution concentration is employed.

As mentioned in Section 5(C), the 100% transmission in AAS or zero fluorescence signal in AFS is set usually while pure solvent or a blank solution whose composition matches that of the samples is nebulised into the flame. If the calibration graph is known to be linear over the desired concentration range it is, in theory, necessary to employ only one standard solution to enable the calibration graph to be constructed; normally, however, it is preferable to employ several standard solutions of different concentrations in order to obtain good precision with practical samples. When the calibration graph is known to be a curve the use of sufficient standard solutions to construct the graph may still permit quite accurate analysis. Under conditions for least photometric error in AAS, i.e. measurement between 0.2 and 0.7 absorbance units, the calibration curve most frequently would cover a 20 to 200-fold increase in concentration above the detection limit. This often results in inconvenience in AAS, as this range above the detection limit corresponds to the use of solutions of relatively high salt content which may cause nebulisation difficulties. Thus in AAS, calibration curves are normally constructed for use between absorbance values of ca. 0.02—0.7. Apart from such considerations as the photometric accuracy, the accuracy of determinations when curved calibration graphs are employed is affected by the actual shape of the curve. The rate of change of analytical signal strength with concentration governs the reliability with which the concentrations of samples producing similar absorbance values may be differentiated. The slope of the curve may vary considerably over the concentration range of interest; depending on the shape of the graph (bending away from, or towards, the concentration axis), the certainty with which it is possible to differentiate between samples producing similar low absorbances may be poor or good compared to the situation at high absorbance values. Some modern atomic absorption spectrophotometers provide for correction of read out so that absorbance values observed appear to follow a linear calibration even though the actual uncorrected calibration data are non-linear.

Frequent checks of the calibration are recommended during operation in both AAS and AFS. Standard solutions should be nebulised

222

with each set of sample solutions to ensure that wavelength drift, fuel flow rate or change in nebuliser performance has not occurred and resulted in inaccurate analyses.

(1) AAS Calibration Graphs

The dependence of the analytical absorption signal on analyte concentration has already been discussed in Section 2(C). With a sharp line source and the measurement of absorbance, i.e. in the case of the majority of practical AAS analyses, the calibration graph of absorbance vs. concentration should be linear over a wide range of concentration. In practice, the source line is never ideally sharp, and this results in a curvature of the graph towards the concentration axis; the greater the line width the greater is the curvature and the lower the concentration at which it becomes significant. When a continuum source is employed there is no simple relationship between absorbance and concentration and the percentage absorption must be measured. As in this case the percentage absorption is directly proportional to concentration at low concentrations, and to the square root of the concentration at higher values, it is best to plot the calibration graph as $\log (\Delta I/I_0)$ vs. log (concentration); the graph has a slope of 1 at low values and of 0.5 at high values (see Section 2(C)(6)). Any tendency for the slope to decrease further at very high concentrations may be reduced by increasing the band-pass of the monochromator as is also decribed in Section 2(C)(6).

The above deviations from linearity predictable from the theoretical basis of the techniques are frequently accompanied by other deviations which may be described as interference effects. As these vary with the nature of the sample rather than the instrumentation used they may often prove more troublesome. These types of interference have been discussed in detail in Section 6, but the principal effects may be repeated in summary. When a non-absorbing spectral line is emitted by the source at a wavelength close to that of the absorption line, a reduction in sensitivity will be observed if this line is transmitted by the monochromator. The effect will progressively become more serious as the absorbance at the analytical line increases; the calibration graph bends towards the concentration axis. When an atom is easily ionised, the number of atoms present in the absorption cell at low sample concentrations will be seriously depleted by the formation of ions. The percentage ionisation decreases with increasing concentration of atoms in the cell; the calibration graph bends *away* from the concentration axis.

(2) AFS Calibration Graphs

As shown in Section 2(D), the theoretical dependence of atomic fluorescence signal strength on the analyte concentration is more complex than that in AAS; fortunately the usual applications of AFS for analytical work allow a fairly simple approach from the practical viewpoint. This is based on the assumption that AFS is most frequently applicable for the measurement of low analyte concentrations in the atom cell, or that the amount of absorption occurring in the atom cell is small. It is then true that the graph of concentration vs. fluorescence signal is linear for both continuum and sharp-line sources. When determinations are attempted at high concentrations (the exact value depends on the element concerned) the calibration graph bends towards the concentration axis. This is partly explained by the theoretical predictions given in Sections 2(D)(4) and 2(D)(5) and partly by the increasingly serious re-absorption of the resonance fluorescence signals as the atom population in the cell increases. In addition to these deviations from linearity explained in Section 2(D) the same type of deviation due to ionisation effects will cause concavity of the calibration graphs in AFS as in AAS.

(F) OTHER CALIBRATION METHODS

Although most routine analytical determinations by AAS and AFS are carried out with calibration graphs as described in Section 5(E)., a number of other simple techniques are available. These methods sometimes show advantages over the conventional calibration procedure for certain analyses and the most useful ones are described below.

(1) Standard Addition Methods

This method may be applied within the range of concentrations corresponding to linear calibration graphs. The signal is measured for the sample alone and from the sample plus aliquots of a solution of the analyte element of known concentration. The main advantage of the method in AAS and AFS is that in effect exact matching of the

224

sample and 'standard' solutions (sample to which known amounts of analyte element are added) is achieved automatically. This is most useful in the analysis of samples where complex matrix element effects are observed on the analytical signal. An analysis may be carried out by this method by making only two measurements, i.e. for the sample alone and for the sample plus one aliquot of standard analyte solution, but three or four measurements (sample plus one, two and three aliquots of standard analyte solution) usually result in greater accuracy. In order to avoid the calculation of volume corrections it is usual to make all the solutions to the same volume before nebulisation into the flame. A graph of absorbance or fluorescence signal vs. concentration of each addition is constructed. Extrapolation of this graph line back to zero signal then enables the analyte concentration of the (diluted) sample solution to be determined.

(2) Dilution Method

This method is an adaptation of the standard addition method which may be used in preference under certain circumstances. A measurement of the analytical signal S_1, from a volume V_1 of a solution containing a known concentration C_1 of the analyte element is made. A volume V_2 of the unknown sample, containing a concentration C_2 of analyte, is then added and a second signal, S_2, is then recorded. The unknown concentration, C_2, is then given by

$$C_2 = \frac{C_1[S_2(V_1 + V_2)(V_2 - S_1 V_1/V_2)]}{S_1}$$

Provided that sufficient accuracy is obtained with only two measurements the method is rapid and easily applied. Smaller volumes of solution of the sample may be used than with the addition method as it is not necessary to record a signal for the sample solution alone. Also with concentrated sample solutions, or at least analyte solutions which produce a large analytical signal, it is not necessary to perform an additional operation to obtain sufficient dilution to reach the linear part of the calibration.

(3) Absolute Analysis by AAS

A technique which would allow absolute analyses by AAS would offer several advantages over the conventional methods utilising calibration against standards. The feasibility of absolute AAS methods has been discussed by Rann [259], de Galan [260], and de Galan and Samaey [261]. Several reports of the measurement of free atom fraction and oscillator strengths by AAS have also been made, and these measurements involve many similar experimental difficulties to absolute analysis [262—265]. There are two main problems associated with absolute analysis by AAS. In the first place it is necessary to have an exact theoretical expression relating absorption measurements which are feasible to the number of absorbing atoms. Secondly, it is necessary to calibrate experimentally the apparatus employed, both in order to evaluate all of the parameters which must be substituted into the basic equation and to relate the number of absorbing atoms to the actual concentration of the analyte sample. Although each of these presents difficulties, the most serious from the viewpoint of the analytical chemist, is the need for an absorption measurement corresponding to a simple accurate expression over a sufficiently wide range of experimental conditions, and the selection of an absorption cell in which the number of absorbing atoms can be readily related to the sample concentration. The simplest expression is that for the integrated absorption coefficient given in Section 2(C)(3). As explained in that section, however, it is difficult to use this to establish a routine analytical technique. A more convenient measurement would be the peak absorbance made with the sharpest possible source line under conditions of low optical density. Even under these limited circumstances, however, the measurement cannot be applied with equal success to all elements; the absorption line used may have a hyperfine structure which is sufficiently wide or well-separated in the source and/or atom cell to cause a serious error in the measurement of the absorbance. The problem has been considered in some detail by L'Vov [266]; he suggested that measurement in a high pressure cell might offer one solution. Under these conditions the Lorentzian broadening would result in a much wider absorption line than that obtained using other absorption cells at atmospheric pressure. The hyperfine structure would then have an insignificant effect on the absorbance measurement for a much wider range of elements, and the errors due to use of a broadened source line would also be reduced.

The ease with which the number of absorbing atoms may be related to the actual sample concentration also varies with the type of

226

absorption cell used and the analyte element. It is usually possible to measure fairly accurately the proportion of the sample which actually reaches a region of the absorption cell where the analyte atoms can give rise to a signal, although the experimental technique may prove more simple for a L'Vov furnace, for example, than for a conventional premixed flame and pneumatic nebuliser. The measurement of the fraction of the sample reaching the region in which absorption can occur which actually forms atoms, however, is more difficult. The most promising solution might be the use of an atom cell in which the free atom fraction is unity for the element concerned. With the atom cells currently available, particularly high temperature flames, this would mean that a useful range of elements would be amenable to absolute analysis. However, after deleting those elements for which the lack of a suitable sharp line source or the effect of hyperfine structure cause additional problems the range of elements becomes much smaller. Also the combined effects of inaccuracies in the various calibrations required and deviations from the theoretical assumptions which must be made are likely to restrict the accuracy of the results severely.

(G) PRECISION AND ACCURACY

The precision of a measurement refers to the reproducibility obtained if the same measurement is repeated several times. The accuracy of a measurement is a measure of the deviation of the measurement from the actual 'correct' value. Thus it is possible to carry out a precise series of determinations by AAS and AFS without obtaining an accurate result, but an accurate determination is unlikely to be obtained with an imprecise measurement.

(1) Precision

The precision of an analysis is usually found by making several determinations and noting the variation from the mean. For most purposes this is best done by calculation of the standard deviation, S, defined as

$$S = \sqrt{\frac{\Sigma(x_i - \bar{x})^2}{n-1}}$$

where x_i is the result of one measurement and $\bar{x}$ the mean of n measurements. The precision may also be quoted in terms of the coefficient of variation, F, which is given simply as

$$F = \frac{100\,S}{\bar{x}}$$

For routine analysis by AAS, F may lie in the range 0.3—1.0%, depending on the type of spectrophotometer in use and the nearness of the sample concentration to the detection limit. However, with modern digital read-out AAS instruments, which are capable of displaying a single reading as the result of an automatic integration of 10 to 100 measurements, it is possible to obtain values of F better than 0.1%.

In general, the precision which can be obtained in AAS and AFS depends on the level of fluctuations in the emission source, atom cell and electronic components of the spectrophotometer. These will govern not only the precision of an isolated measurement but also the precision with which the calibration graph is constructed.

(2) Accuracy

Evaluation of the accuracy of an analysis by AAS or AFS is considerably more time-consuming than an estimate of the precision. If a test only of the accuracy cannot be made using standard analysed samples (such as those available from the National Bureau of Standards or the British Standards Institute) the analytical procedure may be tested by reference of the results to those given by a referee method employing a classical absolute technique or other instrumental method. The recovery of analyte element may also be checked by making standard additions of analyte element to samples and conducting the procedure in the normal fashion.

When good precision is available it is possible to achieve a high level of accuracy in trace analysis by AAS or AFS. Much depends on the recognition of the existence of systematic error when dealing with practical samples. Provided that matrix elements effects on nebulisation and atomisation are recognised and properly compensated for or eliminated by sample pretreatment, the use of AAS or AFS provides results at least as accurate as other instrumental techniques of trace analysis.

6. Interferences

With the exception of a greater freedom from spectral interference effects, the techniques of analytical atomic absorption and fluorescence spectroscopy are subject to the same types of interference as those encountered in flame emission spectroscopy. Thus, when concomitant elements affect an alteration in those physical or chemical properties or processes which control the final population of neutral ground state atoms of the sample in the absorption cell interference with the analytical signal results. The principal types of interferences encountered therefore result from physical effects and from chemical effects in the condensed phase or vapour phase. Each of these types of interference, and the occurrence of spectral interference, will be considered separately.

(A) PHYSICAL INTERFERENCES

The term physical interference in AAS and AFS is usually employed to denote any influence of concomitant materials present with the analyte element in the sample solution on one or more of the physical processes involved during the nebulisation and atomisation, or on one of the physical properties of the sample solution. The principal interferences encountered in both techniques result from the effect of solution viscosity on sample aspiration, and the effect of the vapour pressure, surface tension and temperature of the sample solution on the nebulisation process and solvent evaporation. Physical interference effects may also result from solute evaporation effects and the scattering of incident radiation from the source. Many examples of the occurrence of physical interferences are described in the literature and continue to be reported with increasing frequency. Extensive documentation of these effects will not be attempted here; the nature of the interferences caused by effects on each of the physical processes and properties mentioned will be considered separately and illustrated by reference to some commonly encountered examples where applicable.

(1) Aspiration

When an interference on the observed atomic absorption or fluorescence signal may be traced to an effect on the sample aspiration rate, this is most frequently caused by variation in viscosity of the sample solution. The sample solution is drawn up the capillary in indirect and direct nebulisers by the pressure drop created by the high-velocity gas stream at the capillary orifice. The hydrostatic head (due to the column of liquid lifted) and the viscosity of the sample solution tend to oppose the aspiration. The aspiration pressure usually greatly exceeds the pressure due to the hydrostatic head, however, so that the aspiration rate (sample uptake rate, $cc.min^{-1}$) is not significantly affected by small changes in the hydrostatic head or density of the sample liquid. The force of capillarity also opposes the aspiration, but this effect is also very small and any variation caused by different surface tensions in the sample liquids is unimportant in relation to the aspiration pressure.

When the sample liquid leaves the capillary under conditions of laminar flow, and with low velocity, the sample flow-rate theoretically is proportional to the pressure drop along the capillary and the fourth power of the capillary diameter and inversely proportional to the viscosity of the liquid. The dependence of the flow rate on the fourth power of the capillary diameter has been confirmed experimentally for the nebuliser of a flame photometer [283], and Winefordner and Latz [284] have demonstrated that the flow rate closely obeys the Pouseuille equation for viscous flow for low but not high flow rates. A variation in nebuliser uptake rate caused by change in viscosity of sample solutions is not necessarily accompanied by a proportional change in the atomic absorption or fluorescence signal observed in the flame. Willis [185] has shown that with the indirect nebulisers employed with premixed flames, variation in sample uptake rate may be compensated for by changes in nebulisation efficiency and drop-size distribution which occur simultaneously; the change in the atomic concentration of the analyte element which is produced in the flame may then be quite small. The effect of variation in sample viscosity on the flow rate for total consumption nebuliser—burners may be serious. Because of the large amount of solution entering the flame, much of it as large droplets, substantial temperature changes may occur when the sample uptake rate varies. The residence time of the sample droplets in the flames supported at these burners is also short; if a large number of large droplets enter the flame when the uptake rate is increased the desolvation process may be less efficient. Owing to this temperature effect and the effect on desolvation, when the sample uptake rate is increased substantial-

230

ly, decreased absorption or fluorescence signals may be obtained with this type of burner.

When an interference effect in AAS or AFS can be directly attributed to varying viscosities in the sample solutions it may be overcome by preparation of the standard solutions used to construct the calibration curve at similar viscosity in the same solvent. Alternatively, the method of standard addition to the sample solution may be employed.

(2) Nebulisation

The physical processes which occur during nebulisation have a large influence on the sensitivity and selectivity of flame methods of analysis (see Section 4(E)). Owing to their complexity, however, and the difficulty of designing experiments to alter one parameter affecting the nebulisation while maintaining the other parameters unchanged, nebulisation is perhaps the least well-characterised physical process involved in analytical flame spectroscopy. The drop-size distribution and nebuliser efficiency depend on the nature of the nebulising gas and sample solvent in a manner which may vary from one nebuliser type and geometry to another. The following paragraphs describe briefly the principal effects applicable with indirect nebulisers.

The effect of the surface tension, density and viscosity of the sample solution on the drop-size distribution at a concentric nebuliser may be demonstrated from the empirical equation established by Nukiyama and Tanasawa [285]. This relates the mean droplet diameter in micrometers, D_0, to the properties of the solution aspirated.

$$D_0 = \frac{A_1}{v}\left(\frac{S}{d}\right)^{\frac{1}{2}} + A_2\left(\frac{n}{(Sd)^{\frac{1}{2}}}\right)^{0.45}\left(1000\frac{V_{\text{liq}}}{V_{\text{air}}}\right)^{3/2}$$

where A_1 and A_2 are numerical constants, v is the velocity of the nebulising air at the orifice (m.sec^{-1}), S is the surface tension (dyn.cm^{-1}), d is the density of the sample liquid (g.cm^{-3}), n is the viscosity (dyn.sec.cm^{-2}) and V_{liq} and V_{air} are the velocities of the liquid and air flows (cm^3.sec^{-1}) respectively. This equation, which was established experimentally, is applicable in this form only between certain limits of d, S and n for subsonic velocities with the particular nebuliser employed by Nukiyama and Tanasawa with aqueous solution and air as the nebulising gas. The equation may nevertheless be employed to show the general manner in which mean droplet diameter is controlled by the physical properties of the solution nebulised. For nebulisers in which V_{air}/V_{liq} is greater than about 5000, only the first term of the equation is important; the mean droplet diameter is then determined principally by the air velocity at the orifice, and the density and surface tension of the solution. When V_{air}/V_{liq} is less than about 5000, however, the second term of the equation is significant and the viscosity of the sample becomes an important factor. The mean droplet diameter should decrease with increase in the air velocity at the orifice and with decrease in surface tension and viscosity of the sample solution. As the mean droplet diameter affects the nebuliser efficiency, efficiency of transport of solution through connecting tubing to the burner, desolvation and solute vaporisation in the flame, changes of viscosity and surface tension in the sample solution can result in interference with the analytical signal observed. This property is used to advantage in the use of organic solvents of low viscosity and surface tension and at high vapour pressure to enhance sensitivity in flame methods of analysis.

Solvent evaporation in the nebuliser chamber affects the drop-size distribution of the sample reaching the burner [283,286—288]. The extent of solvent evaporation depends on the drop size and residence time in the nebuliser, the molecular weight of the vapour, the diffusion coefficient and saturation vapour pressure of the solvent in air and the temperature. The smaller droplets may lose their solvent completely by evaporation during their residence time in the nebuliser. As the evaporation proceeds, however, the decreasing drop size and increasing dissolved solid concentration change the effective saturation vapour pressure for the vapour. When a concomitant material causes a substantial change in the drop-size distribution through its effect on the solution viscosity or surface tension, or in the dissolved-solid concentration, it is reasonable to suppose that the change in saturation vapour pressure which occurs may affect the rate of evaporation. This change is probably small, however, and overshadowed by the effect on the analytical signal of the gross changes in drop-size distribution and nebulisation efficiency caused by change in viscosity or surface tension. The extent of evaporation can be increased substantially by preheating the gas used for nebuli-

232

sation [289] or by using a heated nebuliser chamber [290]. After removal of most of the solvent in a heated nebuliser chamber the gases may be cooled to condense and collect the solvent [291]. The fine droplets and dry particles may then pass into the flame and a substantial increase in the sample transported results.

The temperature of the solution aspirated can affect the analytical absorbance or fluorescence signals observed. The viscosity, surface tension and vapour pressure of the solution aspirated are temperature dependent. Change in temperature of the sample solution may therefore affect the uptake rate and drop-size distribution, although a proportional change in analytical signal may not be observed. With indirect nebuliser systems it is generally observed that when sample solution temperatures differ only slightly ($\pm$ 3°C) the errors are insignificant.

(3) Solvent and Solute Vaporisation in Flames

With indirect nebulisers, although the smaller droplets may lose their solvent by evaporation in the nebuliser chamber, the desolvation process is only usually completed when the aerosol enters the flame. The rate of the desolvation process in the flame is controlled by the droplet size, boiling point and specific heat of vaporisation of the solvent, thermal conductivity of the gas and the flame temperature (see Section 4(E)(1)). The process is rapid with well-designed indirect nebulisers and the premixed flames normally employed in AAS. Interference with the analytical signal due to the radiation scattering and loss of atomic concentration which would accompany incomplete desolvation is therefore seldom encountered. The situation is quite different with total consumption nebuliser—burners, however, where incomplete desolvation occurs due to the large number of large droplets introduced which may have only a short residence time in the flame. Parsons and Winefordner [292] reported only 50—90% evaporation for water in such a flame in the part of the flame normally used for analysis.

The dry particles which remain after complete desolvation of the aerosol droplets must be vaporised and dissociated to provide the atomic population for analysis. The size and other physical properties of the particle determine the rate of vaporisation. When the composition of the dry particles varies in the presence of concomitant elements these physical properties may vary and give rise to disturbance of the analytical signal. Thus, for example, when many large particles are formed when a sample solution has a high dissolved-solid content, the rate of vaporisation may be slower than that of the smaller particles produced from pure aqueous solutions of the analyte element. Other interferences, frequently classified as chemical interferences, result from the different physical properties of the dry particles of compounds formed between analyte and concomitant elements in the flame compared to corresponding properties of the particles formed for the analyte element in the absence of concomitant elements. The physical properties of most concern in these cases are the melting and boiling points of the compounds. Further discussion of these effects is presented in Sections 4(E)(2) and 6(B).

(4) High Dissolved-Solid Content Samples and Light Scattering

With sample solutions which contain high concentrations of other dissolved solids, reduction in the analytical atomic absorption signal may be observed compared to that obtained with a pure aqueous solution of the same analyte concentration. The sample uptake rate and nebuliser efficiency is lower owing to the increased viscosity of the solutions, and frequently incrustation of the nebuliser and burner head lead to high signal noise levels. As mentioned above, physical occlusion of analyte in the large particles formed from such solutions on desolvation may also retard the vaporisation process in the flame.

In addition to the interference effects mentioned above in AAS and AFS for samples of high dissolved-solid content, with such solutions problems from scattered radiation may also become significant. Willis [293] first proposed that the scattering of the incident radiation by particles in the flame might explain the small unspecific absorption obtained in the presence of concomitant elements in AAS. For small particles, whose diameter is much smaller than the wavelength of the radiation measured, Rayleigh scattering (proportional to the fourth power of the frequency) should occur. A wavelength dependence of this type was observed by Willis for the nebulisation of concentrated (5%) solutions of simple salts into a low temperature flame. For particles whose diameter is larger than the wavelength of the radiation measured the Mie [294] scattering theory predicts that only a slight dependence of scattering on wavelength should occur. This type of behaviour was observed by Gidley [295]

234

when 10% solutions of the salts of elements which form refractory oxides (titanium, zirconium, and hafnium) were nebulised. Koirtyohann and Pickett [296] calculated the mean size of solid particles formed on desolvation with a burner for premixed flames, and using the Mie theory concluded that the experimentally observed non-specific absorption could not be accounted for wholly on the basis of light scattering. The same authors [297] demonstrated that absorption by gaseous molecular species caused at least part of the observed non-specific absorption. For example, the molecular absorption spectra of CaOH and SrOH have been observed [298] and may give rise to non-specific background absorption effects. Non-specific background absorption, either by scattering or molecular absorption, may be more serious in cool flames than hot flames. Interference by non-specific absorption is normally corrected for by measurement of the absorbance at wavelengths near to the resonance line using a non-absorbing line from the source and subtracting this from the absorbance at the resonance line. Alternatively the absorbance at wavelengths in the neighbourhood of the resonance line may be measured using a continuum source [60] and the value obtained subtracted similarly.

The phenomenon of light scattering by particles in the flame is more important in atomic fluorescence spectroscopy than in atomic absorption spectroscopy. When resonance fluorescence signals are measured, the signal which results from scattering of the primary source radiation by particles in the flame cannot be distinguished from the atomic resonance fluorescence signal. Veillon and co-workers [57], who employed a continuum source for AFS, corrected for scattered radiation by using a scanning monochromator. When step-wise or direct-line fluorescence signals are employed for analysis, scattered radiation from the source may frequently be eliminated by the use of a simple filter in the incident light beam from the source to select only that line used for irradiation. Problems of scattered radiation are more serious with total consumption nebuliser—burner systems than when indirect nebulisers and premixed flames are employed. Ellis and Demers [221], however, have employed the total-consumption burner with a hydrogen-entrained air flame and report less scattering problems with this flame than with the hydrogen—oxygen flame at a similar burner. With premixed air—hydrocarbon and nitrous oxide—acetylene flames very few scattering phenomena have been observed [108,213]. Scattering has only been observed to cause serious interferences in these flames for samples containing high concentrations of matrix elements.

The term chemical interference in AAS and AFS is employed to denote any reaction of concomitant materials present with the analyte element in the sample solution which occurs in the solid or vapour phase to affect the population of free atoms of analyte element formed in the flame. Such reactions in the solid phase are frequently also referred to as *condensed* phase interferences.

(1) Chemical Interferences in the Solid Phase

Non-specific interferences in the solid phase, where no compounds are formed between the analyte and concomitant material in the flame, may occur for any analyte element and are physical rather than chemical in nature. An example of this type of non-specific solid phase interference is the relatively small depressive effect on AAS signals for analyte elements due to occlusion of analyte within solid particles of salts of the matrix element in the flame. Specific interferences in the solid phase, however, are usually due to a reaction of the analyte element or one of its compounds with concomitant material to form a different compound with different thermochemical properties. This type of chemical interference is frequently specific to one analyte element or a small group of elements, e.g. the interference of phosphate on the alkaline earth elements in AAS in air—hydrocarbon flames due to compound formation.

It is the rate of evaporation of the particles of compounds formed between the analyte and concomitant elements compared to the evaporation rate of the analyte particles alone which determines the severity of the chemical interference effect. As described in Section 4(E)(2) it is not essential for the flame temperature to be higher than the melting or boiling points of the compound formed in the dry particle for vaporisation to proceed to completion. The flame temperature and composition, however, do influence the extent of formation and rate of evaporation of solid compounds in the flame, and in high temperature flames the incidence of chemical interference is lower than in cool flames. Owing to this effect of compound formation on the rate of release of vapour of the analyte, the severity

236

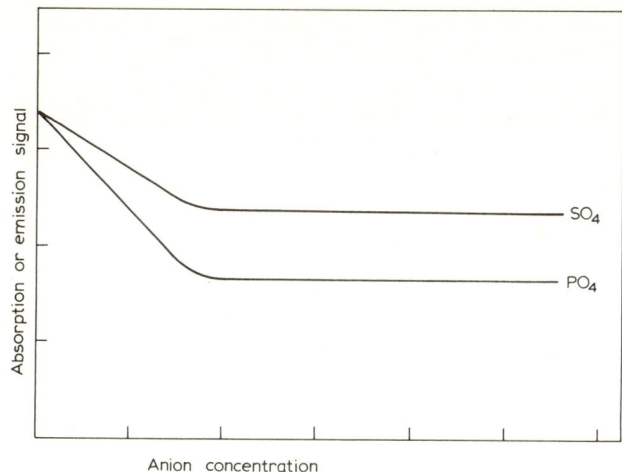

Fig. 34. Typical results for suppression by phosphate or sulphate emission or absorption in air—acetylene flame.

of any chemical interference observed may also depend on the size of the particles formed (and therefore nebuliser performance) and the height of observation of absorption or fluorescence in the flame (i.e. the residence time in the flame).

One of the best known chemical interference effects in analytical flame spectroscopy is the depression of calcium atomic concentration in the presence of phosphate. The effect has been observed and studied, both in emission and absorption spectroscopy, by many workers. A similar effect is also observed for calcium in the presence of sulphuric acid. Figure 34 shows the typical manner in which the calcium signal, observed in emission or absorption at the 422.7 nm line or in emission at the CaOH bond at 554 nm, varies with the concentration of phosphate or sulphate. Similar results may be obtained with premixed hydrocarbon—air flames and with oxy-hydrogen or oxy-acetylene flames supported at total consumption nebuliser—burners. The interference observed is virtually linear with the anion concentration up to a certain atomic concentration ratio of P/Ca, after which there is no further suppression with increasing anion concentration. The P/Ca concentration ratio above which the suppression is constant is not dependent upon the calcium concentration , height of observation in the flame or flame temperature. The relative severity of the interference depends on these factors, however. Values of the P/Ca ratio at the "knee" of the curve are reported in the literature between 0.3 and 1.1. The severity of interference is usually greater with turbulent flames burning at total consumption nebuliser—burners than for premixed flames with indirect nebulisers. The experimental observations may be explained by the postulation that a relatively non-volatile compound is formed between calcium and phosphorus in the flame. The particles of this compound, which probably contains only calcium and phosphorus in association with some oxygen, persist in the flame and evaporate only slowly. The limited volatility of the particles explains why the severity of the interference effect observed is less when the calcium atomic absorption signal is measured higher up in the flame (after a longer residence time) or in hotter flames (more rapid evaporation). As the evaporation rate also depends on the particle size, the observation of less serious interference with premixed flames and indirect nebulisers than at total consumption nebuliser—burners might be accounted for by the smaller drop size and particle size obtained with the former. It has also been observed that the relative depression of the calcium signal by phosphorus is smaller at lower calcium concentrations in the sample solution; this effect leads to curvature of the calibration graph for calcium towards the concentration axis when phosphorus is present. This effect can also be explained due to the relatively greater evaporation rate of the smaller particles of Ca—P—O compound formed at low concentrations than the larger particles formed with more concentrated sample solutions.

As well as chemical interference from concomitant anions on the determination of cations, which are quite prevalent in hydrogen—air and hydrocarbon—air flames used in AAS and AFS, mutual chemical interference between cations may also be encountered. A well-known example lies in the interference of aluminium salts in the determination of the alkaline earth elements and magnesium in hydrocarbon-air flames. The magnitude of the interference observed depends on the anion or acid which accompanies the aluminium. Dickson and Johnson [299] found that the degree of depression of calcium atomic absorption by aluminium varied with the anion present in the order $PO_4^{3-} > SO_4^{2-} > Cl^- > ClO_4^-$. It is most probable that an involatile Al—Ca—O compound is formed, and the extent of its formation depends on the manner in which the aluminium salts with different anions decompose in the flame.

Many other examples of chemical interference for various analyte

238

and concomitant elements have been observed. Gilbert [300] has listed elements which may cause interference on various analyte elements by the formation of mixed oxide species of low volatility. The list includes boron, beryllium, chromium, iron, molybdenum, silicon, titanium, uranium, vanadium, tungsten and rare earth elements.

The occurrence of specific interference between concomitant element and analyte of a chemical nature, rather than a non-specific physical interference in the solid phase, may frequently be recognised from the shape of the graph of the analytical signal vs. concentration of the concomitant element. Where compound formation is prevalent such graphs normally show a pronounced break or knee similar to that described for the calcium and phosphate example. The position of the break on the concentration axis usually remains independent of height of observation, although the depressive effect may be less severe when observations are made high in the flame. Specific chemical interference and a non-specific solid phase interference may occur at the same time. Thus, for example, in addition to the specific effect of aluminium on calcium described above due to formation of a low volatility Al—Ca—O compound, in the presence of a very high concentration of aluminium considerable amounts of aluminium oxide are also formed. In this way the Al—Ca—O compound is bound within relatively large particles of aluminium oxide and the release of calcium atoms to the flame is further impeded. Under these conditions the shape of the analytical signal vs. interfering element concentration graph is modified and a sharp break may no longer be observed. A useful indication of the occurrence of specific chemical interference, rather than a vapour phase interference, can be obtained by experiments conducted using separate nebuliser units for the analyte and concomitant element. When a condensed phase chemical interference exists, separate introduction into the flame in this way of the species involved in compound formation should yield no interference effect. If, on the other hand, the interference is due to a reaction in the vapour phase, the interference effect should still be observable. Experiments of this type have been undertaken by several workers [184,301,302,15] to confirm the occurrence of condensed phase interference of phosphate and other anions on calcium and magnesium in flame emission and absorption spectroscopy.

(2) Elimination of Solid Phase Interferences

The depressive effect of concomitant elements caused by chemical interference in the solid phase may frequently be eliminated, or at least minimised, by judicious choice of operating conditions during the analysis. Where the effect itself may not be eliminated, errors due to the interference may be prevented by suitable preparation of standards and samples.

It is important first to optimise the nebuliser and flame conditions to minimise chemical interference. This is achieved by attention to the following factors.

(a) Particle size

In order to achieve the high rate of particle evaporation the particles should be as small as possible. Careful adjustment of sample uptake rate, use of an organic solvent, and the insertion of barrier surfaces within the nebuliser chamber to assist the formation of as high a proportion of small droplets as possible is thus advisable. Large droplet and particle size accounts for much of the greater severity of condensed phase interference observed with total consumption nebuliser—burners compared to premixed flames with indirect nebulisers. When there is no lack of sensitivity for the detection of the analyte by absorption or fluorescence in the samples to be analysed, dilution of the sample solutions assists the formation of small particles and minimisation of interference.

(b) Height of Observation

When atomic absorption or fluorescence measurements are made as high as possible in the flame commensurate with good stability and low noise levels, the chemical interference effects are minimised. The low volatility compound formed between the analyte and concomitant elements will then have been permitted the longest possible residence time in the flame to release the analyte.

(c) Flame Temperature and Composition

The rate of vaporisation of the compound formed in the flame will depend on the flame temperature and composition. The most serious

240

chemical interferences in the solid phase are usually observed in cooler flames (i.e. air—hydrogen and argon—hydrogen—entrained air and even air—hydrocarbon flames). Even in air—hydrocarbon flames, however, control of fuel: oxidant concentration ratio may result in less severe chemical interference. Slightly fuel-rich air—acetylene flames, in which the low concentration of oxidising species in the interconal zone tends to promote dissociation of molecular compounds, may give rise to slightly less pronounced chemical interference than a fuel-lean flame.

The advantageous effect of the high temperature of the oxy-hydrogen and oxy-acetylene flames for suppression of chemical interference effects is difficult to realise in practice owing to the adverse effects on sample residence time and particle size which accompany their use with total consumption nebuliser—burner units. The use of hot premixed nitrous oxide—acetylene and oxygen—acetylene flames, however, has been shown to possess great advantages from the viewpoint of elimination of chemical interferences. The interference of phosphorus on calcium determinations by AAS may be completely suppressed in these flames [206,209,303]. Similarly the depressive effect of aluminium on magnesium absorption may be eliminated in the nitrous oxide—acetylene flame. These hot flames, particularly when used in the fuel-rich mode, are able to eliminate or substantially reduce most of the chemical interferences observed in cooler flames. This may sometimes only be achieved at the cost of somewhat reduced sensitivity due to ionisation of the analyte atoms at the high flame temperature. Suppression of ionisation, however, is a relatively simple matter (see Section 6(B)(4)), and the use of these hot flames will undoubtedly very soon be the most popular method for the avoidance of chemical interferences.

When it is not possible, through optimisation of the nebuliser and flame conditions described above, to eliminate a particular chemical interference the use of a releasing or protective agent or suitable addition of the interfering element to the standard solutions used to prepare the calibration graph may achieve a similar result.

(3) Use of Releasing and Protective Agents

The term releasing or protective agent is applied to a substance which may restore the AAS or AFS signal for the analyte element in the presence of an interference to the original value obtained in the absence of the interfering element. The substances employed for this purpose are also sometimes called 'buffers'.

Some of the most commonly employed releasing agents are cations with which the interfering species combines in preference to the analyte element. Thus strontium may be used as a releasing agent to prevent the interference of phosphate on the calcium determination. Usually a considerable excess of the releasing agent must be employed. It then consumes the interfering species by formation of a compound of low volatility and allows the analyte to vaporise without interference by compound formation. Similarly, in the determination of magnesium or strontium, calcium may be employed as a releasing agent for the interference from phosphate, silicate or aluminium. David [304] utilised magnesium as a releasing agent to suppress the interference of phosphate, sulphate, aluminium, and silicate on the determination of calcium in plant materials by AAS. Lanthanum is particularly suitable for use as a releasing agent in the determination of the alkaline earth elements as it exhibits no background emission or absorption at their resonance lines in hydrocarbon—air flames. Other rare earths have been employed to suppress the interference of aluminium, phosphate, and sulphate in the determination of strontium [305]. Copper may be used in the determination of platinum by AAS in a low-temperature flame to suppress the interference from elements such as antimony, tungsten, cobalt and nickel [306].

In the determination of the alkaline earth elements and magnesium by AAS the addition of certain substances which are not normally considered to form complexes with elements such as aluminium, iron, phosphate, etc. is nevertheless effective in suppressing their interference. Examples of this type of protecting agent are ethylene glycol, mannitol, dextrose, sucrose and glycerol. These compounds have been found to suppress the interference of phosphate on the determination of calcium and strontium [307,308]. A similar protective action is observed with ammonium chloride which has been employed to eliminate the interference of aluminium in the determination of sodium and chromium [309]; interferences in the determination of molybdenum by AAS are also suppressed by addition of excess NH_4Cl [310]. Chelating agents such as EDTA and 8-hydroxyquinoline have also been found to suppress interferences in the determination of calcium and magnesium. Thus EDTA has been reported to suppress the interference of selenium, tellurium, boron, aluminium, silicate, nitrate, phosphate and sulphate in these determinations [311,312], while 8-hydroxyquinoline has been employed to eliminate aluminium interferences [313,314]. No full explanation for the mechanism of action of these protective agents has been proposed, but it is possible that the substance added is vaporised very rapidly from the particle produced on desolvation to leave a residual sponge-like structure of the compound formed between the analyte and concomitant element. The large surface area of this residual particle would then result in an increase in rate of vaporisation of the analyte element.

(4) Chemical Interferences in the Vapour Phase

Vapour phase interferences result from the influence of concomitant elements on the dissociation and ionisation equilibria which control the concentration of atoms of the analyte element in the flame. Excitation interference, in which concomitant elements affect the analytical signal by disturbance of the equilibria between the electronically excited and unexcited states of the analyte atom, are also considered as vapour phase interferences but may be shown to be insignificant in analytical flame spectroscopy when conditions of thermal equilibrium prevail within the part of the flame employed. As distinct from condensed phase interferences, vapour phase interference is expected to occur even when the analyte and concomitant elements are introduced into the flame via separate nebulisers.

(a) Dissociation Equilibria

An account of the most important dissociation equilibria occurring between analyte atoms and the bulk constituents of the flame gases or other species added to the flame with the analyte has been given in Section 4(E)(3).

When sample solutions of low dissolved-solid content are nebulised there is little tendency for the analyte element to form stable gaseous compounds with the other species (halide, etc.) introduced simultaneously into the flame from the sample. Under these conditions, when an analyte element forms a stable molecule in the gaseous phase this is usually an oxide or hydroxide formed with atomic oxygen or hydroxyl species from the flame gases. The partial pressure in the flame of the flame species involved ([O] or [OH]) is usually several orders of magnitude greater than the partial pressure of the analyte element; the flame species concentration may therefore be regarded as unaffected by the small amount consumed to form the molecular compound with the analyte element. The degree of dissociation attained, and thus the analytical sensitivity in AAS or in AFS, then depends on the equilibrium constant for the dissociation of the molecule. The degree of dissociation therefore remains substantially constant as long as the temperature and flame gas composition are maintained unaltered (by control of gas flow rates). When the partial pressure of the flame gas species involved is high and effectively constant, being stabilised also by its relation to the dissociation equilibria of the bulk constituent gases, little vapour phase interference on the analyte partial pressure caused by disturbance of the dissociation equilibria of its compound can be caused by the introduction of a concomitant element which also forms a gaseous compound in the flame.

When the partial pressure of the flame gas partner (e.g. atomic oxygen) is low, however, and that of the analyte and concomitant element are similar, depletion of the flame species may occur by gaseous compound formation (e.g. oxide) with the concomitant element. This depletion then gives rise to a greater degree of dissociation of the gaseous compound (oxide) formed by the analyte element. This effect then results in an interference with the analytical signal, which increases. This type of enhancement of absorbance in the presence of concomitant elements which form thermally stable oxides has been observed in AAS with fuel-rich nitrous oxide—acetylene flames. Sachdev et al. [315] observed an enhancement of vanadium absorption by the addition of aluminium or titanium which could not be explained by an effect on the ionisation equilibria. Similarly titanium effects an enhancement of aluminium absorption and vice versa [316,317] and yttrium has been reported to enhance sensitivity in the detection of rare earth elements [318].

It is well-established that molecules of the halides of the alkali and alkaline earth metals may be formed in the vapour phase in some flames and give rise to a suppression of the observed analytical signal [319,320]. The formation of halides is most pronounced in cool flames and when high concentrations of halide or halogen acid are present in the sample solution. Mandelshtam [319] explained this behaviour from a consideration of the dissociation equilibria for the various alkali metal halides. The dissociation constants, K_D, for the dissociation

$$MX \rightleftharpoons M + X$$

where $K_D = p_M p_X / p_{MX}$ and p_M, p_X and p_{MX} are partial pressures of alkali metal atom, halogen and alkali metal halide, respectively, are shown in Table 8 for air—acetylene (2500 K) and air—coal gas (2000 K) flames.

In the air—acetylene flame the dissociation of halides is complete even when relatively high concentrations of halide are introduced into the flame with the analyte element. As may be evident from the dissociation constants, however, in cooler flames, such as air—coal gas or air—propane, incomplete dissociation of alkali metal halides may be obtained even at fairly low concentrations. When the concentration of halogen acid or halide in the sample is increased, the dissociation may be suppressed and depression of the analytical signal observed. This explanation may hold for the cooler flames, but does not explain the depressive effect of halogen acids observed in air—acetylene flames. In this hotter flame even considerable increase in halogen concentration should not cause significant metal halide formation. To account for the experimental observations of this depression it is necessary to postulate that the ionisation equilibrium for the alkali metal is disturbed by the addition of the strongly electronegative halogen so that the degree of ionisation of the metal is increased. For further discussion of this effect see ref. 321.

For easily atomised elements other than the alkali metals, vapour phase interferences from halides and acids are usually not encounter-

TABLE 8

Dissociation constants of some alkali metal halides at 2000 K and 2500 K

Compound	K_D (atm)	
	2000 K	2500 K
NaCl	6.3×10^{-6}	2.2×10^{-4}
NaBr	1.1×10^{-4}	1.4×10^{-3}
KCl	2.5×10^{-6}	7.9×10^{-5}
RbCl	6.3×10^{-6}	4.7×10^{-5}
RbBr	4.0×10^{-5}	3.8×10^{-4}

ed. Slight depression of the atomic absorption signals sometimes encountered for elements such as silver, zinc, cadmium, and bismuth in the presence of acids can usually be accounted for by their effect on the sample nebulisation. Few data are available concerning the extent to which true vapour phase interactions are involved in the interference effects observed for different anions in the AAS of elements which form stable oxides. Many of these effects, however, may also be largely due to the effect of the anion on the nebulisation process and the rate of vaporisation of the solute particles in the flame.

(b) Ionisation Equilibria

The effect of the ionisation of species added to flames on the population of free atoms of easily ionised analyte elements has been described in Section 4(E)(4). The equilibria outlined there may be invoked to explain the interference effects from concomitant elements due to their ionisation. The mutual interference effects of the alkali metals in the determination of these elements are probably the most widely encountered ionisation interference effects in flame spectroscopy. As a result of their low ionisation potentials quite pronounced ionisation effects are observed in the determination of the alkali metals in the presence of each other by AAS*. The enhanced absorption observed in the determination of sodium, for example, in the presence of potassium is explained by the suppression of the degree of ionisation of sodium due to the higher electron concentration in the flame when potassium is added to the flame and undergoes ionisation. Few other analyte elements suffer serious ionisation interference in hydrocarbon—air flames; these flames are not hot enough to give rise to significant ionisation of elements other than the alkali metals and alkaline earths. In hotter flames such as oxy-acetylene and nitrous oxide—acetylene, however, some other elements are appreciably ionised. Thus in addition to potassium, sodium, lithium, rubidium, caesium, calcium, barium and strontium, elements with ionisation potentials between 6 and 7.5 eV are also ionised to some extent in these flames. Ionisation effects may therefore result in mutual interference in AAS between elements such as

---

* In a premixed air—acetylene flame at 2500 K aspiration of 10^{-3} M solutions of each of the alkali metals produces a degree of ionisation of ca. 0.95 for caesium, 0.75 for rubidium, 0.6 for potassium, 0.2 for sodium and 0.1 for lithium.

aluminium, vanadium, titanium and some rare earths in these flames. Such inter-element effects due to interaction of the ionisation equilibria for these elements must be carefully distinguished from those due to competitive dissociation equilibria (see Section 6(B)(3)(a)).

The most common methods of overcoming ionisation interference effects in AAS and AFS are to make the concentration of the concomitant element the same in the standards and the analytical sample, or to suppress the analyte ionisation by addition of a high concentration of an easily ionised element to the standards and samples to 'swamp' the effect by providing a very high electron concentration in the flame. In some cases the concomitant elements themselves may be present in sufficiently high concentration to suppress the analyte ionisation completely. The suppression of magnesium ionisation in this way in its determination in silicate rocks by the concomitant elements present has been reported [322].

(C) SPECTRAL INTERFERENCES

The term spectral interference is commonly used in flame emission spectroscopy to denote any effect on the analyte emission signal from emission by concomitant elements in the same wavelength region. In order to achieve minimal spectral interference in flame emission spectroscopy a monochromator of high resolution is required. In AAS and AFS, however, the selective nature of the absorption process involved, and the use of modulated light sources and a.c. detection systems, result generally in greater freedom from spectral interference from concomitants in these techniques than in flame emission spectroscopy. There are, however, some interference effects which occur in these techniques as a result of the emission, absorption or scattering properties of the concomitant elements and these may be termed spectral interferences.

(1) Spectral Interferences in AAS

There are three principal causes of interference with AAS signals which may be considered as spectral interference; spectral 'overlap' and radiation scatter interferences, flame emission interference and interference from unwanted source radiation.

(a) Spectral Overlap and Radiation Scatter Interference

When the wavelength of one of the resonance lines of a concomitant element is almost coincident with the resonance line of the analyte element monitored for its determination by AAS, high and inaccurate results may be obtained in the determination. This effect is a true spectral interference by the atomic absorption process. The severity of the interference effect depends on the wavelength separation and the line widths of the source line and the absorption lines in the flame; when a sharp line from a hollow-cathode lamp is employed the separation of the line widths of the source line of the analyte element and absorption line of the concomitant element must be less than about 0.01 nm for a substantial effect to be observed. Specific examples of this type of interference which have been observed in practice are given in Table 9. Figure 35 shows the results obtained by Fassel and co-workers [323] for the spectral interference of the vanadium absorption line at 308.211 nm on the determination of aluminium by AAS at the 308.215 nm line. This type of interference is usually overcome by selection of an alternative absorbing line of the analyte element for its determination by AAS when the interfering concomitant element is suspected or known to be present in the sample.

A second type of true spectral interference occurs when the concomitant element forms molecules in the vapour phase; the molecular absorption spectra of these molecules may then overlap the analyte resonance line emission from the source. Koirtyohann and Pickett [60] observed molecular absorption by alkali halides in an oxy-

TABLE 9

Spectral line interferences observed in AAS

Analyte, source emission line (nm)	Interferent, flame absorption line (nm)	Wavelength separation (nm)	Ref.
Cu, 324.754	Eu, 324.753	0.001	323
Fe, 271.902	Pt, 271.904	0.002	323
Si, 250.689	V, 250.690	0.001	323
Al, 308.215	V, 308.211	0.004	323
Hg, 253.652	Co, 253.649	0.003	324
Mn, 403.307	Ga, 403.298	0.009	325

248

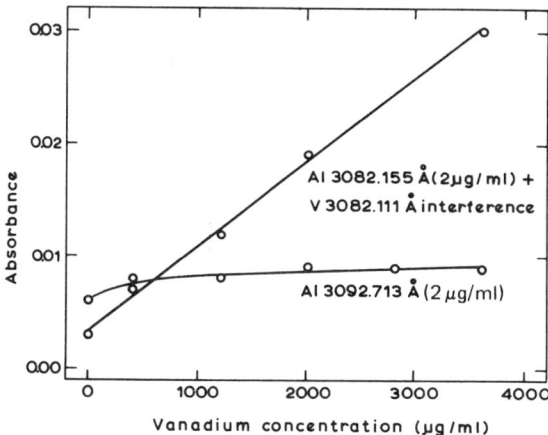

Fig. 35. Spectral interference of vanadium on aluminium absorbance at 308.215 nm (from ref. 323). (Courtesy of Spectrochim. Acta).

hydrogen flame passed into a long tube. These workers have also observed absorption spectra which closely resembled the emission spectra for the alkaline earth oxide and hydroxide molecules. The most serious interference of this type observed was by CaOH absorption in an air—acetylene flame in the vicinity of the barium 553.55 nm resonance line. The absorption from a solution containing 1% calcium was about that expected for 75 ppm barium. Capacho-Delgado and Sprague [298] also observed this effect for calcium and barium; these workers also demonstrated that the interference effect disappears in a nitrous oxide—acetylene flame owing to less extensive formation of calcium oxide. Less serious interferences of the same type have been observed on the absorption signals for lithium at 670.8 nm due to SrO bands, sodium 589.0 nm due to CaO and chromium 357.9 nm due to MgOH. This type of interference due to high background absorption in the presence of concomitant elements may be corrected for by measurement of the absorption at wavelengths near the resonance line using a continuum source, and subtraction of this absorbance from that at the wavelength of the resonance line.

In the presence of high concentrations of dissolved solids, particularly when the matrix element forms a refractory oxide and a low temperature flame is employed, solid, unevaporated particles may be present in the flame gases. Under these conditions a small fraction of the incident radiation from the source may be scattered by the particles and give rise to an apparent high value of absorbance for the analyte element. The severity of interference by scattering in this way depends on the flame type, wavelength of measurement and nature of the sample. Billings [326] and Koirtyohann and Pickett [297] have studied this effect of scattering in detail. These last-mentioned authors have reported that it is not possible to account for the non-specific absorption effects observed simply on the basis of light scattering by particles, and that molecular absorption probably accounts for most of the light losses observed. The use of a continuum source to correct for both effects is recommended [297]. Light scattering is also considered briefly in Section 6(A) on physical interferences.

(b) Flame Emission Interference

Early atomic absorption instrumentation which employed d.c. source and detector operation was prone to serious errors due to detection and amplification of the flame emission at the resonance line wavelength from the analyte element or atomic or molecular emission in the same wavelength region from concomitant elements. This effect was predictably most serious for those elements whose resonance lines used for AAS lie in the visible region of the spectrum. Modern instrumentation which invariably employs a.c. source and detector operation is very much less subject to this type of interference. When the flame emission is intense in the wavelength region to which the monochromator is set a large d.c. signal is received at the photomultiplier. Although this is not amplified by the amplifier, it does register as noise and in extreme cases photomultiplier saturation may result. The effect of flame emission in this way is frequently observed for several elements in the nitrous oxide—acetylene flame due to its high background emission over a wide range of wavelengths in both the visible and ultraviolet regions of the spectrum. When a sample solution contains a high concentration of a concomitant element which produces molecules in the vapour phase whose band emission is in the same wavelength region as the analyte line from the hollow-cathode, loss of analytical precision may result from the increase in the signal noise level which is observed. This effect is minimised by operation at narrow spectral slit width and high lamp operating currents.

250

(c) Unwanted Source Radiation

The wavelength of an atomic line from the filler gas used in the hollow-cathode source, a line from an impurity in the cathode material, or a non-absorbing line of the analyte element itself may lie close to the wavelength of the analyte resonance line from the source. This unwanted radiation is not absorbed by the analyte atoms in the flame but may be transmitted to the detector and amplified at the detector system. Instead of $\log I_0 /I$ being measured as absorbance, therefore, where I_0 and I are the incident and transmitted intensities of the resonance line, $\log (I_0 + i)/(I + i)$ is recorded, where i is the unabsorbed radiation intensity. The overall sensitivity is lowered and non-linear calibration graphs are obtained. This effect is not a true spectral 'interference' as it occurs even in the absence of the concomitant element. The effect is minimised by operation at narrow spectral band-pass.

(2) Spectral Interferences in AFS

At the present stage of development of the technique of AFS no single particular type of instrumental assembly has found widespread usage; both line and continuum sources have been recommended and d.c. and a.c. source and detector systems are both employed. It is to be expected, however, that a.c. operated systems will prevail over d.c. systems except in certain special cases (e.g. with non-flame cells). Problems encountered from flame background emission noise should therefore be no greater in AFS than in AAS, except that smaller signal strengths require detection against this background. In contrast to AAS, in AFS unwanted source radiation does not reach the detector (in the absence of scatter), so that the type of effect outlined in Section 6(C)(1)(c) is seldom encountered.

Spectral interference effects in AFS analogous to those described in Section 6(C)(1)(a) may, however, occur. When a sharp line source, such as a high intensity hollow-cathode lamp, is employed the frequency of occurrence of spectral line overlap interference should be no greater in resonance fluorescence measurements than in AAS. The effect of overlap of the analyte emission line from the source and concomitant element absorption lines in the flame will give rise to an inaccurate AFS signal for the analyte element. This may occur both when resonance and non-resonance fluorescence is measured; a somewhat greater effect may be observed in the case of resonance fluorescence, however, as the radiation emitted as fluorescence at the same wavelength may also be absorbed by atoms of the concomitant element before it leaves the flame. The effect of spectral overlap interference of this type on the signal which registers as atomic fluorescence at the detector is governed by the fluorescence characteristics of the elements concerned; an enhancement or a suppression may occur depending on whether the concomitant element re-emits the absorbed radiation as resonance or non-resonance fluorescence and the quantum efficiency of the fluorescence process compared to that for the fluorescence of the analyte element. The effect of spectral overlap of cadmium 228.802 nm on arsenic 228.812 nm and iodine 206.163 nm on bismuth 206.170 nm [268] and gallium 403.298 nm on manganese 403.307 nm [327] in AFS has been observed. Spectral interference by molecular absorption similar to that observed in AAS should also be encountered in AFS.

When a continuum source, such as a xenon arc, or a line source operated under conditions of high self-absorption, is employed for AFS it is possible for the source to stimulate fluorescence simultaneously from more than one element at wavelengths which are relatively close together. Spectral interference may result under these conditions when the analysing monochromator transmits both the fluorescence line of the analyte element and that of the concomitant element. Control of this type of effect is clearly dependent upon the resolution of the monochromator. Thus the spectral interference in this way by nickel 323.3 nm on copper 324.7 nm, copper 324.7 nm on silver 328.1 nm, nickel 341.5 nm on silver 338.3 nm, cobalt 419.1 nm on indium 410.2 nm, and cobalt 352.7 nm on nickel 352.4 nm have been recorded in work with a 500 W xenon arc source and a grating monochromator of large spectral bandpass [58].

The effect of scattered light from particulate material in the flame is more important in AFS than in AAS. In measurements of resonance fluorescence scatter of incident radiation from the source by particles produced from concomitant elements in the flame may give rise to high results. The measurement of direct-line or stepwise fluorescence, on the other hand, enables any problems from scattered light to be eliminated. An auxiliary filter is then desirable in the incident light beam to prevent illumination of the flame by radiation from the source at the wavelength at which the fluorescence is to be measured. When resonance fluorescence measurements are made using a continuum source, a correction for scattered light may be made by scanning the wavelength region on either side of the resonance line emission and subtracting the observed 'continuum' background caused by scattered light. The problems which are encountered due to scattered light in AFS have not been found to be as extensive as was originally anticipated; with premixed flames and efficient nebuliser units even relatively high concentrations of refractory oxide forming elements may be tolerated, e.g. in an air—acetylene flame in the determination of elements such as iron [274,275], manganese [239], nickel and cobalt [275]. Studies with the use of premixed nitrous oxide—acetylene flames in AFS also reveal relatively little spectral interference by scattering of radiation.

REFERENCES

1. W.H. Wollaston, Phil. Trans. Roy. Soc. London, Ser. A., 92 (1802) 365.
2. D. Brewster, Report of 2nd Meeting of British Association, 1832, p. 320.
3. G. Kirchoff, Phil. Mag., 20 (4) (1860) 1.
4. G. Kirchoff and R. Bunsen, Phil. Mag., 20 (4) (1860) 89.
5. G. Kirchoff and R. Bunsen, Phil. Mag., 22 (4) (1861) 329.
6. T.T. Woodson, Rev. Sci. Instrum., 10 (1939) 308.
7. A. Walsh, Spectrochim. Acta, 7 (1955) 108.
8. C.T.J. Alkemade and J.M.W. Milatz, Appl. Sci. Res. Sect. B, 4 (1955) 289.
9. C.T.J. Alkemade and J.M.W. Milatz, J. Opt. Soc. Amer., 45 (1955) 583.
10. E.L. Nichols and H.L. Howes, Phys. Rev., 23 (1924) 472.
11. C.T.J. Alkemade, Proc. Xth Colloq. Spectroscopicum Intern., Spartan Books, Washington, D.C., 1963.
11a. J.W. Robinson, Anal. Chim. Acta, 24 (1961) 254.
12. J.D. Winefordner and T.J. Vickers, Anal. Chem., 36 (1964) 161.
13. J.D. Winefordner and R.A. Staab, Anal Chem., 36 (1964) 165.
14. J.D. Winefordner and R.A. Staab, Anal. Chem., 36 (1964) 1367.
15. W.T. Elwell and J.F. Gidley, Atomic Absorption Spectrophotometry, 2nd edn., Pergamon, London, 1966.
16. N.P. Ivanov, Atomno-absorbcionnyj Analiz, IREA, Moscow, 1966.
17. B.V. L'Vov, Atomno-absorbcionnyj Spektralnyj Analiz, Izdatelstvo Nauka, Moscow, 1966.
18. J.W. Robinson, Atomic Absorption Spectroscopy, Dekker, New York 1966.
19. J. Ramirez-Munoz, Atomic Absorption Spectroscopy, Elsevier, Amsterdam, 1968.
20. E.E. Angino and G.K. Billings, Atomic Absorption Spectrophotometry in Geology, Elsevier, Amsterdam, 1968.
21. W. Slavin, Atomic Absorption Spectroscopy, Interscience, New York, 1968.

22. I. Rubeska and B. Moldan, Atomic Absorption Spectrophotometry, Iliffe, London, 1969.
23. F. Rousselet, Spectrophotometry by Atomic Absorption Applied to Biology, C.D.V. and S.E.D.E.S., Paris, 1967.
24. R. Herrman, Fortschr. Chem. Forsch., 5 (1966) 515.
25. R. Herrman, Z. Instrumentenk., 75 (1967) 101.
26. A. Hulanicki, Chem. Anal. (Warsaw), 11 (1966) 211.
27. H. Massman, Chimia, 21 (1967) 217.
28. R. Mavrodineanu, Encycl. Ind. Chem. Anal., 1 (1966) 160.
29. A. Petrakev, Khim. Ind. (Sofia), 38 (1966) 256.
30. H. Prugger, Glas. Instrum. Tech., 10 (1966) 610.
31. I. Rubeska, Chem. Listy, 61 (1967) 865.
32. W. Slavin, Appl. Spectrosc., 20 (1966) 281.
33. M. Suzuki and T. Takeuchi, Jap. Anal., 15 (1966) 1003.
34. A. Walsh, XIIth Colloq. Spectroscop. Intern., Hilger and Watts, London, 1965, p. 43.
35. A. Walsh and J.B. Willis, in F.J. Welcher (Ed.), Standard Methods of Chemical Analysis, Vol III, Part A, 6th edn., Van Nostrand, Princeton, New Jersey, 1966, p. 105.
36. H.L. Kahn, J. Chem. Educ., 43 (1) (1966) A7.
37. H.L. Kahn, J. Chem. Educ., 43 (2) (1966) A103.
38. M.L. Girard and F. Rousselet, Ann. Pharm. Fr., 25 (1967) 197.
39. M.L. Girard and F. Rousselet, Ann. Pharm. Fr., 25 (1967) 207.
40. H. Schueller, Landwirt. Forsch., 21 (1967) 93.
41. H.L. Kahn, J. Metals, 18 (1966) 1101.
42. J. Ramirez-Munoz, Ion (Bucaramanga, Columb.), 2 (12) (1966) 15.
43. J.D. Winefordner and J.M. Mansfield, in G.G. Guilbault (Ed.), Fluorescence, Dekker, New York, 1967.
44. J.D. Winefordner and J.M. Mansfield, Appl. Spectrosc. Rev., 1 (1968) 1.
45. T.S. West, in Trace Characterisation, Chemical and Physical, Nat. Bur. Stand. U.S. Monograph 100, U.S. Government Printing Office, Washington, D.C., 1967.
46. R. Mavrodineanu and H. Boiteux, Flame Spectroscopy, Wiley, New York, 1965.
47. M.L. Parsons, W.J. McCarthy and J.D. Winefordner, Appl. Spectrosc., 20 (1966) 223.
48. A.C.G. Mitchell and M.W. Zemansky, Resonance Radiation and Excited Atoms, Cambridge Univ. Press, 1934; reprinted 1961.
49. J.D. Winefordner, M.L. Parsons, J.M. Mansfield and W.J. McCarthy, Spectrochim. Acta, 23B (1967) 37.
50. V.A. Fassel and V.G. Mossotti, Anal. Chem., 35 (1963) 252.
51. N.P. Ivanov and N.A. Kozyreva, Zh. Anal. Khim., 19 (1964) 1266.
52. V.L. Ginsberg and G.P. Satarina, Zavodsk. Lab., 31 (1965) 249.
53. N.P. Ivanov and N.S. Kozyreva, Zavodsk. Lab., 31 (1965) 566.
54. M. Margoshes and M.M. Darr, Nat. Bur. Stand, Tech. Note 272, 1965, p. 18.
55. L. de Galan, W.W. McGee and J.D. Winefordner, Anal. Chim. Acta, 37 (1967) 436.

56. C.W. Frank, W.G. Schrenk and C.E. Meloan, Anal. Chem., 39 (1967) 534.
57. C. Veillon, J.M. Mansfield, M.L. Parsons and J.D. Winefordner, Anal. Chem., 38 (1966) 204.
58. M.S. Cresser and T.S. West, Spectrochim. Acta, 25B (1970) 61.
59. V.A. Fassel, V.G. Mossotti, W.E.L. Grossman and R.N. Kniseley, Spectrochim. Acta, 22 (1966) 347.
60. S.R. Koirtyohann and E.E. Pickett, Anal. Chem., 37 (1965) 601.
61. V.G. Mossotti and V.A. Fassel, Spectrochim. Acta, 20 (1964) 1117.
62. V. Svoboda, Anal. Chem., 40 (1964) 1384.
63. L. Klein, Appl. Opt., 7 (1968) 677.
64. A.V. Sheklein and V.A. Popov, Zh. Nauch. Prikl. Fotogr. Kinematogr., 9 (1964) 192; Chem. Abstr., 61 (1964) 7844h.
65. W.R.S. Garton, W.H. Parkinson and E.M. Reeves, Astrophys. J., 140 (1964) 1269.
66. D.C. Manning and J. Vollmer, At. Absorption Newslett., 6 (1967) 38.
67. G.I. Goodfellow, Appl. Spectrosc., 21 (1967) 39.
68. R.E. Popham and W.G. Schrenk, Appl. Spectrosc., 22 (1968) 192.
69. G. Rossi and N. Omenetto, Talanta, 16 (1969) 263.
70. W.G. Jones and A. Walsh, Spectrochim. Acta, 16 (1960) 249.
71. A. Walsh, L.S.U. Intern. Symp. Modern Methods Anal. Chem., January 1962.
72. W.T. Elwell and J.A.F. Gidley, Anal. Chim. Acta, 24 (1961) 71.
73. H. Massman, Z. Instrumentenk., 71 (1963) 225.
74. L.R.P. Butler and A. Strasheim, Spectrochim. Acta, 21 (1965) 1207.
75. C. Sebens, J. Vollmer and W. Slavin, At. Absorption Newslett., 3 (1964) 165.
76. D.C. Manning, D. Trent and J. Vollmer, At. Absorption Newslett., 4 (1965) 234.
77. J. Vollmer, C. Sebens and W. Slavin, At. Absorption Newslett., 4 (1965) 306.
78. F.J. Fernandez, D.C. Manning and J. Vollmer, At. Absorption Newslett., 8 (1969) 117.
79. J.V. Sullivan and A. Walsh, Spectrochim. Acta, 21 (1965) 721.
80. Z. van Gelder, Appl. Spectrosc., 22 (1968) 581.
81. J.B. Dawson and D.J. Ellis, Spectrochim. Acta, 23A (1967) 565.
82. D.A. Katskuv, G.C. Lebedev and B.V. L'Vov, Zh. Prikl. Spectrosk., 10 (1969) 215.
83. D.G. Mitchell and A. Johansson, Spectrochim. Acta, 25B (1970) 175.
84. S. Tolansky, High Resolution Spectroscopy, Methuen, London, 1947.
85. W.E. Bell, A.L. Bloom and J. Lynch, Rev. Sci. Instrum., 32 (1961) 688.
86. R.J. Atkinson, G.D. Chapman and L. Krause, J. Opt. Soc. Amer., 55 (1965) 1269.
87. N.P. Ivanov, L.V. Minervina, S.V. Baranov, L.G. Pofralidi and I.I. Olikov, Zh. Anal. Khim., 21 (1966) 1129.
88. J.R. Hollahan, J. Chem. Educ., 43 (1966) A401, A497.
89. R.G. Brewer, Rev. Sci. Instrum., 32 (1961) 1356.
90. V.B. Gerard, J. Sci. Instrum., 39 (1962) 217.

91. J.M. Mansfield, J.D. Winefordner and C. Veillon, Anal. Chem., 37 (1965) 1049.
92. J.M. Mansfield, M.P. Bratzel, H.O. Norgordon, D.O. Knapp, K.E. Zacha and J.D. Winefordner, Spectrochim. Acta, 23B (1968) 389.
93. R.M. Dagnall, K.C. Thompson and T.S. West, Anal. Chim. Acta, 36 (1966) 269.
94. R.M. Dagnall and T.S. West, Appl. Opt. 7 (1968) 1287.
95. K.M. Aldous, R.M. Dagnall and T.S. West, Anal. Chim. Acta, 44 (1969) 457.
96. R.F. Browner, R.M. Dagnall and T.S. West, Anal. Chim. Acta, 45 (1969) 163.
97. E. Jacobson and G.R. Harrison, J. Opt. Soc. Amer., 39 (1949) 1054.
98. A.T. Forrester, R.A. Gudmundsen and P.O. Johnson, J. Opt. Soc. Amer., 46 (1956) 339.
99. W.F. Meggers and F.O. Westfall, J. Res. Nat. Bur. Stand. U.S., 44 (1950) 447.
100. G.F. Kirkbright and M. Sargent, Spectrochim. Acta, 25B (1970) 577.
101. R.E. Zacha, M.P. Bratzel and J.D. Winefordner, 40 (1968) 1733.
102. D. Cooke, R.M. Dagnall and T.S. West, Anal. Chim. Acta, 54 (1971) 381.
103. M.D. Silvester and W.J. McCarthy, Spectrochim. Acta, 25B (1970) 229.
104. M.S. Cresser and T.S. West, Anal. Chim. Acta, 51 (1970) 530.
105. B.J. Russell, J.P. Shelton and A. Walsh, Spectrochim. Acta, 8 (1957) 317.
106. J.D. Winefordner and R.A. Staab, Anal. Chem., 36 (1964) 165.
107. G.I. Goodfellow, Anal. Chim. Acta, 35 (1966) 132.
108. R.M. Dagnall, P. Young and T.S. West, Talanta, 13 (1966) 803.
109. R.F. Browner, R.M. Dagnall and T.S. West, Talanta, 16 (1969) 75.
110. R.S. Hobbs, G.F. Kirkbright, M. Sargent and T.S. West, Talanta, 15 (1968) 997.
111. W. Slavin, D.J. Trent and S. Sprague, At. Absorption Newslett., 4 (1965) 180.
112. D.C. Manning, D.J. Trent, J. Sprague and W. Slavin, At. Absorption Newslett., 4 (1965) 255.
113. D.J. Trent, D.C. Manning and W. Slavin, At. Absorption Newslett., 4 (1965) 335.
114. A. Strasheim, Nature, 196 (1964) 1194.
115. A. Strasheim and H.G.C. Human, Spectrochim. Acta, 23B (1968) 265.
116. H.G.C. Human, L.R.P. Butler and A. Strasheim, Analyst, 94 (1969) 81.
117. H.G.C. Human, L.R.P. Butler and A. Strasheim, Paper presented at XVth Colloq. Spectroscop. Intern., Madrid, 1969.
118. S. Greenfield, P.B. Smith, A.E. Breeze and N.M.D. Chilton, Anal. Chim. Acta, 41 (1968) 385.
119. D.C. Manning and W. Slavin, At. Absorption Newslett., (1962) No. 8.
120. R.K. Skogerboe and R.A. Woodriff, Anal. Chem., 35 (1963) 1977.
121. C.S. Rann, Spectrochim. Acta, 23B (1968) 827.
122. G.F. Kirkbright and S.J. Wilson, Analyst, 95 (1970) 833.
123. G.L. Vidale, Space Science Lab. Aerospace Operation, Gen. Elec. T.I.S. Rept. R605D330, 1960.
124. J.P. Mislan, Paper presented at 7th Conf. Anal. Chem. in Nuclear Technol., Gatlinberg, Tenn., 1963.

125. R.D. Hudson, Phys. Rev., 135 (1964) 1212.
126. S.P. Choong and W. Loong-Seng, Nature, 204 (1964) 276.
127. F.S. Tomkins and B. Ercoli, Appl. Opt. 6 (1967) 1299.
128. R. Woodriff, R.W. Stone and A.M. Held, Appl. Spectrosc. 22 (1968) 408.
129. R. Woodriff and G. Ramelow, Spectrochim. Acta, 23B (1968) 665.
130. R. Woodriff and R.W. Stone, Appl. Opt. 7 (1968) 1337.
131. B.V. L'Vov, Spectrochim. Acta, 24B (1969) 153.
132. H. Massman, Spectrochim. Acta, 23B (1968) 215.
133. U. Ulfvarson, Acta Chem. Scand., 21 (1967) 641.
134. H. Brandenburger and H. Bader, Helv. Chim. Acta, 50 (1967) 1409.
135. T.S. West and X.K. Williams, Anal. Chim. Acta, 45 (1969) 27.
136. R.G. Anderson, I.S. Maines and T.S. West, Anal. Chim. Acta, 51 (1970) 355.
137. J.F. Alder and T.S. West, Anal. Chim. Acta, 51 (1970) 365.
138. W.G. Jones and A. Walsh, Spectrochim. Acta, 16 (1960) 249.
139. B.M. Gatehouse and A. Walsh, Spectrochim. Acta, 16 (1960) 602.
140. A. Walsh, Proc. Xth Colloq. Spectroscop. Int. 1962, p. 127.
141. J.A. Goleb and J.K. Brody, Anal. Chim. Acta, 28 (1963) 457.
142. J.A. Goleb, Anal. Chem., 35 (1963) 1978.
143. N.P. Ivanov, M.N. Gusinsky and A.D. Jesikov, Zh. Anal. Khim., 20 (1965) 1133.
144. M. Marinkovic, B. Bojovic and D. Pesic, Proc. XIIIth Colloq. Spectrosc. Int., Hilger and Watts, London, 1967, p. 1181.
145. U.I. Belyaev, L.M. Ivantsov, A.V. Karyakin, P.H. Phi and V.V. Shemet, Zh. Anal. Khim., 23 (1968) 508.
146. T. Kantor and L. Erdey, Spectrochim. Acta, 24B (1969) 283.
147. J.W. Robinson, Anal. Chim. Acta, 27 (1962) 465.
148. J.W. Robinson, Ind. Chem. 38 (1962) 226, 362.
149. R.R. Wendt and V.A. Fassel, Anal. Chem., 37 (1965) 920.
150. R.R. Wendt and V.A. Fassel, Anal. Chem., 38 (1966) 337.
151. K.E. Friend and A.J. Diefenderfer, Anal. Chem., 38 (1966) 1763.
152. C. Veillon and M. Margoshes, Spectrochim. Acta, 23B (1968) 503.
153. S. Greenfield, P.B. Smith, A.E. Breeze and N.M.D. Chilton, Anal. Chim. Acta, 41 (1968) 385.
154. H. Brandenburger, Chimia, 22 (1968) 449.
155. V.G. Mossotti, K. Laqua and W.D. Hagenah, Spectrochim. Acta, 23B (1967) 197.
156. S.D. Rasberry, B.F. Scribner and M. Margoshes, Appl. Opt., 6 (1967) 81, 87.
157. A.V. Karyakin and V.A. Kaigorodov, Zh. Anal. Khim., 23 (1968) 930.
158. B.J. Russell and A. Walsh, Spectrochim. Acta, 15 (1959) 883.
159. J.V. Sullivan and A. Walsh, Appl. Opt. 7 (1968) 1271.
160. J.A. Bowman, J.V. Sullivan and A. Walsh, Spectrochim. Acta, 22 (1966) 205.
161. R.M. Lowe, Spectrochim. Acta, 24B (1969) 191.
162. N.A. Sebestyen, Spectrochim. Acta, 25B (1970) 261.
163. H.L. Kahn, At. Absorption Newslett., 7 (1968) 40.

164. Beckman Instruments, Fullerton, California.
165. A.C. Menzies, Anal. Chem., 32 (1960) 898.
166. J.W. Robinson, Anal. Chem., 33 (1961) 1226.
167. L.R.P. Butler and A. Strasheim, Spectrochim. Acta, 21 (1965) 1207.
168. F.J. Feldman, Anal. Chem., 42 (1970) 719.
169. J.D. Winefordner and T.J. Vickers, Anal. Chem., 36 (1964) 1947.
170. J.D. Winefordner, M.L. Parsons, J.M. Mansfield and W.J. McCarthy, Anal. Chem., 39 (1967) 436.
171. A. Smithells and H. Ingle, Trans. Chem. Soc., 61 (1892) 204.
172. E.A. Boling, Spectrochim. Acta, 22 (1966) 425.
173. P.B. Zeeman and L.R.P. Butler, Appl. Spectrosc., 16 (1962) 120.
174. R.N. Kniseley, A.P. D'Silva and V.A. Fassel, Anal. Chem., 35 (1963) 910.
175. R.N. Kniseley, A.P. D'Silva and V.A. Fassel, Anal. Chem., 36 (1964) 1287.
176. T.G. Cowley, V.A. Fassel and R.N. Kniseley, Spectrochim. Acta, 23B (1968) 771.
177. A. Hell, W.F. Ulrich, N. Shifrin and J. Ramirez-Munoz, Appl. Opt. 7 (1968) 1317.
178. J.A.F. Gidley and J.T. Jones, Analyst, 85 (1960) 248.
179. V.C.O. Schuller and G.S. James, J. S. Afr. Inst. Mining Met., 62 (1962) 786.
180. L.R.P. Butler, J.S. Afr. Inst. Mining Met. 62 (1962) 796.
181. O.E. Clinton, Spectrochim. Acta, 16 (1960) 985.
182. L.R.P. Butler, A.S. Strasheim and E.C. Maskew, J.S. Afr. Inst. Mining Met., 62 (1962) 796.
183. R. Lockyer and G.E. Hames, Analyst, 84 (1959) 385.
184. C.T.J. Alkemade and M.H. Voorhuis, Z. Anal. Chem., 163 (1958) 91.
185. J.B. Willis, Spectrochim. Acta, 23A (1967) 811.
186. E. Beckman and P. Waentig, Z. Phys. Chem., 68 (1910) 385.
187. K.R. May, J. Appl. Phys., 20 (1949) 932.
188. W.H. Walton and W.C. Prewett, Proc. Phys. Soc., B62 (1949) 341.
189. J. Spitz and G. Uny, Appl. Opt. 7 (1968) 1345.
190. J. Stupar and J.B. Dawson, Appl. Opt. 7 (1968) 1351.
191. G.A. Hemsalech, Phil. Mag., 33 (1917) 1.
192. W. Herschel, Z. Phys., 47 (1928) 147.
193. H. Staubel, Naturwissenschaften, 40 (1953) 337.
194. H. Staubel, Mikrochim. Acta, (1955) 329.
195. P.T. Gilbert Jnr., Paper presented at Eastern Analytical Symp., New York, 1960.
196. J.D. Winefordner, J.M. Mansfield and T.J. Vickers, Anal. Chem., 35 (1963) 1607.
197. R. Püschel, L. Simon and R. Herrman, Optik (Stuttgart), 21 (1964) 441.
198. P.T. Gilbert Jnr., A.S.T.M. Spec. Tech. Publ. No. 269, 1959, p. 73.
199. P.T. Gilbert Jnr., Beckman Bull. No. 3063, Fullerton, California, 1961.
200. J.W. Robinson and R.J. Harris, Anal. Chim. Acta, 26 (1962) 439.
201. V.A. Fassel and V.G. Mossotti, Anal. Chem., 35 (1963) 252.
202. W. Slavin and D.C. Manning, Anal. Chem., 35 (1963) 253.
203. F.B. Dowling, C.L. Chakrabarti and G.R. Lyles, Anal. Chim. Acta, 29 (1963) 29, 489; 28 (1963) 392.

204. J.B. Willis, Nature, 207 (1965) 715.
205. G.F. Kirkbright, M.K. Peters and T.S. West, Talanta, 14 (1967) 789.
206. M.D. Amos and J.B. Willis, Spectrochim. Acta, 22 (1966) 1325.
207. G.F. Kirkbright, M.K. Peters, M. Sargent and T.S. West, Talanta, 15 (1968) 663.
208. J.B. Willis, J.O. Rasmuson, R.N. Kniseley and V.A. Fassel, Spectrochim. Acta, 23B (1968) 725.
209. D.C. Manning and L. Capacho-Delgado, Anal. Chim. Acta, 36 (1966) 312.
210. E.E. Pickett and S.R. Koirtyohann, Spectrochim. Acta, 23B (1968) 235.
211. G.F. Kirkbright, M. Sargent and T.S. West, Talanta, 16 (1969) 245.
212. V.G. Mossotti and M. Duggan, Appl. Opt. 7 (1968) 1325.
213. D.N. Hingle, G.F. Kirkbright and T.S. West, Analyst, 93 (1968) 522.
214. R.M. Dagnall, G.F. Kirkbright, T.S. West and R. Wood, Anal. Chim. Acta, 47 (1969) 407.
215. G.F. Kirkbright, R.M. Dagnall, T.S. West and R. Wood, Analyst, 95 (1970) 425.
216. R.M. Dagnall, G.F. Kirkbright, T.S. West and R. Wood, Anal. Chem., 42 (1970) 1029.
217. M.D. Amos and P.E. Thomas, Anal. Chim. Acta, 32 (1965) 139.
218. R.M. Dagnall, K.C. Thompson and T.S. West, Analyst, 93 (1968) 153.
219. J.B. Willis, V.A. Fassel and J.A. Fiorino, Spectrochim. Acta, 24B (1969) 157.
220. D.N. Armentrout, Anal. Chem., 38 (1966) 1235.
221. D.W. Ellis and D.R. Demers, Anal. Chim. Acta, 38 (1966) 1945.
222. H.L. Kahn and J.E. Schallis, At. Absorption Newslett., 7 (1968) 5.
223. T.C. Rains, in J.A. Dean and T.C. Rains (Eds.), Flame Emission and Atomic Absorption Spectrometry, Dekker, 1969, p. 363.
224. C. Feldman and R. Dhumwad, Proc. 6th Conf. Anal. Chem. Nucl. Technol., Gatlinberg, Tenn., U.S.A.E.C., TID-7655, 1963.
225. K. Fuwa and B.L. Vallee, Anal. Chem., 35 (1963) 942.
226. S.R. Koirtyohann and C. Feldman, in J.E. Forrete (Ed.), Developments in Applied Spectroscopy, Vol. 3, Plenum Press, New York, 1964.
227. E.J. Agazzi, Anal. Chem., 37 (1965) 365.
228. J. Stupar, B. Podobnik and J. Korosin, Croat. Chim. Acta, 37 (1965) 141.
229. J. Stupar, Mikrochim. Acta, (1966) 722.
230. Ju.V. Zeljukova and N.S. Poluetkov, Zh. Anal. Khim., 18 (1963) 435.
231. I. Rubeska, J. Stupar, At. Absorption Newslett., 5 (1966) 69.
232. G.F. Kirkbright and T.S. West, Appl. Opt., 7 (1968) 1305.
233. G.F. Kirkbright, A. Semb and T.S. West, Talanta, 14 (1967) 1011.
234. G.F. Kirkbright, A. Semb and T.S. West, Talanta, 15 (1968) 441.
235. D.N. Hingle, G.F. Kirkbright and T.S. West, Analyst, 94 (1969) 864.
236. R.S. Hobbs, G.F. Kirkbright and T.S. West, Analyst, 94 (1969) 554.
237. G.F. Kirkbright, M. Sargent and T.S. West, At. Absorption Newslett., 8 (1969) 34.
238. G.F. Kirkbright, M. Sargent and T.S. West, Talanta, 16 (1969) 1467.
239. L. Ebdon, G.F. Kirkbright and T.S. West, Talanta, 17 (1970) 965.
240. G.M. Hieftje and H.V. Malmstadt, Anal. Chem., 40 (1968) 1860.

241. F.A. Williams, Proc. 8th Symp. Combustion, Pasadena, California, 1960, Williams and Wilkins, Baltimore, U.S.A., 1962, p. 50.
242. C.T.J. Alkemade, in J.A. Dean and T.C. Rains (Eds.), Flame Emission and Atomic Absorption Spectrometry, Vol. 1., Dekker, New York, 1969.
243. W.H. Foster Jnr. and D.N. Hume, Anal. Chem., 31 (1959) 2028.
244. M.R. Baker, K. Fuwa, R.E. Thiers and B.L. Vallee, J. Opt. Soc. Amer., 48 (1958) 576.
245. J. Janin and A. Bouvier, Spectrochim. Acta, 20 (1964) 1787.
246. J.A. Dean, Rec. Chem. Progr., 22 (1961) 179.
247. L. Brewer and R.F. Porter, J. Chem. Phys., 22 (1954) 1867.
248. D.J. Halls and A. Townshend, Anal. Chim. Acta, 36 (1966) 278.
249. D.J. David, Spectrochim. Acta, 20 (1964) 1185.
250. J.V. Vejc and L.V. Gurvitch, Opt. Spektrosk., 2 (1957) 274.
251. S. De Jaegere, J. Deckers and A. Van Tiggelen, VIIIth Intern. Comb. Symp. 1960, Williams and Wilkins, Baltimore, 1962, p. 155.
252. P.F. Knewstubb and T.M. Sugden, Research (London) 9 (8) (1956) S32.
253. B. Lewis and G. Von Elbe, Combustion, Flames and Explosions in Gases, Academic Press Inc., New York, 1951.
254. M.N. Saha and H.K. Saha, A Treatise on Modern Physics, Vol. 1, Allahabad and Calcutta, 1934.
255. D.C. Manning and L. Capacho-Delgado, Anal. Chim. Acta, 36 (1966) 312.
256. J.B. Willis, Appl. Opt., 7 (1968) 1295.
257. C.S. Rann and A.N. Hambly, Anal. Chem., 37 (1965) 879.
258. R.F. Browner, R.M. Dagnall and T.S. West, Anal. Chim. Acta, 50 (1970) 375.
259. C.S. Rann, Spectrochim. Acta, 23B (1968) 827.
260. L. de Galan, Spectrochim. Acta, 24B (1969) 629.
261. L. de Galan and G.F. Samaey, Anal. Chim. Acta, 50 (1970) 39.
262. B.V. L'Vov, Spectrochim. Acta, 17 (1961) 761.
263. S.R. Koirtyohann and E.E. Pickett, Paper presented at XIIIth Colloq. Spectroscop. Int., Ottawa, 1967.
264. L. de Galan and J.D. Winefordner, J. Quant. Spectrosc. Radiat. Transfer, 7 (1967) 251.
265. L. de Galan and G.F. Samaey, Spectrochim. Acta, 25B (1970) 245.
266. B.V. L'Vov, Ind. Lab., 28 (1963) 987.
267. R.M. Dagnall, K.C. Thompson and T.S. West, Talanta, 15 (1968) 677.
268. R.M. Dagnall, K.C. Thompson and T.S. West, Talanta, 14 (1967) 1467.
269. R.S. Hobbs, G.F. Kirkbright and T.S. West, unpublished work.
270. D.R. Demers and D.W. Ellis, Anal. Chem., 40 (1968) 860.
271. B. Fleet, K.V. Liberty and T.S. West, Anal. Chim. Acta, 45 (1969) 205.
272. D.C. Manning and P. Heneage, At. Absorption Newslett., 6 (1967) 124.
273. J.I. Dinnin, Anal. Chem., 39 (1967) 1491.
274. L. Ebdon, G.F. Kirkbright and T.S. West, Anal. Chim. Acta, 47 (1969) 563.
275. J. Matousek and V. Sychra, Anal. Chem., 41 (1969) 518.
276. D.C. Manning and P. Heneage, At. Absorption Newslett., 7 (1968) 80.
277. T.J. Vickers and S.P. Merrick, Talanta, 15 (1968) 873.

260

278. T.S. West and X.K. Williams, Anal. Chim. Acta, 42 (1968) 29.
279. R.M. Dagnall, K.C. Thompson and T.S. West, Talanta, 14 (1967) 1151.
280. R.M. Dagnall, K.C. Thompson and T.S. West, Talanta, 14 (1967) 557.
281. R.F. Browner, R.M. Dagnall and T.S. West, Anal. Chim. Acta, 46 (1969) 207.
282. J. Matousek and V. Sychra, Proc. Int. AAS Conf., Sheffield, 1969.
283. R.E. Bernstein, S. Afr. J. Med. Sci., 20 (1955) 57.
284. J.D. Winefordner and H.W. Latz, Anal. Chem., 33 (1961) 1727.
285. S. Nukiyama and Y. Tanasawa, Transl. Repts. 1—6, Trans. Soc. Mech. Eng. Jap., 4, 5, 6, (1938—40), Defence Research Board, Dept. of Nat. Defence, Ottawa, Canada.
286. C.Th.J. Alkemade, Ph.D. Thesis, University of Utrecht 1954, Excelsior, The Hague, Holland.
287. R.D. Caton Jnr. and R.W. Bremner, Anal. Chem., 26 (1954) 805.
288. P. Porter and G. Wyld, Anal. Chem., 27 (1955) 733.
289. R.A.G. Rawson, Analyst, 91 (1966) 630.
290. C.A. Dubbs, Anal. Chem., 24 (1952) 1654.
291. A. Hell, 5th Austr. Spectrosc. Conf., Perth, June, 1965.
292. M.P. Parsons and J.D. Winefordner, Anal. Chem., 38 (1966) 1593.
293. J.B. Willis, in D. Glick (Ed.), Methods of Biochemical Analysis, Vol. XI, Wiley, New York, 1963.
294. Mie, Ann. der Physik, 25 (1908) 377.
295. J.A.F. Gidley, Limitations of Detection in Spectrochemical Analysis, Hilger and Watts Ltd., London, 1964, p. 25.
296. S.R. Koirtyohann and E.E. Pickett, Anal. Chem., 38 (1966) 1087.
297. S.R. Koirtyohann and E.E. Pickett, Anal. Chem., 38 (1966) 585.
298. L. Capacho-Delgado and S. Sprague, At. Absorption Newslett., 4 (1965) 363.
299. R.E. Dickson and C.M. Johnson, Appl. Spectrosc., 20 (1966) 214.
300. P.T. Gilbert Jnr., Analysis Instrumentation 1964, Proc. Xth Nat. Anal. Inst. Symp., San Francisco 1964, Plenum Press, New York, 1964.
301. S. Fukishima, Mikrochim. Acta, (1959) 596.
302. W. Schuhknecht and H. Schinkel, Z. Anal. Chem., 162 (1958) 266.
303. V.A. Fassel and D.A. Becker, Anal. Chem., 41 (1969) 1522.
304. D.J. David, Analyst, 84 (1959) 536.
305. J.I. Dinnin, Anal. Chem., 32 (1960) 1475.
306. A. Strasheim and G.J. Wessels, Appl. Spectrosc., 17 (1963) 65.
307. C.A. Baker and F.W.J. Garton, UKAEA, AERE-R-3490, 1961.
308. T.C. Rains, H.E. Zittel and M. Ferguson, Talanta, 10 (1963) 367.
309. L. Barnes Jnr., Anal. Chem., 38 (1966) 1085.
310. D.J. David, Analyst, 93 (1968) 79.
311. A.C. West and W.D. Cooke, Anal. Chem., 32 (1960) 1471.
312. T.V. Ramakrishna, J.W. Robinson and P.W. West, Anal. Chim. Acta, 36 (1966) 57.
313. F.J. Wallace, Analyst, 88 (1963) 259.
314. M. Yanagisawa, M. Suzuki and T. Takeuchi, Talanta, 14 (1967) 933.
315. S.L. Sachdev, J.W. Robinson and P.W. West, Anal. Chim. Acta, 37 (1967) 12.

316. T.V. Ramakrishna, P.W. West and J.W. Robinson, Anal. Chim. Acta, 39 (1967) 81.
317. J.B. Headridge and D.P. Hubbard, Anal. Chim. Acta, 37 (1967) 151.
318. R.J. Jaworowski, R.P. Weberling and D.J. Bracco, Anal. Chim. Acta, 37 (1967) 284.
319. S.L. Mandelshtam, C. R. Acad. Sci. URSS, 22 (1939) 403.
320. T.F. Borovik-Romanova, Zh. Anal. Khim., 16 (1961) 664.321.
321. I. Rubeska, in J.A. Dean and T.C. Rains (Eds.), Flame Emission and Atomic Absorption Spectrometry, Dekker, New York, 1969.
322. R.W. Nesbitt, Anal. Chim. Acta, 35 (1966) 413.
323. V.A. Fassel, J.O. Rasmuson and T.C. Cowley, Spectrochim. Acta, 23B (1968) 579.
324. D.C. Manning and F. Fernandez, At. Absorption Newslett., 7 (1968) 24.
325. J.E. Allan, Spectrochim. Acta, 24B (1969) 13.
326. G.K. Billings, At. Absorption Newslett., 4 (1965) 357.
327. L. Ebdon, G.F. Kirkbright and T.S. West, unpublished work.

Chapter 3

Diffuse reflectance spectroscopy

R.W. FREI, M.M. FRODYMA and V.T. LIEU

1. Introduction

In transmission spectroscopy the spectral properties of a solution are determined by measuring the monochromatic light transmitted by the solution. By taking advantage of the transparent nature of the medium under investigation, it is possible to measure the absorption of dissolved substances as a function of wavelength. Difficulties appear, however, when an attempt is made to obtain similar information with turbid and colloidal systems where light-scattering phenomena result in substantial energy losses; and the technique becomes entirely unsuitable when one wishes to determine the absorption spectra of substances adsorbed on solid surfaces. It is in situations of this last type that reflectance spectroscopy can make a unique contribution.

Unlike transmission spectroscopy, reflectance spectroscopy is still very much in the development stage. The technique is not, however, as new as one might expect. This is particularly true of diffuse reflectance spectroscopy. Diffuse reflectance, in contrast to specular reflectance, found early application in industries, such as the paper, paint, dye, textile, printing and ceramic industries, where the measurement of color is important in routine quality control processes [1]. Relatively sophisticated filter-type reflectometers, such as those designed by Taylor [2] and Benford [3], began to make their appearance soon after the first World War. Not long thereafter Hardy [4] developed the first recording reflectometer. All these instruments had one design feature in common, in that they made use of an integrating sphere, or "Ulbrichtkugel" [5], which permitted the collection of diffusely reflected light but excluded the undesirable specular component. It was not until after 1960 [6] that there was much appreciation of the potential of specular reflectance techniques as an analytical tool. This aspect of reflectance spectroscopy will not be included in this discussion.

Other than the acceptance accorded by industry there has until recently been little interest in the application of the diffuse reflectance technique to the solution of analytical and chemical constitution problems. This has changed with the ready availability of diffuse reflectance attachments for practically all commercial spectrophotometers and the literature in the field is experiencing a period of rapid growth. One of the most prominent contributors has been Kortüm, who has written an excellent monograph on the subject that is available in an English translation [7]. Another comprehensive treatment of the subject of reflectance spectroscopy is one written by Wendlandt and Hecht [8]. Both texts include a consideration of specular reflectance techniques. Also noteworthy are the published proceedings of the American Chemical Society Symposium on Reflectance Spectroscopy which was held in 1967 [9].

2. Theory

(A) KUBELKA—MUNK EQUATION

The radiation reflected from a finely divided solid consists of a regular, or specular, reflection component and a diffuse reflection component. The former arises from reflection occurring at the surface of the system with no transmission through it, while the latter results from radiation that has penetrated and subsequently reappeared at the surface of the system following partial absorption and multiple scattering within the system. Because of the complementary nature of these two components, the elimination or minimization of regular reflection is an important consideration in the design of instruments and the preparation of samples for the measurement of diffuse reflectance.

The most general theory treating diffuse reflection and the transmission of light-scattering layers is that developed by Kubelka and Munk [10,11]. For an infinitely thick opaque layer (achieved in practice with a layer thickness of a few mm), the Kubelka—Munk

264

equation may be written in the form

$$F(R_\infty) \equiv (1 - R_\infty)^2 / 2R_\infty = k/s \qquad (1)$$

where R_∞ is the diffuse reflectance of the layer relative to a non- or low-absorbing standard such as magnesium oxide, k is the molar absorption coefficient of the sample, and s is the scattering coefficient. If s remains constant, a linear relationship should be observed between $F(R_\infty)$ and k. This has been confirmed for weakly absorbing materials where the contribution of regular reflectance is small [12].

When the reflectance of a sample diluted with a non- or low-absorbing powder is measured against the pure powder, the absorption coefficient, k, may be replaced by the product 2.303 ϵc

$$F(R_\infty) \equiv (1 - R_\infty)^2 / 2R_\infty = 2.303 \; \epsilon c/s \qquad (2)$$

where ϵ is the extinction coefficient and c is the molar concentration [13]. Since $F(R_\infty)$ is proportional to the molar concentration under constant experimental conditions, the Kubelka—Munk equation in

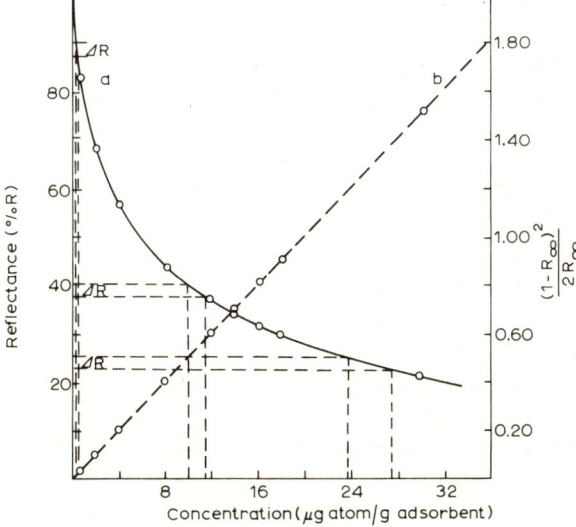

Fig. 1. (a) ——, Percent reflectance (%R) as a function of the concentration of an adsorbed species. (b) ---, Kubelka—Munk value as a function of the concentration of an adsorbed species.

References pp. 345—354

this form is comparable to the Beer—Lambert law of transmission spectroscopy. At high enough dilutions, the regular reflection from the sample approximates to that from the standard and is thus cancelled out in any comparison measurement. For a system which conforms to the Kubelka—Munk equation the relationship among the concentration of an adsorbed species, the percent reflectance, and the Kubelka—Munk value are illustrated in Fig. 1. If, even at high dilutions, deviations from the linear relationship indicated by curve (b) are observed, these, as in the case of true deviations from the Beer—Lambert law, may be taken as an indication of the occurrence of associative or dissociative processes at the boundaries of the particles constituting the sample.

(B) OPTIMUM CONCENTRATION RANGE FOR ANALYSIS

Even with the systems which conform to the Kubelka—Munk equation, reflectance analysis is of limited use at high and low concentrations. With high concentrations of absorbing material, so little radiation is reflected that the sensitivity of the spectrophotometer becomes inadequate. At the other end of the concentration spectrum, the instrument reading error becomes disproportionately large compared to the quantity being measured. The situation can perhaps be visualized more readily with the aid of curve (a) in Fig. 1, which represents percent reflectance ($\%R$) as a function of concentration. When an arbitrary error amounting to ΔR is plotted at 25, 40 and 90%R, it is readily apparent that the corresponding absolute error in terms of concentration is greatest at 25%R. This, in turn, leads to a large relative error in the determination of the concentration. At the other extreme, 90%R, the absolute concentration error, though smallest of the three, approaches in magnitude the quantity being measured. This, too, leads to a large relative error in the determination. It is obvious that there must be some intermediate point, say 40%R, where these two effects can be balanced in such a way as to reduce the relative error to a minimum.

For systems exhibiting no deviation from the Kubelka—Munk equation, the optimum conditions for maximum accuracy can be deduced by computing the relative error dc/c [14]. Thus eqn. (2) can be written in the form

$$c = k'(1 - R_\infty)^2 / 2R_\infty \tag{3}$$

266

where k' is a constant equal to $s/2.303\epsilon$. The error in c is

$$\mathrm{d}c = k'(R_\infty^2 - 1)\mathrm{d}R_\infty/2R_\infty^2 \qquad (4)$$

and the relative error in c is

$$\mathrm{d}c/c = (R_\infty + 1)\mathrm{d}R_\infty/(R_\infty - 1)R_\infty \qquad (5)$$

Assuming a reading error amounting to one reflectance unit, that is $\mathrm{d}R_\infty = 0.01$

$$(\mathrm{d}c/c) \times 100 = (R_\infty + 1)/(R_\infty - 1)R_\infty = \% \text{ relative error in } c \qquad (6)$$

To determine the value of R_∞ which will minimize the relative error in c, $\mathrm{d}(\% \text{ error in } c)/\mathrm{d}R_\infty$ is equated to zero. The positive solution of the resulting equation

$$R_\infty^2 + 2R_\infty - 1 = 0 \qquad (7)$$

indicates that the minimum relative error in c, or the optimum value for reflectance measurements, occurs at a reflectance value of 0.414, which corresponds to a reflectance reading of 41.4%R. This is presented graphically in Fig. 2, where the relative % error, computed with the use of eqn. (6), is plotted as a function of the % reflectance. As is shown there, the minimum in the resulting curve corresponds to the 41.4%R value obtained by employing eqn. (7).

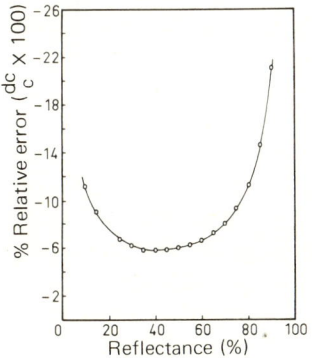

Fig. 2. Percent error, calculated from eqn. (6), as a function of percent reflectance.

References pp. 345—354

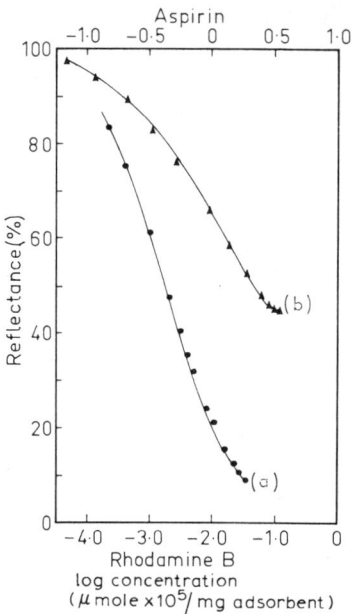

Fig. 3. Percent reflectance as a function of the logarithm of concentration [14].
▲, Non-conforming system (aspirin); ●, Ideal system (Rhodamine B).

Another approach to the selection of the optimum range for reflectance analysis is that suggested by Ringbom [15], and later Ayres
[16], who evaluated relative error and defined the suitable range for
absorption analysis by plotting the absorption (1—transmittance)
against the logarithm of the concentration. When this approach is
employed in conjunction with the diffuse reflectance technique,
plots such as those depicted in Fig. 3 result. Curve (a) was obtained
with an ideal system; curve (b) typifies systems that do not conform
to the Kubelka—Munk equation. If a sufficiently wide range of concentrations is included, the general form of the curves obtained is the
same, with an inflection point occurring at 41.4%R with ideal systems and at some other value for those that are non-ideal. The optimum range for analyses corresponds to that portion of each curve
exhibiting the greatest slope. Since a considerable section of each of
the curves in Fig. 3 is also linear about this point, it is apparent that
good accuracy can be expected over a wide concentration range. The

268

maximum accuracy can be estimated with the use of the equation

$$\frac{dc}{c} \times 100 = 2.303 d(\log c) \times 100 = \frac{2.303 d(\log c)}{dR} \times 100 dR \tag{8}$$

Assuming a constant reading error of 1% R, i.e., $dR = 0.01$

% relative error $= 2.303 \; d(\log c)/dR \tag{9}$

For the two systems under consideration, the optimum range for analysis can be arrived at from a consideration of the curves depicted in Fig. 3 and the % relative error resulting from a reading error of 1%R can be computed with the use of eqn. (9) and the slope of the appropriate curve, $\Delta \% R/\Delta \log c$. When this is done, the data presented in Table 1 are obtained.

TABLE 1

Optimum range for analysis and % relative error

Curve	System	Optimum range, (% R)	Relative error in concentration/ 1% reading error
a	Ideal	20—65	– 6.0
b	Non-ideal	55—85	– 6.0

It appears that the minimum error in diffuse reflectance spectrophotometric analysis is approximately —6% per 1%R reading error, regardless of whether the system in question conforms to the Kubelka—Munk equation or not. This value can obviously be decreased by reducing the reading error. Although the reading error can be reduced to 0.5%R without too much difficulty, it would be unrealistic to expect a precision better than 0.1—0.2%R and, therefore, a smaller minimum error than —1 to —2%.

The Ringbom method has the advantage of not only making available the optimum range and maximum accuracy, but also of providing a plot which is usable as a calibration curve. The rather large difference between the optimum range computed for an ideal system (20—65%R) and that found for a non-ideal system (55—85% in this instance) is noteworthy. In view of this difference it is suggested that, when the optimum range is not known, reflectance values lying in the upper portion of the optimum range, as computed for an ideal system, be used for analyses.

As may be seen from the above, the usual methods for the measurement of diffuse reflectance leave much to be desired with respect to accuracy when one is operating near the ends of the reflectance scale. It is often possible, however, to reduce considerably the error resulting from instrumental uncertainties by modifying the way in which the measurement is carried out. The techniques employed for this purpose, which are analogous to those used in transmission spectrophotometry [17], are differential in nature and can be categorized as (1) low-reflectance methods, useful with more concentrated samples, and (2) high-reflectance methods, useful with samples of very low concentration.

(1) Low-reflectance method

In the usual procedure for the measurement of reflectance, two preliminary instrumental adjustments must be made to avail oneself of the full reflectance scale. These involve setting the instrument to read 0 with the photocell in darkness and 100 when the photocell is exposed to the light reflected from a non- or low-absorbing standard. In the low-reflectance method, this procedure is altered so that the latter scale setting is made with the use of a standard whose concentration is somewhat less than that of the analytical sample. The zero adjustment is made in the usual way. This procedure effectively expands the reflectance scale and thus makes possible a more accurate reading of a reflectance value.

The behavior of an ideal system under the above conditions can be deduced from the Kubelka—Munk equation. If the relationship between R_{so}, the reflectance of the differential standard relative to a non- or low-absorbing standard, and C_s, the concentration of the differential standard, is expressed in the form of eqn. (3) and then subtracted from a similar expression for the relationship between R_{xo}, the reflectance of the analytical sample relative to a non- or low-absorbing standard, and C_x, the concentration of the analytical sample, the difference can, following substitution and rearrangement,

270

be put in the form [18]

$$C_x = \frac{k'(1 - \sigma R_{xs})^2}{2\sigma R_{xs}} - k'h \tag{10}$$

where R_{xs} is the reflectance of the analytical sample relative to the differential standard and σ and h are constants equal, respectively, to R_{so} and $[(1-\sigma)^2/2\sigma] - C_s/k'$. As indicated, C_x would bear a straight-line relationship to $(1-\sigma R_{xs})^2/2\sigma R_{xs}$.

Labinowich [19] employed this approach in an effort to increase the accuracy of a procedure designed to determine by diffuse reflectance the amount of copper concentrated on a chromatographic column. Some of his results are presented in Fig. 4, which depicts the percent error arising from a reading error of 1%R (estimated graphically) as a function of the reflectance for various concentrations of copper(I)—neocuproine complex adsorbed on magnesium silicate. A consideration of these results reveals that it is not only possible to extend the optimum range for analysis to higher reflectance values

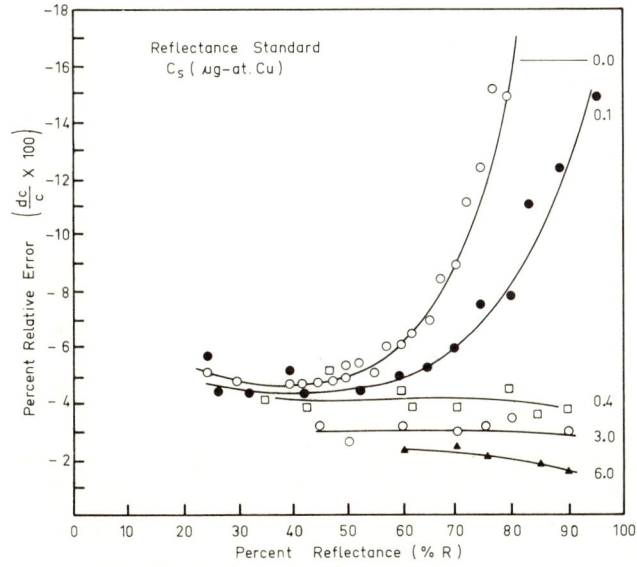

Fig. 4. Percent relative error arising from a reading error of 1%R, estimated graphically, as a function of the reflectance for various concentrations of copper (I)—neocuproine complex adsorbed on magnesium silicate [19].

References pp. 345—354

by increasing the concentration of the differential standard, but also to reduce the percent error itself, with the greatest reductions occurring as the concentration of the differential standard approaches that of the analytical sample. It has been demonstrated experimentally that the low-reflectance method is capable of providing analytical results whose accuracy nears that obtainable by classical volumetric or gravimetric procedures [18].

(2) High-reflectance method

This technique is particularly useful in situations where the concentration of the absorbing species is so low that the reflectance value obtained in the ordinary way lies above the optimum reflectance range for analysis. In this case, expansion of the scale is achieved by setting the instrument to read 0 when the photocell is exposed to light reflected from a reference sample somewhat more concentrated than the sample being analyzed; the full scale adjustment is made in the ordinary way. Usually the zero adjustment is made with the use of the dark-current control, while the full-scale setting is achieved by varying the sensitivity and slit width. With some instruments adjustment of the full-scale point may alter the zero setting, and vice versa, so that it is necessary to employ a series of successive approximations to achieve the desired result.

Under the conditions obtaining for the high-reflectance method, the reflectance of a sample relative to the differential standard is given by the expression [20]

$$R_{xs} = (R_{xo} - R_{so})/(1 - R_{so}) \qquad (11)$$

where R_{xs}, R_{xo} and R_{so} have the same meaning as they do in the derivation of eqn. (10). Following rearrangement, eqn. (11) can be combined with eqn. (3) to give

$$C_x = k'[(1-\sigma)(1-R_{xs})]^2/[2R_{xs}(1-\sigma) + \sigma] \qquad (12)$$

where δ is a constant equal to R_{so}, and C_x is the concentration of the analytical sample. If the concentration of the differential standard is such that the incident light is completely absorbed, i.e., $R_{so} = 0$, eqn. (12) reverts to eqn. (3), the more common form of the Kubelka—Munk equation.

272

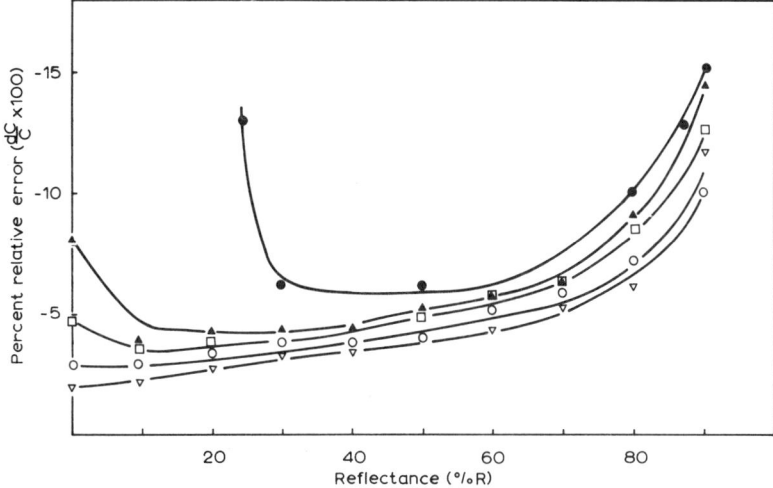

Fig. 5. Percent relative error arising from a reading error of 1%R, estimated graphically, as a function of the reflectance for various concentrations of nickel dimethylglyoxime adsorbed on cellulose [20]. Concentration of differential standard, μg Ni/40mg cellulose: ●, photocell in darkness; ▲, 107; □, 83; ○, 59; △, 5.9.

Zaye [21] made use of this technique in the determination of trace amounts of inorganic cations resolved on chromatoplates and found it led to increased accuracy. As may be seen from Fig. 5, which presents the percent error arising from a reading error of 1%R (estimated graphically) as a function of the reflectance for various concentrations of nickel dimethylglyoxime complex adsorbed on cellulose, the use of the high-reflectance method not only extends the optimum range for analysis toward regions of lower concentration but also reduces the relative error in the process.

Implicit in the discussion of both differential methods has been the assumption that the reading error associated with the differential reflectance measurement is independent of both the reflectance of the sample relative to the differential standard and the differential standard itself. This assumption has been verified experimentally [19,20].

(D) MULTICOMPONENT SYSTEMS

In situations involving a finely divided solid mixture of n light-absorbing components whose reflectance functions are additive, the Kubelka–Munk function $F(R_\infty)$ can also be adapted for simultaneous analysis [22]. The function of the total reflectance $R_{\infty T}$ of the mixture at some wavelength i may then be represented as the sum of all of the individual reflectance functions, or

$$F(R_{\infty T})_i = \sum_{j=i}^{n} \tau_{ij}\, c_j \tag{13}$$

where j refers to the components and τ is the slope of the Kubelka–Munk plot of $F(R_\infty)$ as a function of concentration. Equation (13) can be put in a more explicit form by writing as many equations as there are components in the mixture, or

$$F(R_{\infty T})_a = \tau_{a1} c_1 + \tau_{a2} c_2 + \dots + \tau_{an} c_n \ , \tag{14}$$

$$F(R_{\infty T})_b = \tau_{b1} c_1 + \tau_{b2} c_2 + \dots + \tau_{bn} c_n \ , \tag{15}$$

and so forth. These equations are, of course, valid only for those concentration ranges where adherence to the Kubelka–Munk equation is observed. At higher concentrations interferences due to saturation of the first monomolecular adsorption layer lead to marked deviations from linearity [23] which, in extreme cases, result in calibration curves asymptotic to the horizontal axis. The useful concentration range can, however, be extended by the use of such semi-empirical relationships as

$$(1 - R_\infty)^2 / 2R_\infty = k \log C \tag{16}$$

which made possible almost a fivefold extension in one situation [24]. Other functions which tend to give linear calibration curves under suitable conditions have been suggested by Lermond and Rogers [25].

Frei et al. [22] validated the above approach in a study of the simultaneous determination of binary mixtures of dyes adsorbed on silica gel. The spectra of the individual dyes, as well as spectra of sample mixtures of the dye pairs (Orange G–Crystal Violet and Fuchsin–Brilliant Green) are presented in Figs. 6 and 7, respectively. That the reflectance functions of the dyes are additive was indicated by the fact that spectra for the dye pairs arrived at by the addition of

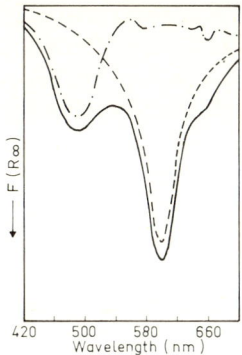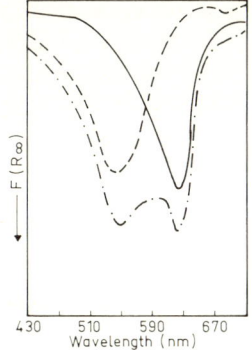

Fig. 6. Reflectance spectra of Orange G, — · — ; Crystal Violet, ---; and a mixture of the two, ——— [22].

Fig. 7. Reflectance spectra of Fuchsin, ---; Brilliant Green, ———; and a mixture of the two, — · — [22].

the individual reflectance functions coincided closely with those obtained experimentally. The reliability of the approach in the hands of inexperienced analysts was tested by employing it in an introductory chemical analysis course to determine the composition of dye pair "unknowns" [26]. In the case of four samples involving the Orange G—Crystal Violet pair, an average deviation from the true value of 1.9% was obtained for Orange G and of 3.3% for Crystal Violet. The comparable figures for Fuchsin and Brilliant Green were 2.1% and 3.1%, respectively. The average standard deviation achieved with four sets of four samples each was ± 2.4%. As expected, the precision was somewhat poorer than that usually attained in single-component analysis and was limited by sample-preparation processes.

The procedures used in these multicomponent analyses, such as the selection of the proper wavelengths for optimum results, were similar to those employed in transmission spectrophotometry.

(E) DIFFUSE REFLECTANCE SPECTRA

(1) Presentation of spectra

There is a variety of ways in which diffuse reflectance spectra can be presented. One is to plot log $F(R_\infty)$ as a function of the wavelength or the wave number. This type of relationship is known as the "typical color curve" of a sample and facilitates the comparison of spectra in identification procedures. The reason for this becomes apparent if one considers the expression that results when one takes the logarithm of eqn. (2), or

$$\log F(R_\infty) = \log \epsilon + \log 2.303c/s \tag{17}$$

Since the scattering coefficient, s, is nearly independent of wavelength, plots of the type being discussed should correspond to the absorption spectra obtained by transmission measurements except for a displacement along the ordinate axis equivalent to the expression $\log 2.303 c/s$. It is then a simple matter to establish quickly the identity or non-conformity of samples of different concentrations because their spectra can be superimposed merely by displacing them along the ordinate axis.

More frequently, however, reflectance spectra are presented in the forms $\log (1/R_\infty)$, or R_∞, versus wavelength or wave numbers. These correspond, respectively, to the absorbance and transmittance curves of transmission spectroscopy. In fact, the quantity $\log (1/R_\infty)$ is often referred to as the "apparent absorbance".

(2) Elimination of regular reflectance

It has already been pointed out that reflected radiation consists of two components: regular and diffuse reflectance. Since the contribution of the regular reflectance distorts diffuse reflectance spectra, which are usually less structured than transmission spectra, it is essential that this interference be reduced as much as possible. This can be achieved by diluting the light-absorbing species with a non- or low-absorbing powder. The results obtained by applying such a procedure in conjunction with the reflection spectrum of anthraquinone are depicted in Fig. 8 [13]. In this case, dried sodium chloride was used both as a diluent and as a reflectance standard. As is shown there, the two main absorption peaks at 29,600 and 39,000 cm^{-1} become sharper with increased dilution. At sufficiently low concentrations, the spectrum becomes independent of concentration and approaches in appearance that of the transmission spectrum obtained with the use of a dilute ethanolic solution of anthraquinone. It is also noteworthy that the spectrum of pure anthraquinone bears little

276

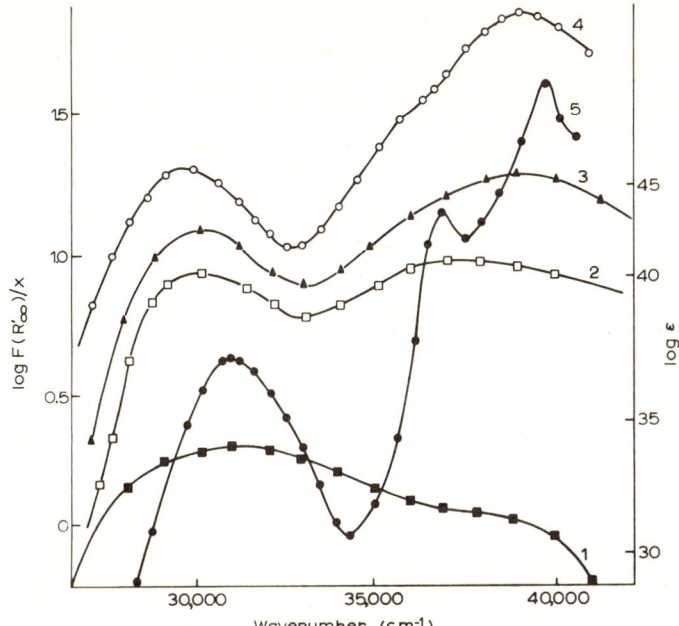

Fig. 8. Reflectance spectra of anthraquinone. (1) Undiluted; (2) adsorbed on NaCl, $x = 1.26 \times 10^{-2}$; (3) $x = 5.0 \times 10^{-3}$; (4) $x = 1.9 \times 10^{-4}$; (5) transmission spectrum in diluted ethanolic solution [13].

resemblance to the spectrum of the sodium chloride-diluted anthraquinone or of its alcoholic solution.

This dilution technique has an added advantage when used with highly light-absorbing substances in that it permits one to work in the optimum reflectance range ($20-65\%R$) for analysis.

(3) Effect of adsorbent

The selection of a diluent in diffuse reflectance spectroscopy affects the reflectance spectrum in very much the same way as the selection of a solvent in transmission spectroscopy affects the transmittance spectrum. For samples that are sufficiently dilute so that the species under investigation cannot achieve a monomolecular saturation of the diluent (in practice this generally is achieved in the 10^{-3} to 10^{-5} mole fraction range), the spectrum obtained is that of the species of interest as modified by the contribution of the diluent.

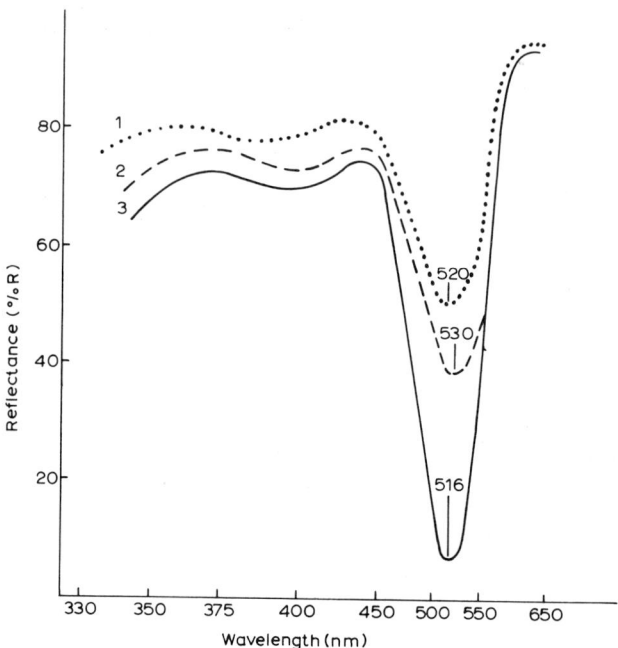

Fig. 9. Reflectance spectra of Eosine B adsorbed on (1) silica gel (Merck thin-layer chromatography grade); (2) filter paper (Whatman No. 42); and (3) alumina (Merck thin-layer chromatography grade) [27].

An indication of the extent of this effect is provided by Fig. 9, which contrasts the reflectance spectra of eosine B adsorbed on filter paper, alumina, and silica gel [27].

When compared with transmittance spectra, reflectance spectra tend to have broader peaks which are displaced. Vibrational structure, if present at all, is strongly suppressed [13,28,29]. Curves 4 and 5 of Fig. 8 provide an example of the contrast between reflectance and transmittance spectra. In cases where there is a strong interaction between the diluent and the species under investigation, and chemisorption is involved, the reflectance spectra obtained are significantly different from those of the free molecules [23,30—32].

(4) Moisture effect

The presence of moisture tends to increase the extinction of a sample diluted with a low- or non-absorbing powder, with the result that samples appear darker when moist. Kortüm ascribed this to the dependence of the scattering coefficient, s, on the ratio of the refractive indices of the powder, n, and of the surrounding medium, n_0 [33]. The displacement of air by water in moist samples reduces the ratio n/n_0, which, in turn, reduces s. As indicated by the expression $F(R_\infty) = k/s$, any reduction in the scattering coefficient gives rise to greater extinction.

The presence of moisture may lead to the development of vibrational structure in the spectrum of a substance. For example, whereas the reflectance spectrum of anthracene adsorbed on absolutely dry sodium chloride exhibits little vibrational structure, vibrational structure emerges when moisture is admitted [13].

(5) Effect of particle size

With weakly light-absorbing material, or with strongly light-absorbing material that has been diluted with a non- or low-absorbing powder, the observed extinction of a sample decreases with decreasing particle size. Thus, as it becomes more finely divided, $CuSO_4 \cdot 5H_2O$ appears lighter due to the inability of incident radiation to penetrate the system as deeply as before because of more efficient scattering by the smaller particles [12].

(6) Qualitative analysis

Notwithstanding the many parameters that influence reflectance spectra, it is possible through the selection of an appropriate set of experimental conditions to use such spectra for identification purposes. Particularly useful is the application of this technique to the in situ identification of substances following their resolution by paper or thin-layer chromatography. Such an approach is not only applicable to substances that are colored or that have characteristic ultraviolet spectra, but also to colorless substances which react with a chromogenic reagent to generate a color. For example, Frei et al., with the use of a modified spray reagent, were able to identify eighteen amino acids resolved on chromatoplates by means of reflectance spectroscopy employed in conjunction with R_f values and visual observation [34]. This procedure enabled one- and two-dimensional chromatograms to be read rapidly and accurately without requiring the conditions necessary for the attainment of reproducible R_f values.

3. Instrumentation

(A) THE INTEGRATING SPHERE

The integrating sphere is the best known, and theoretically best understood, device for the collection of diffuse reflectance. The annular, ellipsoidal mirror is used much less often for this purpose. There are also attachments where reflected radiation is measured by a photo-device without the preliminary collection of such radiation. Examples of all three types are included in a subsequent discussion of commercial instruments available for the measurement of diffuse reflectance. The integrating sphere is particularly useful in situations where a large sample area is illuminated in a single operation rather than those which involve the use of an integrative scanning spot or slit procedure, such as is customarily employed in chromatographic applications [35]. The earliest descriptions of this optical device, which consists of a sphere or hemisphere whose inner wall is coated with material capable of reflecting radiation efficiently in the wavelength of interest, are those of Sumpner [36] and Ulbricht [5].

Various modes of application as well as a large number of experimental arrangements of optical spheres are known. Two basic designs are depicted in Fig. 10. Type (a), which makes use of the substitution method of measuring reflectance, is commonly employed in the design of reflectance attachments for single-beam spectrophotometers. With such an attachment, the diffuse reflectance of a sample is

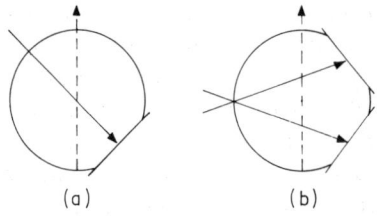

(a)   (b)

Fig. 10. Diagrammatic presentation of the (a) substitution and (b) comparison methods of employing integrating spheres in the measurement of reflectance [8].

measured relative to a suitable standard by substituting the referen↓ material for the sample at the sample port and repeating the measurement under otherwise identical conditions. Type (b), which makes use of the comparison method, permits the simultaneous irradiation of both sample and reference standard and is usually employed in the design of attachments for double-beam recording spectrophotometers. In both (a) and (b) the solid arrows represent incident radiation entering the sphere from an external source and falling on sample or standard surfaces, which are indicated as linear components of the sphere wall. The dashed arrows represent radiation measured by an external photometer following its reflection by the sphere wall and subsequent emergence through an aperture. With samples having a relatively fine texture, the angle of irradiation does not materially affect the amount of diffuse reflectance that is measured [37]. With those possessing a definite structure, however, directional irradiation produces shadow effects and diffuse irradiation can only be achieved with the use of optical spheres of the type depicted in Fig. 10 but with reversed optical paths. In this arrangement an external polychromatic source provides the irradiation energy and the reflected light is then passed through a monochromator, with the results of the measurement being unaltered [38—40]. A hemisphere [41] or circumferential mirror system can also be utilized for the same purpose.

The principles governing the radiation flux distribution in an integrating sphere can best be illustrated by the summation method for infinite series of interreflected rays [42—44]. Assuming a sample having a reflecting capacity equivalent to R_s, is irradiated at a given angle with a radiation flux of intensity S, the flux intensity received by a detector through an aperture of area $\Delta\omega$ can be represented by the expression

$$B_{1s} = R_s S \, \Delta\omega \tag{18}$$

When the same sample is mounted in the sample port of the integrating sphere so that it becomes an integral part of its inner wall, the flux intensity is amplified by multiple reflections occurring at the surface of the wall and can then be defined by the equation

$$B_s = R_s S \, \Delta\omega \, R_w \cdot 1/(1 - R_{sp}) \tag{19}$$

where R_{sp} is the average reflecting power of the sphere and R_w is the reflectance of the material used to coat the sphere walls. Since R_{sp} is usually about 0.95 and since materials selected for coating spheres have reflectance values approaching unity, it is not uncommon to observe a 20-fold amplification in the radiation flux as a result of multiple sphere reflection. The term $1/(1-R_{sp})$ is often referred to as the "sphere efficiency factor".

The average reflecting power of the sphere can be obtained by means of the relationship

$$R_{sp} = (A - \Sigma a)R_w + a_s R_s/A \qquad (20)$$

where A is the total area of the inner surface of the sphere, a_s is the surface area of the sample, and a includes the areas of the incident and emergent apertures. In the substitution mode, the situation that exists when the sample is replaced by a reference standard is described by equations analogous to (19) and (20), or

$$B_r = R_r S \Delta \omega R_w \, 1/(1 - R_{sp'}) \qquad (21)$$

and

$$R_{sp'} = (A - \Sigma a)R_w + a_s R_r/A \qquad (22)$$

where the quantities relating to the reference standard are identified by the subscript r. The relative difference between R_s and R_r determines the average reflecting power of the sphere. This difference becomes negligible, i.e. $R_s \cong R_r$, as a_s becomes quite small, and $R_{sp} \triangleq R_{sp'}$.

Equations (19) and (21) can be combined to give the expression

$$B_s/B_r = R_s/R_r \cdot (1 - R_{sp'})/(1 - R_{sp}) \qquad (23)$$

which indicates that the flux intensity ratios obtained by the substitution method differ from the ratios of the reflecting powers of the sample and the reference standard by the factor

$$\alpha \equiv (1 - R_{sp'})/(1 - R_{sp}) \qquad (24)$$

This factor, which is generally referred to as the "sphere error", can become quite significant [44]. The reflectance of the sample relative to the standard is given by an expression resulting from the combina-

282

tion of eqns. (23) and (24), or

$$R' \equiv R_s/R_r = (B_s/B_r)(1/\alpha) \qquad (25)$$

In general it can be said that the sphere error increases with decreasing values for the reflectance of the sample, R_s, and with increasing values for the ratio of the surface area of the sample, a_s, to the total inner surface area of sphere, A. While the sphere error can be reduced by decreasing a_s at the same time as A is increased, there is a practical limit to the extent that the size of the sphere can be increased because this also results in a decrease in the values obtained for R_r. Although there are ways of looking at the behavior of radiant fluxes in optical spheres other than the one given above [45—52], there is general concurrence in the conclusions reached.

Excellent reviews and discussions of the theory and sources of error in the use of integrating spheres have been provided by Kortüm [7] and Wendlandt and Hecht [8].

(B) REFERENCE STANDARDS AND SPHERE COATING MATERIALS

In 1931 the Commission Internationale de l'Eclairage (CIE) [53] recommended that magnesium oxide serve as the reflectance standard for photometric and colorimetric measurements, with the reflectance at 457 nm of magnesium oxide surfaces prepared under specified conditions being assigned the value of 100%. This assignment is obviously an arbitrary one since the actual reflectance of smoked magnesium oxide surfaces is lower, with reported values differing by as much as 2% [54]. This has been attributed to various factors such as difference in methods employed to measure reflectance [49]; variations in layer thickness [55], particle size [7,52], and density of packing [56]; and differences in methods used in preparing a surface and aging [57]. The difficulty of controlling such a set of experimental variables to the degree necessary if magnesium oxide is to be an absolutely reliable standard is obvious.

The layer thickness is a particularly critical factor, as it has been shown that up to 8 mm of smoked magnesium oxide still possess some transparency [55]. Opinions on the best method for the coating of spheres or plate standards with magnesium oxide differ widely. Middleton and Sanders [57] have found that a coating obtained by burning magnesium ribbon has advantages over one obtained by burning magnesium turnings in a silicate dish, as recommended by early workers [58]. The same investigators also reported that the magnesium oxide underwent a relatively rapid aging process, which they ascribed, in part, to the presence of impurities and, in part, to the decomposition by ultraviolet radiation of the nitride produced during the smoking process [57,59]. The effect of compaction of powdered samples was investigated by Schatz [56], who made the general observation that weak absorbers, such as reference materials, show a decrease, and strong absorbers an increase, in reflectance with increased pressure.

The use of barium sulfate in place of magnesium oxide has been suggested repeatedly [60—62]. It not only is non-transparent in layer thicknesses of about 1 mm [54,63] but also is stable toward aging, undergoing a change in reflectance of less than 0.5% after 595 days [64]. Unfortunately, while good precision is attained with the use of barium sulfate by individual investigators, the results obtained differ considerably from one investigator to another [54]. Magnesium carbonate, whose reflectance properties are similar to those of magnesium oxide has also found some use as a reference material [54,65]. Its main advantage is the convenient block form in which it is made available by many instrument manufacturers. For practical applications, other more durable materials such as glasses (didymium, Vitrolite, opal glasses and the like) [66—69] or porcelain enamel [54] are often preferred. Vitrolite, which is a white structured glass, is available from the U.S. Bureau of Standards [70]. Carl Zeiss employs porcelain material in the built-in swivel standard of its chromatogram spectrophotometer and provides a white glass standard with a treated surface for its "Elrepho" reflectometer. Incidentally, Zeiss also manufactures a powder press which makes possible the preparation of magnesium oxide and barium sulfate standards having a reproducibility better than ± 0.1% [71,72].

The reference standard often has to be selected to fit a particular application. In determinations requiring only relative measurements, a wide range of standards is available. For example, in chromatographic work the plate or column material, such as alumina, cellulose and silica gel, often serves this purpose [73]. Diffuse reflectance spectra obtained for these as well as other materials relative to magnesium oxide are depicted in Fig. 11.

In 1959 the CIE recommended that the "perfect diffuser" be adopted as the reference standard for reflectance measurements of

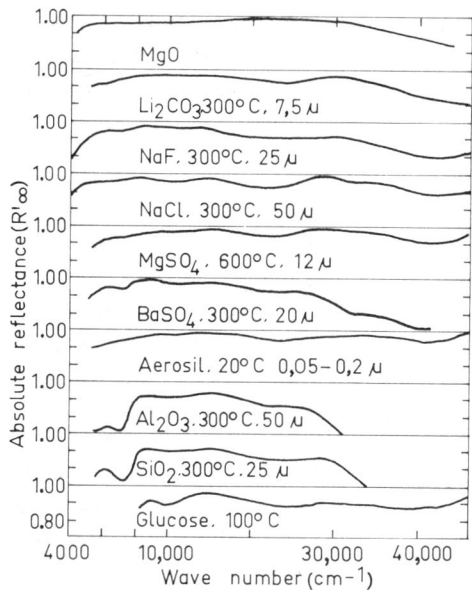

Fig. 11. Absolute diffuse reflectance of some white standards. Drying temperatures and, where available, maximum grain size are also indicated [13].

opaque specimens [74]. Although this recommendation was not implemented immediately, it was agreed at the 16th session of the CIE held in Washington in 1967 that the ideal diffuser should supersede magnesium oxide as the reference standard from January 1, 1969 [75]. Despite the fact that relative reflectances now have to be converted to absolute values, this new precept will undoubtedly gain more general acceptance as new standard materials and improved automatic calculation techniques become available.

Obviously some of the materials utilized as reference standards are also employed as coatings in integrating spheres, since in both functions similar properties are required. Magnesium oxide and barium sulfate are the most widely used substances, even though recently there has been a marked shift toward the use of barium sulfate in commercial instruments. Several methods exist for coating of optical spheres [55,58,61,76]. These involve application either by suitable smoking procedures or in the form of paints. Water glass or plastic materials are often used as additives to increase the mechanical stability of the coatings. Although detailed coating procedures are usually provided by the manufacturer, reflectance attachments which have become commercially available recently have coatings so stable and resistant to aging that they rarely have to be recoated by the user.

(C) FILTER INSTRUMENTS

As in transmission spectroscopy, the first instrument constructed for the measurement of reflected light employed filters to obtain light of a defined wavelength range [2]. Most of the many commercial and non-commercial filter instruments that have been designed since that time have found use primarily in the paint, ceramics, paper and textile industries for the definition and matching of colors and for the investigation of such properties as whiteness, brightness, and the like. While the majority have been equipped with an integrating sphere for the collection of diffuse radiation [77—80], some employ such devices as light pipes, hemispheres, and lens and mirror systems for this purpose. A wide range of commercial instruments is available [81], some specifically designed for use in the industries mentioned earlier.

Typical of the filter reflectometers in current use is the Elrepho (Electric Reflectance Photometer) colormeter manufactured by Carl Zeiss, Oberkochen, West Germany. The optical arrangement of this instrument, which is discussed critically elsewhere [82,83], is presented in Fig. 12. The sample, A, mounted at the sample port of the integrating sphere, is irradiated in a diffuse manner via the optical sphere by two incandescent lamps, which are not shown in this illus-

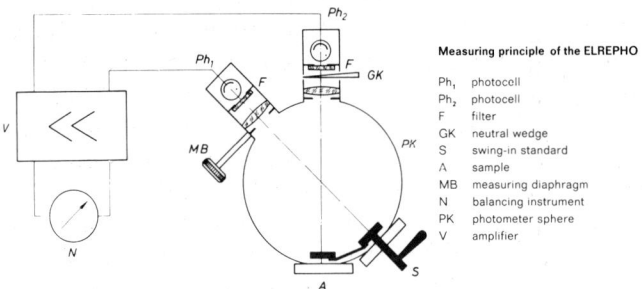

Fig. 12. Optical schematic for the Elrepho colormeter.

tration. Direct irradiation of the samples is prevented by the use of baffles. A built-in, swivel standard, S, is irradiated by the same system. Reflected light strikes the photocells, Ph_1 and Ph_2, at zero degree angle to the reference and sample surfaces, respectively. A neutral wedge, GK, in the measuring beam and an adjustable diaphragm, MB, in the reference beam compensate the two resultant opposing photo currents. The difference between the two currents is amplified at V and indicated by a balancing instrument, N. An automatic modification of the Elrepho which provides an instant readout of color coordinates is also available from Zeiss under the listing RFC 3 colormeter. The useful spectral range is the same as for the manual instrument (400—700 nm), and the reflectance geometries d/8 and 45/0 (diffuse incident radiation, 8° reflected beam, and 45° incident beam, 0° reflected beam) are available.

The Pretema Spectromat FS 38 (Pretema Ltd., Birmensdorf-Zürich, Switzerland) is another filter reflectometer that employs a double-beam arrangement in conjunction with an integrating sphere. Optical geometries d/0 and 45/0 are available, and a small built-in digital computer is programmed to provide a direct readout of CIE color coordinates in printed or digital form. Like the Zeiss RFC 3, this instrument finds wide application in industries where large numbers of color comparisons must be made in connection with quality and production control. Another filter instrument is the IDL Color-Eye (Industrial Development Laboratories, Inc., Attleboro, Mass., U.S.A.) which has been described in detail by Van den Akker et al. [84]. Although this reflectometer also utilizes an integrating sphere, the simultaneous irradiation of the sample and reference standard is achieved with a rotating mirror that directs the polychromatic incident beam alternately to the sample and reference ports. The instrument permits absolute or differential measurements of diffuse and specular reflectance. The Colorede is a semiautomatic modification of the Color-Eye, available from the same manufacturer.

A range of instruments with somewhat different design features is available from Hunter Associates Laboratory, Inc. (Fairfax, Va., U.S.A.). The modern Hunter instruments have developed from some of the earliest known filter-type reflectometers designed and described by Hunter in 1934 [85] and 1940 [86,87]. All of the Hunter Color and Color Difference meters that are currently available operate on the single-beam principle (substitution mode). The solid state wired electronic unit is the same for all D25 models, containing one of three different optical units to give the models D25A, D23M/L and D25P. With the D25A, whose reflectance geometry is 45/0, the diffusely reflected polychromatic light is collected by means of a light pipe and directed toward a set of 3 phototubes. Four broadband spectral filters, tristimulus X, Y, Z amber and X blue, are positioned in front of the detectors so that the photocurrents can be converted to color coordinates. The optical head D25P has an 8 in. diameter optical sphere with the diffuse radiation being viewed through the top of the sphere with the same set of phototubes and filters employed in the D25A. Although the specular reflectance can be excluded in the D25P (as it is in the D25A), with the former unit it is also possible to measure total reflectance by swinging the source or the light sphere about 8° off their common optical axis (8/t). The usual optical geometry for this unit is 0/d. The optical head D25M/L utilizes a rather unusual design. A circular mirror system consisting of faceted annular rings serves to irradiate rough-textured samples, such as textiles, in a perfectly diffuse manner. Light reflected from the sample is viewed at zero degree angle and directed through 4 spectral filters to a set of 4 phototubes. All D25 models are available with a built-in computer for the computation of color coordinates according to the ASTM method.

A few filter-type instruments are also discussed in a later section dealing with recording instruments used for in situ chromatographic work.

(D) MONOCHROMATOR INSTRUMENTS

The design of non-commercial reflectance units equipped with monochromators instead of filters has been discussed by a number of investigators [65,88—90]. The first commercial reflectance spectrophotometer, The General Electric Hardy Reflectometer (General Electric Co., West Lynn, Mass., U.S.A.), was made available in 1958 [91,92]. Since that time, attachments designed for use with general purpose spectrophotometers already in existence have become available for almost every commercial spectrophotometer on the market [81].

One of the best known and most widely used spectrophotometers in North America is the Beckman Model DU (Beckman Instruments, Inc., Fullerton, Cal., U.S.A.). This single-beam instrument can be equipped in minutes with the diffuse reflectance attachment depicted

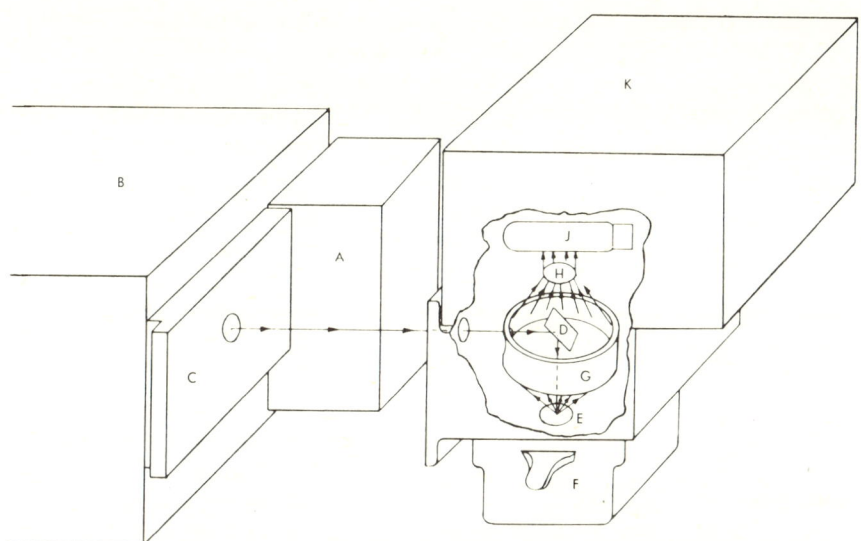

Fig. 13. Diagrammatic representation of the diffuse reflectance attachment for the Beckman Model DU spectrophotometer.

in Fig. 13. A light beam from source A passes through the monochromator, B, to enter the reflectance unit after emerging from a port in the mounting block, C. Thereupon the beam is directed by a deflection mirror, D, toward the surface of the sample, E, at a zero degree angle of incidence. The sample drawer, F, is designed to hold a sample and a reference standard, with the operator positioning one or the other under the incident beam by moving the drawer along a track. The diffusely reflected light is collected by a circular, ellipsoidal mirror, G, at angles in the 35—55 degree range, focused upon a frosted quartz diffusing screen, H, and then impinges upon a phototube, J. The attachment can also be employed in conjunction with the DU-2 model, as shown in Fig. 14.

A schematic diagram of the reflectance accessory for the DK-2 spectrophotometer is given in Fig. 15. This attachment uses an integrating sphere in which both sample and reference are irradiated simultaneously with monochromatic light. An oscillating mirror produces the double-beam effect by alternately deflecting the beam to the sample and reference ports. When the sample is positioned normally, the incident radiation strikes the sample surface at a zero degree angle and the specular reflectance portion is attenuated. Total

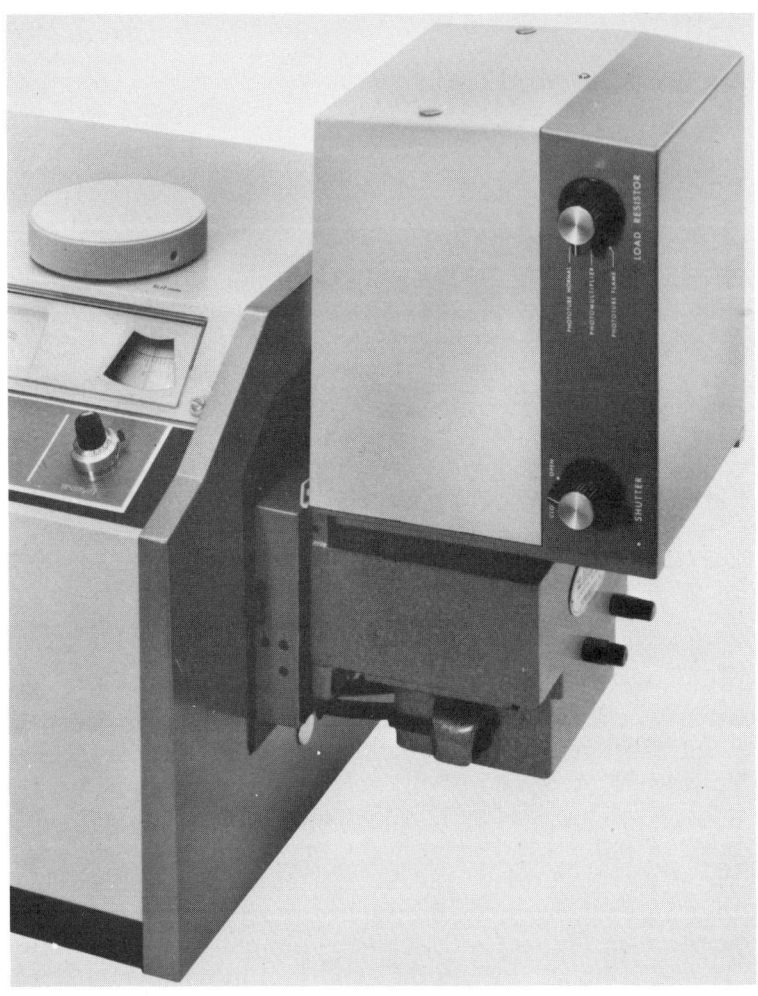

Fig. 14. Reflectance attachment mounted on a Beckman Model DU-2 spectro-photometer.

reflectance can be measured by mounting specimens at a 5° angle to the incident radiation. The spectral range for these two modes of operation is 210—2700 nm, with a lead sulfide detector being em-

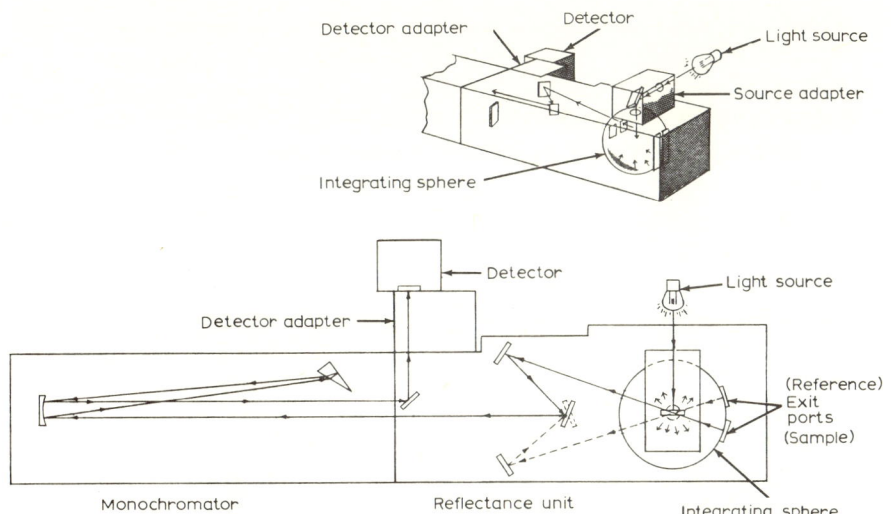

Fig. 15. Optical diagram for the Beckman Model DK-2 spectrophotometer fitted with a diffuse reflectance attachment (optical geometry $d/0$).

ployed in the near infrared region. With samples having a rough texture, the specimens are irradiated diffusely via the sphere by reversing the positions of the light source and the detector.

Diffuse reflectance attachments for the Bausch & Lomb Spectronics 20 and 505 (Bausch and Lomb, Inc., Rochester, N.Y., U.S.A.) have likewise gained wide acceptance. A schematic diagram of the attachment used with the Model 20 is presented in Fig. 16. Monochromatic light passing through the lens—filter system falls on the

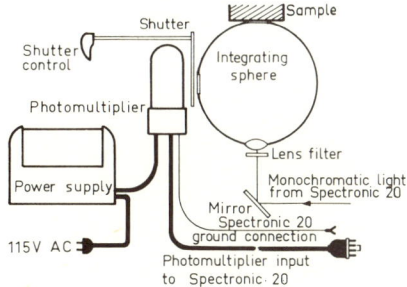

Fig. 16. Schematic diagram of the Bausch and Lomb Spectronic 20 fitted with a diffuse reflectance attachment.

sample or reference positioned at the top of an integrating sphere and the diffusely reflected light is detected by a phototube located at a right angle to the surface of the sample. The monochromator consists of a replica grating having a 20 nm band width. The Model 505 attachment also makes use of an integrating sphere, but operates in the double-beam mode. The sample and reference materials, which are mounted vertically at equatorial ports, are irradiated diffusely via the integrating sphere by a light source situated at the top of the sphere. Light traps located at right angles to the irradiated surfaces are used to eliminate specular reflectance. The standard attachment can be utilized in the spectral region 400—700 nm, with another sphere, coated with a barium sulfate paint, being available if an extension of the range into the ultraviolet is desired.

Three different attachments are available for the Cary Model 14 and Model 15 spectrophotometers (Cary Instruments, Monrovia, Cal., U.S.A.). The one for which an optical schematic is depicted in Fig. 17 makes use of an integrating sphere. Both sample and reference are irradiated at a zero degree angle of incidence by a chopped, monochromatic beam, and the radiation reflected from their surfaces is collected by the sphere and measured by a phototube positioned at the top of the sphere. As with the Beckman DK-2, the optical path can be reversed. This model is also available with a circular mirror in

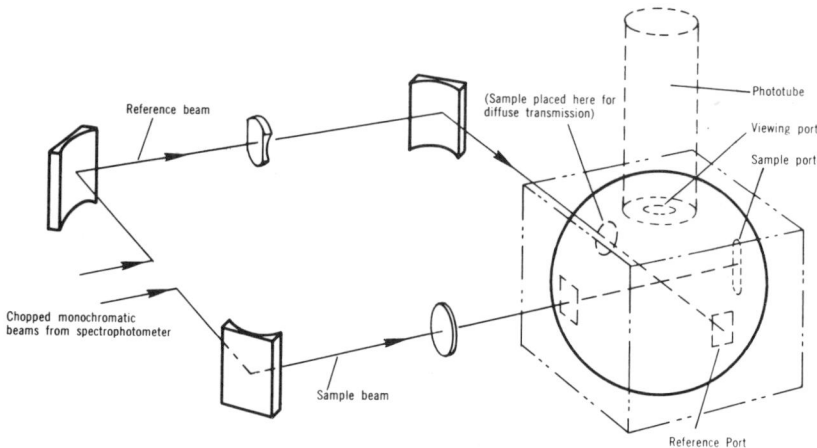

Fig. 17. Optical scheme for the integrating sphere attachment for the Cary Models 14 and 15 spectrophotometers.

292

place of the integrating sphere. The second type of accessory uses a sample sphere and a light-integrating reference box so that the sample and reference beams are completely separate. Unlike the first attachment, which can only be used with the Model 14 spectrophotometer, the second can be employed with either the Model 14 or Model 15. The third and most recent reflectance attachment developed by Cary is one intended for use with its Models 14, 14R and 14RI. Using an altered integrating sphere design, it can be employed in the 220—2500 nm spectral range. The working principle of this particular attachment has been discussed by Hammond and Nimeroff [93]. Cary Instruments also makes available a computer which can be used in conjunction with this instrumental set-up for the computation of tristimulus values in color measurements.

Perkin—Elmer (Perkin—Elmer Corp., Norwalk, Conn., U.S.A.) has designed an attachment for use with its spectrophotometers Models 350 and 450 which has separate spheres for the sample and reference standard, each with its own end-window photomultiplier tube. This dual sphere design, which has been described by Anacreon and Noble [94], makes provision for the inclusion or rejection of specular radiation by the selection of an appropriate aperture cover. Although the manufacture of this particular accessory will be discontinued with the withdrawal of Models 350 and 450 from the market, the Hitachi Perkin—Elmer Model 139 spectrophotometer will still be available with an integrating sphere-type diffuse reflectance attachment.

The Unicam (Pye—Unicam Instruments, Ltd., Cambridge, England) spectrophotometers Models SP500, SP700, and SP800 can be provided with diffuse reflectance attachments Models SP540, SP735, and SP890, respectively. All three accessories have the optical geometry 0/d and make use of ellipsoidal mirrors for the collection of diffusely reflected radiation. Model SP540 is employed in the single-beam mode in very much the same way as is the attachment for the Beckman Model DU described earlier. Models SP735 and SP890, which have been designed for use with double-beam instruments, are quite similar except that the sample is mounted horizontally in the former and vertically in the latter.

A versatile range of reflectance instruments is available from Carl Zeiss, with three reflectance attachments being listed for use with the single-beam PMQII spectrophotometer alone. With the Model RA2 attachment the sample is irradiated with monochromatic light (200—600 nm wavelength range) at an angle of $45°$, and the reflected

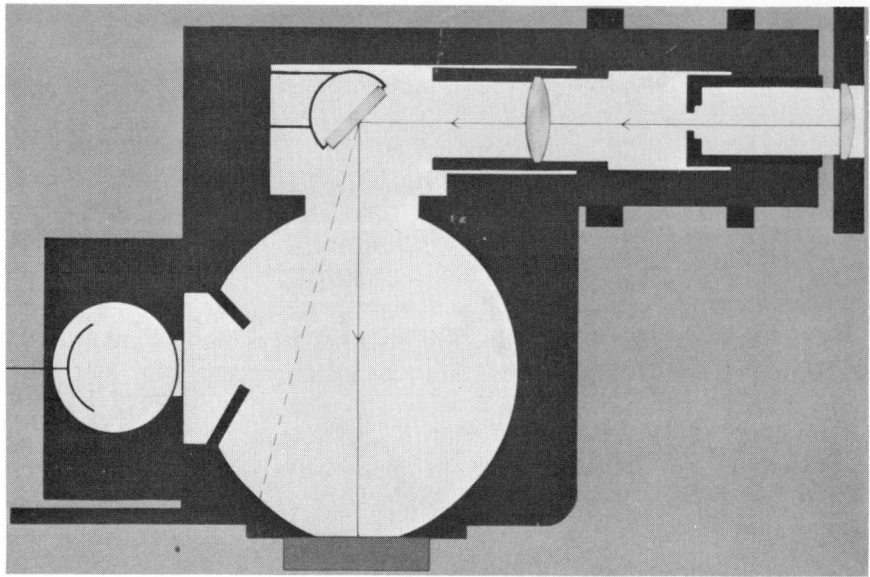

Fig. 18. Optical diagram for the Zeiss Model RA3 reflectance accessory (optical geometry 0/d).

radiation is measured with a phototube positioned at right angles to the sample surface. Since the 45/0 geometry of the RA2 accessory leads to systematic errors when used with rough-textured samples [48,95], such as textiles or coarse-grained powders, the model RA3 accessory is recommended if such samples are to be dealt with. The RA3 can not only be employed in the more conventional 0/d mode depicted schematically in Fig. 18, but it can, in a matter of minutes, also be converted to operation in the d/0 mode by reversing its optical path. This guarantees nearly ideal diffuse illumination of the sample via the integrating sphere of the attachment and elimination of shadow effects. The light reflected from the surface of the sample then passes through a monochromator to the detector. The third attachment available for use with the PMQII has been designed to accomodate samples having a surface area as small as 0.1 mm^2 and is, therefore, well suited for microanalytical work. It can be employed in the 250—2500 nm wavelength range and has been provided with a swivel standard for the setting of the reference value without necessitating removal of the sample during the measuring operation.

An attachment is also available for the double-beam recording spectrophotometer, Zeiss Model DMR21. Like its predecessor, the RA20, which was designed for use with the Model RPQ20A spectrophotometer, it operates with an integrating sphere and an 8/d optical geometry.

A real advance in the field of diffuse reflectance spectroscopy and color measurement has been achieved with the Zeiss Model DMC recording and digital spectrophotometer which is shown in Fig. 19. As shown in Fig. 20, which presents an optical diagram of the instru-

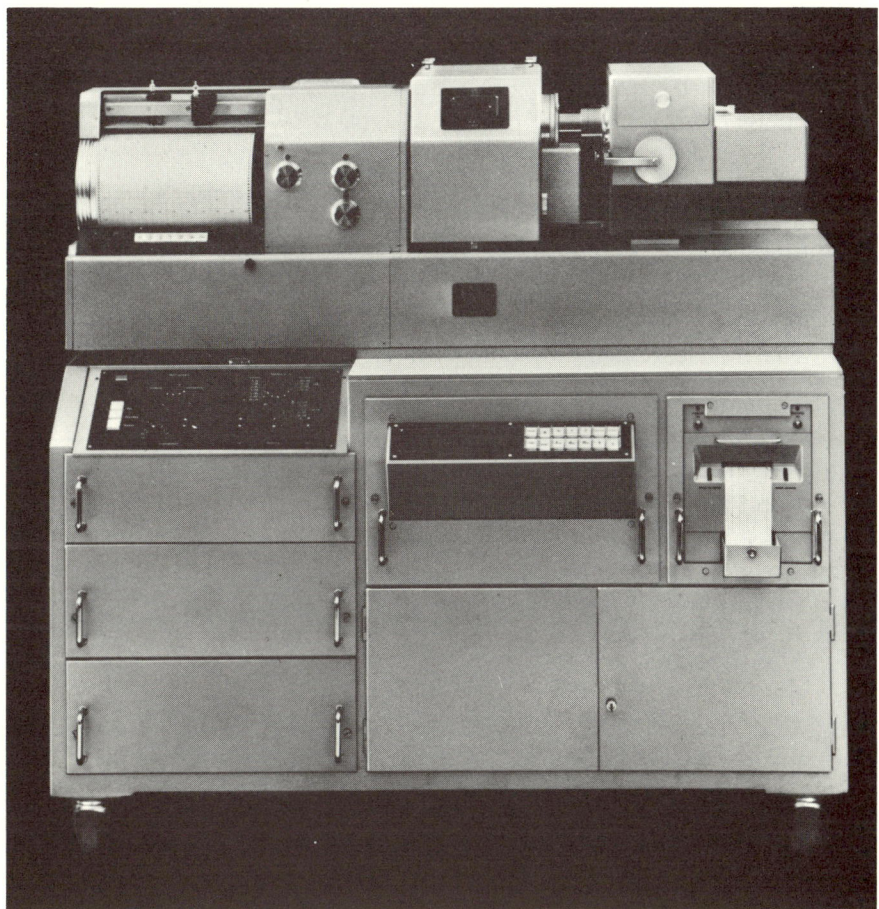

Fig. 19. Zeiss Model DMC25 color spectrophotometer.

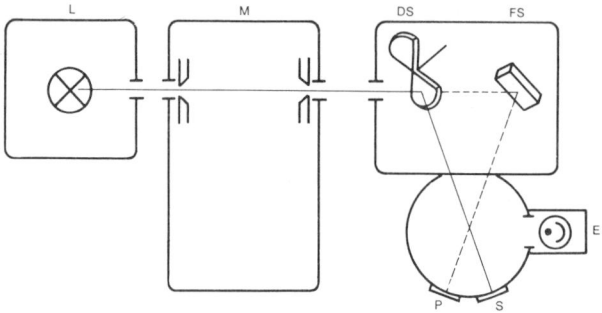

Fig. 20. Schematic diagram of the Zeiss Model DMC25 color spectrophotometer.

ment, the light from L that passes through the monochromator, M, is alternately deflected and transmitted by a mirror—chopper, DS, to the sample, P, and the standard, S, with the diffusely reflected light being monitored by the detector, E. The optical path can be reversed by moving the detector to position L and replacing it with a xenon or tungsten lamp so that the sample and reference material are diffusely irradiated with polychromatic light. A twin analog recorder makes it possible to simultaneously record %R and the Kubelka—Munk function, or transmittance and absorbance, with the measurements being printed out at preselected wavelength intervals. Color coordinates x, y and z provided by a built-in digital computer are made available to the operator in digital form either in print or on a screen. The operator has his choice of two spectral ranges (200—650 nm, and 380—2500 nm), three reflectance geometries (d/8, 8/d and 45/0), and two different sizes of integrating spheres (100 mm and 130 mm diameter). Surfaces of samples can not only be varied from 50 × 38.5 mm to 15 × 20 mm, but can even be reduced further if required. Finally, differential spectroscopy is encouraged by the availability of seven measuring ranges for the measurement of diffuse reflectance.

Additional reflectance measuring devices have been discussed by Judd and Wyszecki [1] and Wendlandt and Hecht [8].

(E) SAMPLE HOLDERS

(1) Commercial and macro cells

No single cell can hold equally well the wide variety of samples

296

subjected to study by diffuse reflectance spectroscopy. Solid samples possessing a definite shape usually present no problems since they can be clamped in a vertical or horizontal position, face-up or face-down, by means of spring-loaded cups. Liquids and powders, however, present some difficulties. Although in some cases transmission cells have been used in reflectance work to hold powders and slurries, problems arise in packing, emptying and cleaning the cells. If the sample is to be positioned horizontally and face-up, small cups, beakers or planchets may serve as cells.

Few commercial sample holders are available for powdered materials. Bausch and Lomb, Zeiss, Hitachi and others sell windowless macro cells which can be packed with the use of special powder presses, but they can be used only with relatively adhesive powders.

Several noncommercial sample holders have been described by Tonnquist [96] and a low-cost cell intended to hold slurries and powders for reflectance measurements in the ultraviolet, visible and near infrared regions of the spectrum has been designed by Barnes et al. [97]. The latter cell, which has been used extensively for studies of adsorption phenomena by spectral reflectance, consists of a Bakelite base and a Bakelite cover plate into which a circular quartz window has been cemented with epoxy resin. The sample is contained in a planchet positioned between the cover plate and the base in a circular well, 0.16 cm deep and 3.2 cm in diameter.

(2) Semimicro cells

A common drawback of the cells discussed so far is the large amount of material required for their packing. Depending on the nature of the material, this ranges from a few grams for the commercial sample holders to 0.5—1.0 g for the Barnes cell. Three types of cells have been developed by Frodyma et al. to circumvent this difficulty: (A) a glass-window cell [24]; (B) a quartz-window cell [98]; and (C) a windowless cell [99]. Cell A, which is depicted in Fig. 21, consists of white paperboard to which a microscope cover glass has been affixed with masking tape. The paperboard is cut to a size which permits its introduction into the sample holder of the specific apparatus being used. In the case of the sample drawer of the Beckman DU reflectance attachment, for which the cell had originally been designed, this would be 3 × 4 cm. In assembling the cell, the analytical sample (approximately 40mg) is compressed between the

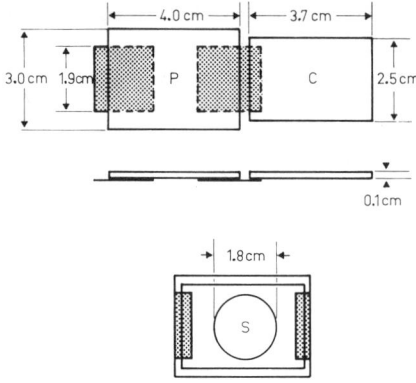

Fig. 21. Dimensions of the elements of a simple semimicro cell for the measurement of diffuse reflectance and sketch of assembled cell. P, backing paper; C, microscope cover glass; and S, sample [24].

cover glass and the paperboard until a uniform layer of densely packed powder is obtained, with care being taken so that no discontinuities appear in the surface. The main advantages of this cell are its low cost and simplicity, which permits its manufacture in the large quantities needed for investigations involving dilution series. It can, of course, be used only in the visible and near infrared region of the spectrum.

Cells B and C are essentially modifications of cell A intended primarily for use in the ultraviolet, although they can also be used in the visible and near infrared. Cell B consists of a circular quartz disk, which has a diameter of 22mm, superimposed on a 40 × 40 × 1mm piece of white paperboard. The disk is held in place by means of a 40 × 40 × 3mm plastic plate affixed to the backing paper with masking tape. Cell C is made by affixing a plastic plate with a circular opening to a piece of backing paper having the same area dimensions. This cell is packed by introducing the sample into the opening and then compressing it with a tamp of appropriate size.

(3) Variable temperature cells

Valuable information can often be obtained from reflectance spectra recorded at elevated temperatures. Special sample holders are required for this purpose, either to maintain the sample at a particu-

298

lar temperature, usually between 100 and 300°C, or to gradually increase the temperature of the sample at a uniform rate. One of the earlier holders devised which made provision for the controlled heating of the sample was one described by Asmussen and Andersen [100], which consisted of a cell located at the top of a cylindrical brass block. A small bulb located in the base of the cylinder regulated the temperature of the sample, which was measured with a thermocouple. A number of cells have also been described by Wendlandt and his coworkers [101]. The first of these [102] involved a cylindrical aluminum block in which was recessed a sample well, 25mm in diameter and 1mm deep. Sample temperatures were measured with one thermocouple, while a second served as a monitor to operate the temperature programmer for the 35 W cartridge heater employed to heat the block. A modification of this assembly for use in dynamic reflectance spectroscopy has been described by Wendlandt and George [103]. Most recent [9] is the cell devised for use with small samples, with which the sample is placed on a strip of glass-fiber cloth, covered with a cover glass, and then clamped to a small aluminum block. This is heated internally by a circular heater element. The entire assembly is mounted on a 5 × 5 cm piece of transite.

A cell intended for low-temperature work has been designed by Symons and Trevalion [104] for use with the Unicam model SP500 spectrophotometer. The sample holders are made of copper rods which have been recessed to accept powdered materials. The rods are cooled by passing liquid nitrogen through them by means of copper tubing.

(F) INSTRUMENTS FOR IN SITU MEASUREMENTS OF CHROMATOGRAMS

Although the direct scanning of paper and thin-layer chromatograms by transmitting light through them and measuring the absorbance has been carried out for some time, the application of the diffuse reflectance technique to this particular analytical situation has been attempted only recently owing to the lack of suitable instrumentation. Klaus [105,106] and DeGallan et al. [107] have described the construction of accessories suitable for use with the Zeiss Model RA3 reflectance attachment for the scanning of chromatoplates. Other non-commercial instrument modifications for the in situ reflectance measurement of chromatograms have been discussed by Gordon [108] and Hamman and Martin [109]. Beroza et al. [110] have recently devised an inexpensive instrument for the automatic recording of spectral reflectance from thin-layer chromatograms. This device, which makes use of fiber optics, can be used in the single- as well as the double-beam mode, and, with suitable modification, in the ultraviolet as well as the visible region of the spectrum.

(1) Commercial single-beam instruments

One of the first commercially available instruments designed for the direct scanning of thin-layer plates by reflectance spectroscopy was the Zeiss Chromatogram Spectrophotometer, which has been described in several publications [111—116]. In this instrument the chromatogram is illuminated with monochromatic light at a zero angle of incidence and the diffuse radiation is measured at a 45° angle. A light pipe is used for the collection of the radiation, with specular radiation escaping through the opening in the pipe that admits the incident beam. Adjustments that can be made by the operator to satisfy a given experimental situation include varying the movement of the mechanical stage on which the chromatograms are placed, the positioning of the measuring head, and the dimensions of the scanning spot. A built-in standard, which can be swung into the light path, makes it possible to calibrate test values without disturbing the analytical sample. Transmission measurements through the chromatogram are made possible by locating the detector underneath the sample table. With some minor adjustments the instrument can also be used for in situ fluorescence measurements. The working range for reflectance is 200—2500 nm.

Recently Camag Ltd. (Muttenz, Switzerland) has begun production of its Camag-Z-Scanner, which can be used for the direct measurement of reflectance and fluorescence from thin-layer and paper chromatograms. This attachment can be fitted to the Zeiss PMQII spectrophotometer as shown in Fig. 22. In this arrangement the monochromatic light beam strikes the vertically positioned thin-layer plate, or paper, at zero degree angle, and a lens—photodetector assembly picks up the diffuse radiation at angle 45°. The optical path can be reversed, if need be, and the specular reflectance can be excluded either way. The attachment, which has a working range of 220—750 nm, makes provision for the scanning of chromatograms at

Fig. 22. The Camag-Z-Scanner attachment mounted on the Zeiss PMQII spectrophotometer.

two speeds, the manual recording of spectra, and the variation of scanning slits up to 16mm maximum width.

The Leitz Mikrophotometer (E. Leitz, Inc., New York, N.Y., U.S.A.) is another relatively low-cost instrument available for direct reflectance and transmission work with chromatograms. Equipped with a recorder and a motor-driven microscope table, it can make single-beam scans over a distance of 7 cm without readjustment. Interference filters are used for the production of the monochromatic beam, and the reflectance geometry is 45/0. Analytical applications of the instrument are described in the literature [117,118].

Another single-beam instrument, the Vitatron Densitometer TLD 100 (Vitatron, Ltd., Dieren, The Netherlands), has been designed for the in situ measurement of reflectance as well as transmittance and fluorescence. An interesting feature of this instrument is the "flying spot" scanning device pioneered by Goldman and Goodall

[119,120]. This involves the measurement of a chromatographic spot by an oscillating movement of the scanning table in one direction while, at the same time, the plate is moving at a uniform speed in a direction at right angles to the oscillating motion. The integrated values for the optical density measured over the stroke length of the latter movement, which is somewhat larger than the spot diameter, are registered on a chart recorder and the corresponding peak areas are computed with an electronic integrator. This approach is particularly suited to the measurement of irregularly shaped spots, since the spot areas under investigation at a given moment are so small that the material distribution shows up homogeneously. Monochromatic light is obtained with the use of a set of filters and the reflectance geometry is 45/0. Unfortunately, at present, the reflectance mode can only be employed in the visible region and the instrument is not suitable for the measurement of spectra.

Thin-layer chromatography (TLC) attachments are available for three well-known spectro-fluorometers, Farrand's MK1 (Farrand Optical Co., Inc., New York, N.Y., U.S.A.), Aminco-Bowman (American Instrument Co., Inc., Silver Spring, Md., U.S.A.), and Perkin—Elmer's MPF2A. All three can additionally be used in the reflectance mode.

(2) Commercial double-beam instruments

Several instruments are now available for the in situ evaluation of chromatograms by the double-beam mode. The Joyce Loebl Chromoscan (Joyce Loebl and Co., Ltd., Gateshead, England) with chromatogram accessory was one of the earliest on the market. It was originally designed to measure transmission and fluorescence, but it can now be utilized for the direct measurement of spectral reflectance. It is a null-point instrument in which a grey wedge is balanced against the radiation reflected from the sample in a 0/45 mode. The chromatogram attachment, a close-up of which may be seen in Fig. 23, has its own optical system. Radiation provided by a quartz iodide tungsten lamp is rendered monochromatic by a set of interference filters. Although UV lamps can be placed alongside the tungsten lamp in a dual lamphouse which permits a rapid change of light sources, at the moment reflectance work can only be carried out in the visible region. With the use of an accessory, plates up to 20 × 20 cm can be accommodated, and scan speeds relative to the

302

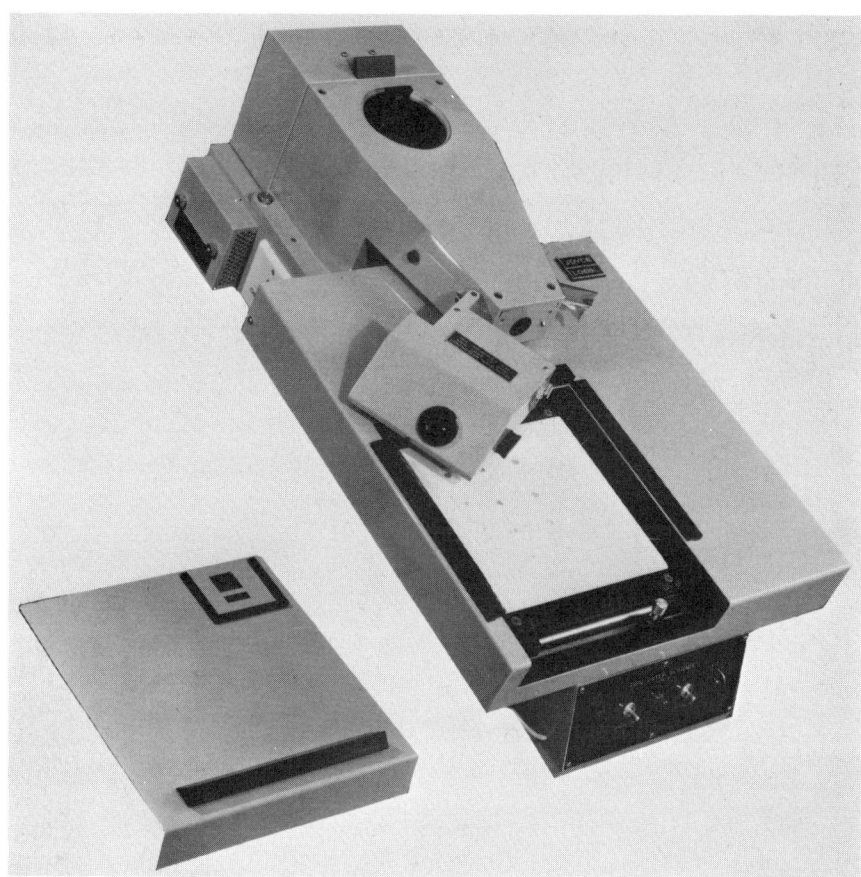

Fig. 23. Closeup of the Joyce Loebl thin-layer chromatogram scanning attachment.

recording drum can be varied. Goldman and Goodall [35,119] have modified the Chromoscan densitometer to permit a detailed comparison of transmission and reflectance measurements on chromatograms.

Farrand's versatile VIS—UV Chromatogram Analyzer, which is pictured in Fig. 24, can be used for reflectance work in both the visible and ultraviolet regions. In this instrument a sliver of monochromatic light, whose length lies at right angles to the direction of scanning, illuminates the plate at zero degree angle so that the spot

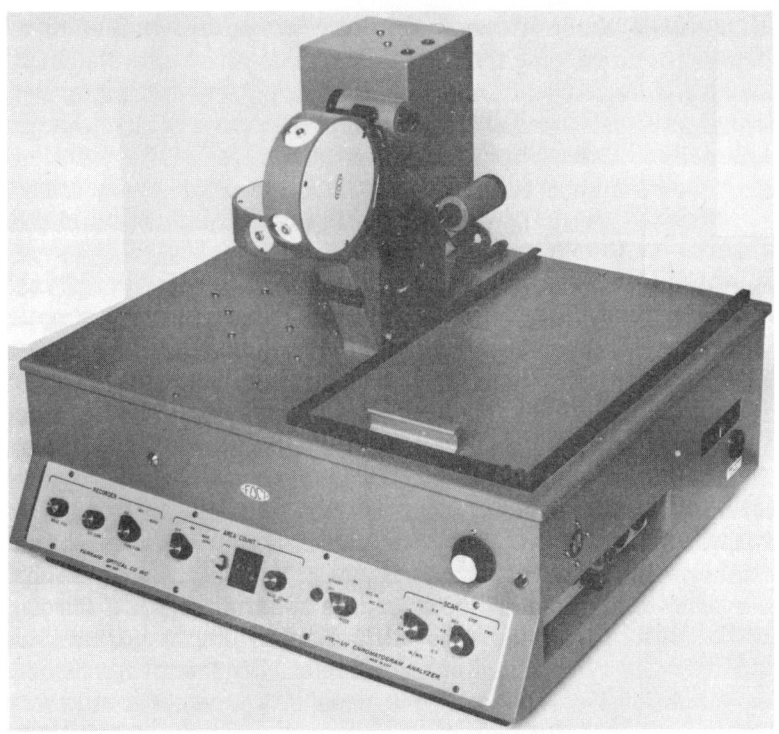

Fig. 24. The Ferrand VIS—UV Chromatogram Analyzer.

being surveyed, as well as some adsorbent on each side of it, are irradiated. Reflectance of the central segment of the sliver, which impinges on the spot, is viewed only by analyzer optics at a 25° angle, while the reflectance of each terminal segment of the sliver is viewed by a reference channel at 45° and the two resulting signals are averaged. The ratio of the sample signal and the average reference signal is then recorded on a conventional strip-chart recorder. This rather unique operation serves to cancel out variations in plate background. The scanning table, which accommodates plates up to 20 × 20 cm, permits manual adjustment along one axis and the choice of nine different scan rates along the other.

Two densitometers, models SD 2000 and SD 3000, marketed by Schoeffel Instrument Corp. (Westwood, N.J., U.S.A.) have recently been modified for use in reflectance spectroscopy. Although both operate on similar principles, the SD 2000 is a lower-cost filter instrument intended for routine use, particularly in medical laboratories, while the SD 3000, with its flexibility of operation, is capable of serving as a research instrument. The latter instrument has a reflectance geometry of 0/45, with the radiation reflected from the sample area and the adsorbent material being monitored by separate phototubes. Both signals are automatically converted into an absorbance reading, i.e. the logarithm of the reference signal/sample signal ratio. To facilitate reflectance measurements, a linear reciprocal ratio can also be provided on request. The instrument can also be operated in the single-beam mode for in situ fluorescence and transmittance work on paper, TLC and electropherograms, and the scanning table, which accommodates plates as large as 20 X 20 cm, can be operated at any one of six speeds.

The Nester-Faust Manufacturing Corp. (Newark, Del., U.S.A.) has recently developed a dual-beam chromatogram scanner, the Uniscan 900, which can be used for reflectance work in the visible region. In this instrument the sample and reference beams are generated by two separate energy sources instead of by splitting a single beam, as is done in most other dual-beam instruments. The Uniscan, whose reflectance geometry is 45/0, can also be employed in the single-beam mode in situations where it is not possible to scan the plate surface along with the sample surface. An unusual feature is the use of a mobile scanning head instead of a moving table to vary scanning speed. Various readout accessories are available.

Available from the American Instrument Company is an instrument specifically designed for the direct measurement on chromatograms of reflectance, transmittance, fluorescence and fluorescence quenching. It can be operated in both the single- and double-beam mode and in the ultraviolet and visible regions of the spectrum.

A fuller description of many of the above instruments is available in a recent paper by Lefar and Lewis [120].

4. Applications

The following list of applications of spectral reflectance is not meant to be exhaustive. Rather, it is intended simply to be indicative of the wide variety of situations in which the technique has been employed with profit.

Since color manifests itself in the form of reflected light emanating from the object under observation, it is not surprising that spectral reflectance techniques have played a major role in the field of color technology.

(1) Principles and methods of color measurement

Although a number of color order systems have been developed [121,122], only the CIE system, which is also known as the ICI (International Commission on Illumination) system and which is by far the most widely used [53,123,124], will be considered in this discussion. Information about the others may be obtained from the many authoritative sources available in this field [1,125—129]. The CIE system is based on the concepts of the three-dimensional nature of color and of the additivity of spectrum colors. Grassmann's law [130], which states that the color-matching functions or tristimulus values of the spectrum colors can be calculated for any specified set of primaries, indicates that colors can be specified in terms of three numbers, or tristimulus values, representing relative amounts added from three primary sources. For the sake of convenience, the three tristimulus values are designated X, Y and Z. Y stands for the light response, or brightness or luminosity, while X and Z describe that aspect of color which permits its identification with various regions of the spectrum, or hue, and which determines its excitation purity, or saturation.

The trichromatic equation for tristimulus values can be put in the form

$$\text{Color } (C) \equiv x(X) + y(Y) + z(Z) \tag{26}$$

where x, y and z represent the chromaticity coordinates of C, obtained with the use of the following set of equations

$$x = X/(X+Y+Z) , \qquad y = Y/(X+Y+Z) , \qquad z = Z/(X+Y+Z) \tag{27}$$

Since the sum of the chromaticity coordinates equals unity, two coordinates suffice to describe a color, and a chromaticity diagram, such as that depicted in Fig. 25, can present colors in terms of their x

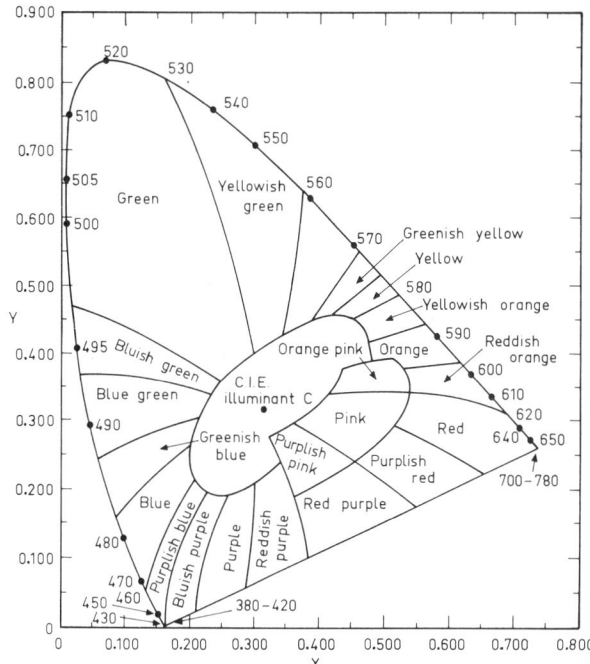

Fig. 25. Chromaticity diagram [140].

and y coordinates. This pair of coordinates, which describes hue and saturation, is known as chromaticity.

The unambiguous designation of color also requires a clear specification of the nature of the illumination. In 1931, CIE recommended the use of three standard illuminants, identified as A, B and C. Illuminant A is representative of the light cast by a gas-filled incandescent lamp operated at 2854 K; B is representative of sunlight at noon; and C is representative of daylight from an overcast sky. Illuminants B and C are obtained by employing the lamp used for A in conjunction with a filter [131] designed to give color temperatures approximating 5000 and 6800 K, respectively. Color temperature may be defined as the temperature at which the color of a black-body radiator matches that of the illuminant [132].

The above, in brief, is the basis of the three approaches that have been used within the CIE system for the measurement of color. All three suffer to an extent from the lack of sensitivity and precision in the instrumentation and the inability to simulate real or standardized conditions. The equi-contrast method depends on the additive mixing of instrumentally measurable amounts of three primaries (e.g. red, green and blue) followed by the visual comparison of the resultant mix, first versus a reference standard (white light), and then versus the color being measured. Several instruments which make use of this approach have been described in the literature [122,133,134], including one [135] which utilizes as many as six primary sources in the light-mixing process.

The trichromatic method represents an attempt to circumvent the limitations imposed by the failings of the human eye, which in this instant is replaced by a suitable photodetector with light filters (tristimulus filters) positioned in front of it to simulate standard observer conditions. The standard observer, according to the CIE, is an attempt to define the color perception of an individual with normal vision in terms of a tristimulus system. This arrangement is generally employed with the red, green and blue regions of the spectrum and can be used for the direct measurement of tristimulus values [136,137] provided that the photocell response is approximately proportional to some linear combination of the distribution curves of these tristimulus filters in the visible spectral region. In accordance with Grassmann's law, it should then be possible to determine whether any two light beams have the same color.

The actual determination is made with a photoelectric tristimulus colorimeter fitted with a CIE illuminant (see pp. 286 ff). Measurements relative to a standard, which is usually magnesium oxide or barium sulfate are taken through each of the filters and the readings thus obtained are then converted to tristimulus values by means of factors provided by the manufacturer along with the filter specifications. Typical of such conversion factors is the set provided with the Elrepho instrument: $X = 0.782R_x + 0.198R_z$; $Y = R_y$; and $Z = 1.181R_z$, where R_x, R_y and R_z are the readings obtained with the use of the three filters. The tristimulus values can then be converted to chromaticity coordinates according to eqn. (27) for entry on a chromaticity diagram similar to that shown in Fig. 25.

The accuracy of this approach depends to a large extent on how closely the observer combination of photocell and tristimulus filters can simulate the characteristics of the CIE standard observer. The precision that is attained by operators that have undergone no special training is comparable to the best that can be achieved visually, while the rapidity of the method exceeds that of both the equicontrast and the spectrophotometric methods.

If greater accuracy is desired in color measurement, the third, or spectrophotometric method, is usually the one that is employed. As the name implies, the method makes use of a spectrophotometer fitted with a reflectance attachment (see pp.288 ff) to measure the reflectance spectrum of the substance under investigation so that it may serve as the basis for the calculation of the parameters employed in the specification of color. Theoretically the tristimulus values should be arrived at by the integration of such expressions as

$$X = \int_0^\infty \bar{x}_\lambda R_\lambda E_\lambda \, d\lambda \,, \quad Y = \int_0^\infty \bar{y}_\lambda R_\lambda E_\lambda \, d\lambda \,, \quad Z = \int_0^\infty \bar{z}_\lambda R_\lambda E_\lambda \, d\lambda \quad (28)$$

where R_λ is the reflectance at wavelength λ, E_λ is a distribution function for the illuminant being employed, and $\bar{x}_\lambda$, $\bar{y}_\lambda$, and $\bar{z}_\lambda$ represent the tristimulus values of the color resulting from a unit density of radiant flux having an infinitely narrow wavelength range, $d\lambda$. Such an approach would be very tedious and, in practice, less time-consuming, approximate methods are used. One of these involves the subdivision of the visible spectrum into a set of finite, equal intervals, say 5 or 10 nm, with the integrals then being approximated by the sums. When such a method is employed, reflectance measurements are taken at fixed intervals and the measured value at a given wavelength is multiplied by an appropriate distribution coefficient. Tables of such distribution coefficients weighted by the energy values for illuminants A, B and C are available [138,139]. The computed values are then added to give the tristimulus values for the object of interest, i.e., ΣX, ΣY and ΣZ and the chromaticity coordinates are calculated with the use of eqns. (27). A detailed discussion of this as well as other methods has been provided by Judd [140].

The use of small digital computers for computing operations such as those discussed above not only makes approximation procedures unnecessary but also increases the number of samples that can be handled. As indicated earlier (see pp. 288 ff), there is already a number of instruments on the market which employ built-in computers which can instantly read out color coordinates.

(2) Pigments

Although the pigment industry still depends on professional color matchers and comparators for color mixing and matching, spectral reflectance techniques are gaining increasing acceptance, particularly in cases where pigments are being mixed with large proportions of white or yellow materials. In situations such as these the Kubelka—Munk function is valid and the absorption coefficients of the components of pigment mixtures commonly used in color formulations are additive at any given wavelength (see pp. 273 ff). With samples having an opaque surface, such as powders, and an infinite layer thickness, the Kubelka—Munk function can be used in the form $(1-R)^2/2R$. With paints, linoleums, plastics and other materials in which the pigment particles are incorporated in a continuous medium, however, a correction factor has to be introduced to take into account specular and internal reflections occurring at the air—medium interface [141,142]. Deviations from the Kubelka—Munk law also occur when colored pigments are present in high concentration. In such instances either a diluent is employed, or a semi-empirical function is introduced which will extend the linear range of the concentration—reflectance relationship.

Duncan [143—145] and others [146,147] have devised a number of procedures for pigment analysis by the diffuse reflectance technique, and computers have been specially designed for this purpose so as to reduce the time required for analysis [148—150]. Such investigations have not only been carried out in the visible, but also in the ultraviolet [151,152] and near infrared regions of the spectrum [88,153]. The technique has also been employed for the prediction of colors obtainable by the mixing of paint pigments [154], and the investigation of the hiding power, opacity and aging of paints [155—158].

Use has also been made of spectral reflectance in the identification of pigments [143]. With binary mixtures the simple inspection of spectra usually suffices for positive identification. For more complex systems absorption coefficients, K, are obtained at specified wavelength intervals and used as the basis of an identification system. It is customary in color technology to report these data in the form of ratios of the K values at different wavelengths relative to the K value at a particular wavelength, i.e. K/K_{425}. These values are then tabulated for comparison with similar values obtained for pigments under investigation. Qualitative studies of this sort have been reported for yellow pigments, the green colors of copper phthalocyanine—benzidine yellow mixtures [159], and azo pigments [160]. Ulrich et al.

310

[161] have developed a method for differentiating between pumpkin and saffron, two pigments which are difficult to distinguish from each other. Reflectance work in the ultraviolet has been employed for the investigation of white pigments [146,162—164].

(3) Biological materials

A reflectance method, called reflection oximetry, has been developed for the measurement of the oxygen saturation of human blood [165]. With samples having a thickness of at least 3 mm, the relationship observed between the oxygen saturation, OS, and the reflected light, $I_r(\lambda)$, at wavelength λ can be expressed as

$$OS = A + [B/I_r(\lambda)] \tag{29}$$

where A and B are constants dependent on the optical geometry of the instrument, the wavelength, and the intensity of irradiation [166]. Polanyi and Hehir [167] have devised a reflection oximeter for use with an absolute reflection method for the determination of OS.

Other biological applications of an unusual nature have also been reported. Lubnow [168] studied the possibility of classifying birds having melanine as a coloring matter in their plumage on the basis of reflectance data, while other investigators [169,170] have carried out reflectance spectroscopic studies on hummingbird feathers. Derksen and Monahan [171] measured the light reflected from human skin in the visible and near infrared, and Luckiesh et al. [172] employed spectral reflectance to study the tanning of human skin as a result of irradiation with ultraviolet light. Still other investigators have reported on reflectance studies of foliage and leaves [173,174].

Opportunities for the utilization of spectral reflectance in dealing with biological problems are for the most part limited only by the ability of the investigator to devise suitable experimental procedures for its application.

(4) Building materials

Spectral reflectance techniques have found wide use in the construction industry for the quality comparison and matching of building materials, and for the investigation of lighting conditions. For example, Moon [175,176], employing a Hardy spectroreflectometer, has studied the reflectance spectra of a wide variety of building materials, ranging from cork and oak flooring, carpets, asphalt tiles, and linoleum to schoolroom materials, such as blackboards, wall paints and window blinds.

For ceramic tiles and glossy paints, gloss properties as well as color factors are important, as a large proportion of the reflected light may be given off in the form of specular radiation. Accordingly, illumination engineers make use of total reflection properties. Taylor [151] and others [171], working in the visible and ultraviolet, reported on the reflection properties and reflectance curves for metal surfaces, wood and white plaster. Reflectance data are also available for a variety of ceramic materials [177—180], sanitary ware [181] and chrome magnesite brick [182].

(5) Food analysis

In recent years food laboratories and the food processing industry have resorted to diffuse reflectance spectroscopy in an attempt to make the comparison of color changes in foods less subjective.

Guerrant [183], for example, used reflectance spectra to follow color changes occurring in foodstuffs that had been stored under deep-freeze conditions (+10°, 0° and —20°F) over a period of 12 months. The changes observed, which seemed to be related to variations in the ascorbic acid content, are typified by the reflectance curves obtained for peaches as shown in Fig. 26.

Naughton et al. [184,185] employed the spectral reflectance technique in studying the "greening" phenomenon observed with tuna fish flesh following its precooking in the canning process. Various heme pigments were investigated and the "greening" effect was traced to an anomalous hemeprotein oxidative process. Similar studies [186,187] were made of the heme pigments in fresh beef, with reflectance spectra being used to determine the relative proportions of myoglobin, oxymyoglobin, and metmyoglobin present. The reflectance technique was also employed to study color changes resulting from the irradiation of meat and meat extract samples with gamma rays [188], and to follow changes caused by the cooking and pickling of meats [189].

A group of instruments were tested and rated as to their suitability for the measurement of reflectance spectra of canned tomato juice

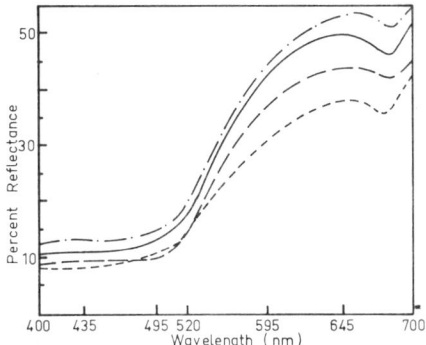

Fig. 26. Light reflectance of peaches. — · —, Before freezing; ——, stored for 12 months at −20°F; − −, stored for 12 months at 0°F; ----, stored for 12 months at 10°F [183].

[190], and a reflectance scale for the grading of tomato purees was devised by Yeatman et al. [191]. Results in the latter study were obtained with the use of a tomato reflectance colorimeter designed by Hunter and Yeatman [192].

Some other investigations which employed the diffuse reflectance technique include ones that involved the grading of sugars [193], the browning of milk on heating [194], and the darkening effect of the roasting process on peanut butter [195]. Diffuse reflectance has also been used in conjunction with paper [196] and thin-layer [27] chromatography in the analysis of food dyes.

(6) Geological materials

Until recently, because of the lack of suitable instrumentation and of an awareness among geologists of the potential of the technique, relatively little use was made of spectral reflectance in investigations dealing with geological materials. One such was a study by Piller [197], which concerned itself with the measurement of color during microscopic ore examinations. The work was carried out with a Zeiss microscope reflectometer designed by Piller himself. After discussing the possible use of the CIE color system for the classification and identification of ores and minerals, he concluded that the specification of such colors according to Helmholtz [125] was more desirable since it achieves a closer correspondence to hue and saturation. There have also been a number of discussions [198—201] concerning errors encountered in microscopic reflectance measurements. In no case were systematic errors found to exceed 1%.

A novel application of the diffuse reflectance technique was its use in the study of desert surfaces [202]. A special instrument, which permitted measurements of extended areas, had to be constructed for this purpose. Measurements were made from heights of 1.5—2 m above the surface, and, in some cases, from a helicopter at an altitude of 200 m. The results indicated that for the desert surfaces studied the reflectance in the blue portion of the spectrum was significantly less than in the red. Varying the height of observation between 2 and 200 meters did not materially affect the reflectance data.

(7) Paper and pulp

The literature concerning the application of diffuse reflectance to paper and pulp materials is extensive because the measurement of color, whiteness, brightness and gloss are important to the paper manufacturing process. An early study was one by Taylor [151], which dealt with the reflection factors for a number of commercial newspapers. Subsequently the reflectance spectra of colored papers [203,204] and wallpaper [176] have also been measured.

Luner and Chen [205] have discussed the use of reflectance methods in the control of the whiteness and brightness of wood pulp during the paper making process. Although the whiteness of the end product is an important indicator of its quality, differences in whiteness are not solely attributable to differences in brightness, but also to real differences in color. In a detailed study undertaken by Vaeck [206,207], it was shown that a touch of blue suggests a higher quality of white whereas any other hue would be taken as an indication of lower quality. It has recently been proposed that the color coordinates of white colors be arranged according to a scale which would take into account individual differences in color perception among observers [208].

Stenius [209,210] has provided a critical discussion of the parameters in the measurement of the whiteness and brightness of papers. Many of the principles involved have equal validity in the textile field, where optical brightners are widely used. The selection of proper reference standards is particularly important in the measurement of samples having a low color saturation, and Budde and Chapman [211] have recently discussed some of the problems relat-

314

ing to the calibration of standards for "absolute brightness" determinations. There has also been a number of studies relating to the validity of the Kubelka—Munk function when used in the measurement of reflectance from paper [212—214].

(8) Pharmaceuticals

Considerable use of diffuse reflectance spectroscopy has been made by the pharmaceutical industry in studying the effects of aging, illumination, temperature and humidity on tablets, powders, creams and emulsions [215], as well as in production control. Typical of these investigations is the series of studies carried out by Lochman et al. [216—219] on the light-fastness of water-soluble dyes employed for the coloring of tablets. Some of the results of these studies are presented in Fig. 27, which shows the effect of normal and excessive illumination upon such a dye, FD&C Blue No. 1, over periods of up to 84 days. The results thus obtained served as the basis for the calculation of rate constants for the fading process. Everhard and Goodhart [220] carried the investigation another step forward by

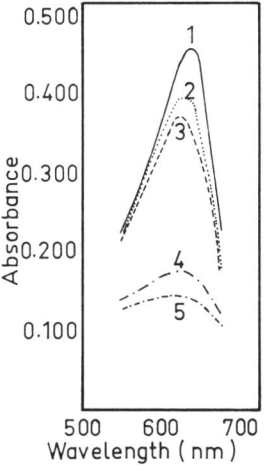

Fig. 27. Reflectance spectra of FD&C Blue No. 1 after different intervals of storage under normal and exaggerated illumination. (1) Initial spectrum; (2) after 42 days under normal illumination; (3) after 84 days under normal illumination; (4) after 42 days under exaggerated illumination; (5) after 84 days under exaggerated illumination [217].

References pp. 345—354

proposing a first-order rate equation which made it possible to estimate the fading that would occur under normal illumination on the basis of results obtained under conditions of high intensity irradiation. Other common applications are typified by two investigations dealing with spectrophotometric reflectance methods for the matching of the colors of solid dosage forms [221,222].

(9) Textiles

One of the earliest applications of spectral reflectance to textile technology was that of Hardy [4], who measured the reflectance spectrum of a piece of green silk in demonstrating the performance of an instrument which he had designed and built and which was the forerunner of General Electric's Hardy recording reflectometer. Since that time there have been numerous publications dealing with color measurements of dyed fabrics by this technique. A comprehensive treatment of this particular aspect of textile technology has been provided by Müller-Gerbes [223].

Lermond and Rogers [25] were the first to advocate the use of differential reflectance spectroscopy (see pp. 270 ff) for use with fabrics colored with strongly absorbing dyes. They also suggested a number of methods for expressing the reflectance as a function of the concentration of the dye. Plots of %R or the Kubelka—Munk function versus concentration resulted in smooth curves suitable for most analytical purposes. Straight line plots were obtained for the concentrations studied when they were expressed in the logarithmic form.

The technique has been employed in investigating the mechanism of the dyeing process [224], color changes resulting from irradiation with light [225], and the influence of elevated temperatures, humidity and aging on numerous fabrics [223]. Fourt and Sookne [226] studied the influence of the angle of incidence upon the measured reflectance of some cotton yarns and found the optical geometry as well as the orientation of the yarn relative to the plane of light to be critical. As mentioned earlier, diffuse illumination via an integrating sphere or ellipsoidal mirror is often preferable for use with samples having pronounced structural characteristics.

Laundering processes and the effectiveness of soap and detergents have also been studied with the use of reflectance measurements [227—229]. Whiteness resulting from the action of optical brighteners often employed in modern detergents can be measured by means of a reflectance spectroscopic method devised by Vaeck [206]. Interference by fluorescent components present in the system is avoided by the use of suitable filters.

(B) THEORETICAL APPLICATIONS

(1) Surface phenomena

Diffuse reflectance spectroscopy is presently the only means available in the ultraviolet, visible, and near infrared for the quantitative investigation of substances adsorbed on a solid surface. The only other alternative for the study of surface phenomena, such as the chemisorption of ethylene on alumina [230], is the transmittance of thin layers of powdered material onto which the species of interest has been adsorbed, and this has been shown to be effective only in the mid-infrared portion of the spectrum.

(a) Sample preparation

Sample preparation in the investigation of surface phenomena is extremely critical. One factor which has considerable influence on reflectance spectra is the particle size of the adsorbent [28]. For this reason, it is essential that the grinding and sifting operations in the preparation of the adsorbent be standardized. Kortüm and his co-workers, who have investigated this adsorbent particle-size effect exhaustively, advocate the use of mortars or ball mills for grinding. Porcelain grinders with two large or six small porcelain balls were found to be suitable for most purposes. Agate grinders with agate or hard metal balls were recommended for harder materials. The same type of grinding equipment should be employed throughout any given dilution series. To insure sample homogeneity, the grinding operation should be carried out for at least six hours, with sixteen hours being preferable. Errors arising from the abrasion of the grinding equipment have been found to be negligible relative to the accuracy provided by the reflectance technique.

The nature of the sample surface also affects reflectance spectra. Surfaces should be smooth and have no gloss, particularly if directional illumination is being employed. These conditions are usually satisfied by using windowed sample cells (see pp. 296 ff) at a zero degree angle of incidence. If the sample holder has no cover, a glass or metal tamp should be used to compress the sample until its surface is smooth and even. The pressure applied during this operation should also be standardized to the extent possible. Layer thicknesses of 2—5 mm are usually adequate to satisfy the infinite layer thickness requirement.

Reference standards, which often consist of the pure adsorbent or diluent used in the particular experiment, are treated in the same manner as the analytical samples. In fact, the use of a twin mill for the simultaneous grinding of reference and sample is highly recommended.

Because the highly polar water molecules compete for the available adsorption sites of the sample material in investigations of this type, it is imperative that moisture be excluded as much as possible during the sample preparation. This can be done by carrying out the entire operation in a glove box whose internal pressure is maintained at a slightly positive value by the introduction of a stream of air that has been passed through a bed of silica gel or molecular sieves. Samples can then be transferred elsewhere by means of desiccators.

While the above procedures are satisfactory for carrying out the mixing or dilution of solid substances, it is sometimes desired to adsorb the species of interest from the gas phase or from solution. If one is dealing with substances that are gaseous or that have a fairly high vapor pressure, the former can be managed quite easily with the use of a closed system, such as that employed by Kortüm and Koffer [231]. Although adsorption from solution is applicable in a wider variety of situations, it poses more problems from the experimental standpoint. In this approach the solvent is often removed by vacuum stripping, or, if it is volatile enough, by simply allowing it to evaporate while the adsorbent—solution mixture is standing in a dry box. While homogeneous mixing of two phases can thus be achieved without recourse to tedious grinding operations which might have undesirable side-effects, the adsorbent in all the methods described must still be prepared in such a way as to have a uniform and definite particle size.

It should be mentioned that the preparation of samples for routine analytical purposes need not be as rigorous as the above.

(b) Adsorbent—adsorbate interaction

Since it was first noted that many compounds underwent a color

318

change following adsorption on active surfaces [232—234], there have been numerous investigations which have employed diffuse reflectance spectroscopy as a means of studying adsorbent—adsorbate interactions.

Kortüm, who has made many significant contributions in the field, and his coworkers [23] made one such study of the system p-dimethylaminoazobenzene (DMAB) adsorbed on barium sulfate which had been carefully heated to eliminate any adsorbed moisture. The reflectance spectra that were obtained for varying amounts of DMAB adsorbed on dry barium sulfate are depicted in Fig. 28. As is shown there, the red band (approximately 19,000 cm^{-1}), attributed to the chemisorption of a mono-molecular layer of DMAB, grows with increased concentration of DMAB until a limiting value is reached. Before this point is reached, however, a shoulder in the yellow region (approximately 24,000 cm^{-1}) appears and continues to grow even after the red band stops. This is ascribed to the physical adsorption of second and subsequent layers of DMAB molecules which commences before the chemisorption process occurring at the adsorbent surface is completed. Viewed with the naked eye, the yellow DMAB turns red upon adsorption, but reverts to its original yellow color

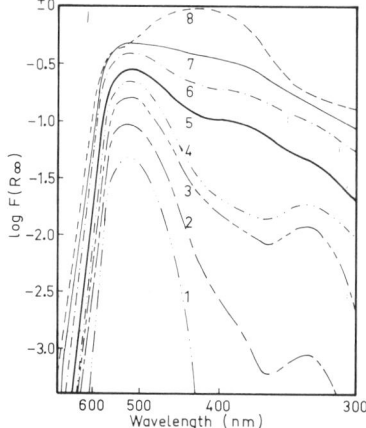 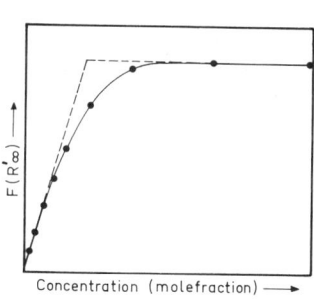

Fig. 28. Reflectance spectra of p-dimethylaminoazobenzene adsorbed on dry BaSO$_4$ at different concentrations (10^{-4} mole fraction) [23]. (1) 1.40; (2) 2.42; (3) 4.07; (4) 6.69; (5) 10.35; (6) 15.72; (7) 25.31; (8) 44.12.

Fig. 29. Adsorption isotherm of p-dimethylaminoazobenzene adsorbed on dry BaSO$_4$ [23].

when the DMAB—BaSO$_4$ mixture is exposed to moisture. This is attributed to the displacement of the DMAB molecules from the adsorption sites by the water molecules. The process is reversible, and can, in this case, be reversed by simply placing the mixture in a vacuum desiccator with calcium chloride. The adsorption isotherm shown in Fig. 29 resulted when the values of $F(R'_\infty)$ for the band associated with the chemisorbed species were plotted versus the mole fraction of the adsorbed DMAB. As indicated there, complete chemisorption is achieved at the limiting value 1.4×10^{-3}. The chemisorption process itself was explained by Kortüm in terms of a Lewis acid—base reaction, with the adsorbent acting as the acid.

Another interesting study of adsorbent—adsorbate interaction was carried out by Kortüm and Vogel [30] and involved the adsorption of malachite green-o-carboxylic acid lactone on alkali halides. A greenish-blue color develops when the colorless lactone is adsorbed on the carefully dried adsorbent, and the color can be discharged by the exposure of the mixture to water vapor. The generation of the color is ascribed to the cleavage of the lactone ring, with the surface interaction resulting in a lowering of the activation energy for the process. Here, again, it is concluded that the chemisorption is due to a Lewis acid—base reaction, except that the cations of the adsorbent lattice act as the acid, while the carboxyl groups of zwitterion that is formed when the lactone ring is cleaved act as the base. Support for this explanation is provided by the infrared spectrum of the adsorbed lactone which exhibits the —COO$^-$ band [235].

Reversible ring cleavages of spiranes were also found to occur at room temperature when adsorbed on dried magnesium oxide and sodium chloride [236]. This is just one of many investigations of photochemical reactions of adsorbed substances by Kortüm and his group. Another has involved a kinetic study of the reversible photochemical conversion of colorless, adsorbed 2-(2'4'-dinitrobenzyl)-pyridine to a blue quinoid-type structure [237]. In the dark, the adsorbed reactant returns to its original colorless form, but so much more slowly than occurs in solution that the rate of change can be measured by reflectance spectroscopy. Not only are the activitation energies for some reactions lowered by adsorption, as in the case of the spiranes, but also there are many situations where the influence of the adsorbent affects the actual course of the reaction. For example, anthracene, which in solution dimerizes to dianthracene under ultraviolet irradiation, is oxidized to anthraquinone in the presence

320

of oxygen when adsorbed on magnesium oxide, silica and alkali halides, with the rate being dependent on the nature of the adsorbent [32]. When the adsorbent is alumina and when the mixture is exposed to ultraviolet light, there is also a further conversion of the anthraquinone to quinizarin. On silica gel, the reaction is slower, and on potassium chloride it does not proceed at all. Diffuse reflectance spectroscopy has also been employed in studying the thermochromic properties of ethylenes adsorbed on magnesium oxide [238] and the thermo- and piezochromic properties of adsorbed bis(thiaxanthylene), biflavylene and dimethyldiacridene [238,239].

There has been a variety of efforts to relate changes in reflectance spectra with the nature and extent of adsorbent—adsorbate interactions. Schwab et al. [240,241] found it possible to arrange adsorbents in a series based on the magnitude of the bathochromic shift induced in the principal absorption bands of a group of eight adsorbed dyes. The adsorbents studied, arranged in order of increasing effectiveness in bringing about this shift, were: quartz powder; air-dried alumina; air-dried silica; alumina dried at 200°; silica dried at 200°; alumina dried at 900°; HCl-treated silica; pure solid adsorbent; HCl-treated alumina; α-alumina; and bentonite. While this arrangement can be looked at as one in which the adsorbents are listed in order of increasing acidity, with the adsorbent—adsorbate interaction being essentially an acid—base reaction, it is not quite clear in all instances whether the acidity in question is due to active hydroxyl groupings or incomplete coordination.

Others have attempted to use polarization theory as the basis for explaining shifts of absorption maxima observed in reflectance spectra, in very much the same way as solute—solvent interactions are interpreted in transmission spectroscopy. Zeitlin et al. [242] employed such an approach in their studies involving a variety of mercury(II) compounds which had been mixed with such adsorbents as alumina, silica gel and sodium fluoride. As expected on the basis of polarizability, bathochromic shifts were observed with the silica gel and alumina mixtures which were not observed with sodium fluoride, and bathochromic shifts observed as a function of the anion were found to be $I > Br > Cl$ and $S > O$. The behaviour of the mercuric iodide—alumina system, however, was anomalous. The spectral change in this case was ascribed to the formation of a yellow modification of mercuric iodide which is normally metastable at room temperature but which is stable in the adsorbed state [243,244]. Kortüm, however, who found the reflectance spectrum of adsorbed mercuric iodide to be similar to the transmittance spectrum of mercuric iodide in concentrated solutions of potassium iodide, has proposed the formation of tetrahedral mercury complex with one of the active adsorbent sites displacing an I^- ligand [31]. Zeitlin and his coworkers have also attempted to extend the polarization approach to organic systems with studies of the three isomeric mononitrophenols adsorbed on starch, talcum and alkali carbonates [245]; of o-nitrophenol adsorbed on alkaline earth oxides [246]; of nitroanilines adsorbed on alkaline earth oxides [247]; and of benzophenone, p-dimethylaminobenzaldehyde and Michler's ketone adsorbed on silica gel and on acidic, neutral and basic alumina [248,249].

Kortüm and his group have also employed diffuse reflectance in their study of the formation of "charge-transfer" bonds [250] resulting from the adsorbent acting as an electron donor and the adsorbate as the acceptor, or vice versa. For example, the spectra obtained for iodine or bromine vapor adsorbed on dried alkali halides corresponded closely with transmittance spectra obtained for solutions of $I_2 X^-$ and $Br_2 X^-$ complexes, where X^- can be I^-, Br^-, or Cl^- [251]. Similarly the adsorption of iodine or bromine on highly dried alumina, aluminium hydroxide, magnesium oxide and calcium oxide resulted in reflectance spectra which were assigned to analogous complexes $I_2 OH^-$ and $Br_2 OH^-$ [252]. Reflectance spectral data obtained for iodine adsorbed on dried aerosil, however, indicated that in that particular case the iodine is only physically adsorbed [231]. On the basis of reflectance data, Frei and his coworkers [253,254] have suggested that "charge-transfer" bonding also occurs when the copper, nickel and cobalt chelates of pyridine-2-aldehyde-2-quinolylhydrazone are adsorbed on cellulose and starch.

Benzene adsorbed on a mixed silica—alumina catalyst under vacuum conditions has been studied by means of diffuse reflectance in the near infrared [255,256]. The spectral information indicates the existence of a redox equilibrium mixture involving the species $C_6 H_6^+$ and $C_6 H_6 H^+$, as well as the presence of physically adsorbed benzene. Numerous other aromatic hydrocarbons have been observed to behave similarly.

(c) Determination of surface area of powders

By making use of color changes resulting from adsorbent—adsorbate interactions, it is possible to determine the surface areas of the adsorbents employed with an accuracy comparable to that provided by the BET method. Kortüm and Oelkrug [29] used Langmuir adsorption isotherms for malachite green-o-carboxylic acid lactone adsorbed on moisture-free sodium chloride of different grain sizes as the basis for computing the available surface of the sodium chloride adsorbent. Since the chemisorptive process can only occur in the monolayer at the surface of the adsorbent, the limiting value as determined from the isotherm provides a relative indication of the active surface area of the adsorbent. The isotherms were obtained in essentially the same manner as that described earlier (see p. 319) for the p-dimethylaminoazobenzene—barium sulfate system. The surface area ratio found for two sodium chloride samples with the use of this method was 1.12, which compares favorably with the 1.11 value obtained using the BET method with nitrogen adsorption. Such an approach should be capable of providing a more realistic value for the available surface areas of chromatographic adsorbents than is currently available by other means.

(d) Determination of equilibrium constants

Deviations from the Kubelka—Munk equation can be employed in certain situations to determine equilibrium constants for the association of molecular species adsorbed on solid surfaces in very much the same way that deviations from the Bouguer—Lambert Law are used in solution chemistry. Braun and Kortüm [257] used this approach to study the hexamethylbenzene-s-trinitrobenzene molecular complex adsorbed on silica gel. Freshly deposited magnesium oxide was used as the reference standard so that the absolute reflectance could be calculated. The dissociation constant for the complex was then determined by iteration procedures [258] from known mole fractions and measured $F(R_\infty)$ values. For such a dissociative equilibrium to be established, the "charge-transfer" forces between the adsorbent and the individual components of the complex must be of the same order of magnitude as the bonding forces between the components themselves. Such was not the case for pyrene-s-trinitrobenzene adsorbed on sodium chloride, where no dissociation was observed [259].

Attempts to determine dissociation constants at elevated temperatures have until now not met with much success because of the difficulty encountered in controlling the moisture content of the analytical samples [260].

(2) The measurement of reflectance at other than ambient temperatures

With few exceptions the reflectance measurements in the studies discussed so far have been made at ambient temperatures. By carrying out such measurements at elevated or reduced temperatures, however, additional applications of the diffuse reflectance technique become possible, such as the monitoring of thermally-induced reactions and the study of crystal structure in terms of the ligand field theory.

(a) Measurements at elevated temperatures

Although there have been some earlier studies [31,261], the majority of the contributions in this field have been those of Wendlandt and his coworkers, who have also provided surveys of recent developments [101,262,263]. Wendlandt has categorized the methods used into "high-temperature reflectance spectroscopy" (HTRS) [102], in which the reflectance spectra are measured under isothermal conditions, and "dynamic reflectance spectroscopy" (DRS) [264], in which the reflectance at a fixed wavelength is measured continuously over a temperature range.

While neither of these approaches (which can be employed in the ultraviolet, visible and near infrared) provides information that cannot be arrived at by the application of other thermal techniques, such as thermogravimetry, differential thermal analysis, and the like, they serve as valuable complements to the more established techniques. While the DRS procedure offers the advantage of being unaffected by any weight losses and thermodynamic effects that occur during the measurement, spurious results are obtained when it is used to determine transition temperatures unless the temperature is increased very slowly [103]. Sample holders and ancillary equipment suitable for use in high temperature work have been described earlier

324

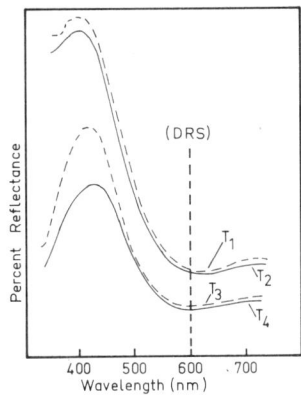
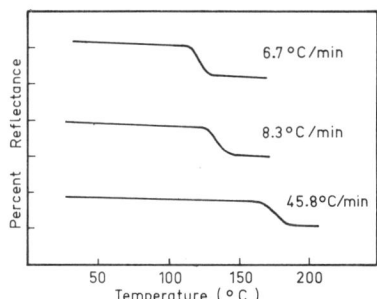

Fig. 30. HTRS curves of $[Cu(en)(H_2O)_2]\,SO_4$ at T_1 (25°); T_2 (75°); T_3 (150°); T_4 (180°) [9].

Fig. 31. DRS curves of $[Cu(en)(H_2O)_2]\,SO_4$ at 600 nm [9].

(see p.298). Procedures for the preparation of samples for analysis are essentially the same as those employed in the study of surface phenomena (see pp.317 ff).

Typical of the studies suitable for the HTRS and DRS treatment is one involving the deaquation of $[Cu(en)(H_2O)_2]SO_4$ [265]. The HTRS curves obtained at 4 different temperatures are depicted in Fig. 30. As shown there, the absorption maximum located at 625 nm at 25° and 75° undergoes a hypsochromic shift to 575 nm when the sample is heated to 150° and 180°. These data indicate that the process whereby the coordinated water is expelled occurs within the 75—150°C range. The DRS technique can then be utilized to locate more precisely the transition temperature. The DRS curves obtained for the same complex at three different heating rates, which are presented in Fig. 31 [263], emphasize the dependence of the observed transition temperature on the rate at which the temperature of the sample is increased. The 115—130° value obtained by using the slowest rate of 6.7°/min is the one that most closely coincides with the HTRS data. Other deaquation processes have been similarly investigated including the chromium(III) and cobalt(III) complexes of the type $[M(NH_3)_5 H_2O]X_3$, where M may be either cation and X may be Cl^-, Br^-, I^- or NO_3^- [266].

Wendlandt and his coworkers have also employed these high tem-

References pp. 345—354

perature reflectance techniques to study the reaction between $CoCl_2.6H_2O$ and KCl [267]; the changes induced in the structure of $CoCl_2.6H_2O$ by the application of heat [268]; the conversion of bis-pyridine cobalt(II) chloride from its violet, octahedral, α-form to its blue, tetrahedral, β-form [269]; and the thermochromic transitions of $Cu_2[HgI_4]$, $Ag_2[HgI_4]$, and AgI [103]. Thermochromic reactions have also been investigated by Hatfield et al. [270], who studied bis(N,N-diethylethylenediamine) copper(II) perchlorate, which at $50°C$ undergoes a color change from red to blue. The two early investigations mentioned above involved the conversion of red mercuric iodide into its yellow modification at $127°C$ [31], and an isothermal, kinetic study of the reaction between iron(III) oxide and calcium oxide [261].

(b) Measurements at reduced temperatures

There has been a number of studies in which the reflectance measurements have been made at reduced temperatures [12,271—274]. It has been found that spectra recorded under these conditions, often at liquid nitrogen temperatures, have sharper structural features than similar spectra recorded at ambient temperatures. This has been attributed to some extent to the elimination of "hot bands" arising from transitions involving vibrationally excited ground states. Such studies, which present no particular practical difficulties, have proved themselves to be useful in the interpretation of complex spectra in terms of the ligand field theory. A cell suitable for use at reduced temperatures has been described earlier (see p.299).

(3) Inorganic systems

In addition to the many studies of inorganic substances listed in the preceding sections, as well as in the subsequent one dealing with chromatographic applications, spectral reflectance has also found application to problems more specifically related to physical-inorganic chemistry [275]. These often involve ascertaining whether the species present in a solution is identical with that in a solid crystallized from that solution, or obtaining the absorption spectrum of a compound that is either unstable in solution or insoluble.

Because of the many advantages afforded by the application of reflectance spectroscopy to ligand field theory, there have been nu-

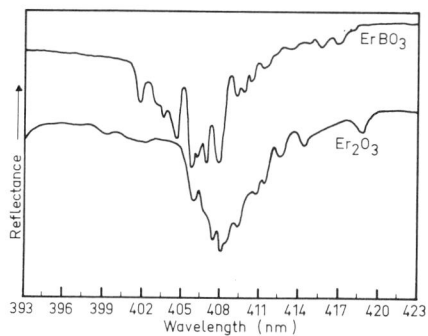

Fig. 32. Reflectance spectra of $ErBO_3$ and Er_2O_3 [280].

merous contributions in this field [276]. Typical of these have been the studies of transition metal complexes in the solid state conducted by Anysas and Companion [277], Asmussen and Bostrup [278], and Clark [272]. White [279] and Ropp [280] have obtained the reflectance spectra of rare earth oxides and other rare earth compounds in the solid state. An indication of the degree of resolution attainable by this means in the visible and ultraviolet is provided by Fig. 32, which depicts the reflectance spectra of $ErBO_3$ and Er_2O_3. With such high resolution, it is possible, for example, to identify optically narrow bands as ones arising from unperturbed $4f$ transitions. Loh [281] has carried out similar investigations employing single crystals rather than powders.

There have been several [282—284] reflectance studies of uranium oxides in the visible and ultraviolet. Changes of the oxygen content were shown up in the reflectance spectra of bulk samples in the $UO_{2.66}$—$UO_{2.00}$ range.

Working in the ultraviolet, Griffiths et al. [285,286] were able to obtain reflectance spectra for a wide variety of compounds, including iodides, iodates, periodates, dithionates, chromates, and oxides. The reflectance technique was found to be particularly useful with the ozonide, superoxide and peroxide of sodium, which are difficult to obtain in single crystal form because of their lack of stability. Kortüm and Herzog [287] have used ultraviolet reflectance spectroscopy as the basis of a method for the determination of rutile in mixtures with a second modification of titanium dioxide, anatase. The ultraviolet spectral reflectance technique has also been employed in a number of investigations of inorganic phosphors, including studies of the effect of copper and lead content on the reflectance of zinc sulfide phosphors [288], and of thallium content on that of potassium bromide phosphors [289]. Studies have also been made of potassium chloride phosphors with thallium [289,290], and of calcium silicate and sodium chloride phosphors with both lead and manganese [290].

Goulden [291], working in the near infrared region, has measured the reflectance spectrum of $CaSO_4.2H_2O$.

Using such powdered mixtures as iron(III) oxide—barium sulfate, Lermond and Rogers [25] explored the potential of the reflectance technique for use with difficultly soluble substances, and found the differential approach (see p. 270) essential in the analysis of samples having a low reflectance. Other powdered mixtures that have been studied by means of diffuse reflectance have been iron(III)oxide diluted with magnesium carbonate [292]; various combinations of lead(II)oxide, silver oxide and zinc oxide diluted with either silica gel or magnesium oxide [293]; and such pairs as aluminium and titanium, silica and aluminium and zirconium oxide and alumina [294].

There have been several assessments of the analytical utility of measuring the reflectance of substances adsorbed on paper. Fischer and Vratny [292] investigated, among other compounds, $Cu(NH_3)_4SO_4$ adsorbed on filter paper strips. In their study of the copper complex of α-benzoinoxime adsorbed on paper, Winslow and Liebhafsky [294] reported that transmission measurements through the paper provided a better signal-to-concentration relationship than did reflectance readings. Ayres [295], in his study of a variety of materials adsorbed on paper, found a linear relationship in the 0.5—5 μg range between the concentration of copper(II), nickel and iron(III) and the peak heights of the reflectance curves obtained for their adsorbed sulfides. Another reflectance study dealing with metal sulfides is that of Takagi et al. [296]. A procedure involving the reflectance measurement of colloidal ferric hydroxide collected on a membrane filter was devised by Mizuniwa et al. [297] for the estimation of the total iron content in boiler water. Palalau [298] also employed the reflectance technique in the determination of mercury in air by measuring the amount of the yellow-orange copper(I) iodide complex of mercury formed on filter paper. Ermolenko et al. [299] employed diffuse reflectance to detect nickel and manganese after concentration on cation-exchange paper and treatment with appropriate chelating reagents.

328

(4) Chromatography

The use of spectral reflectance for the analysis of the components of a mixture following its chromatographic resolution has proven to be quite fruitful, particularly in the case of thin-layer chromatography.

(a) General experimental procedure

Chromatographic resolution. The utilization of diffuse reflectance in conjunction with chromatography does not necessitate any changes in established chromatographic practices for the resolution of mixtures. Information regarding the latter can be obtained from any of the many standard references available in the field.

Detection of resolved substances. Much the same can be said for the spraying procedures that are employed for the detection of resolved colorless substances. For quantitative reflectance work, however, reactions induced by chromogenic sprays must not only be rapid and reproducible, but also must result in the formation of an indicator product which will be stable over at least the time span required for the analysis. Obviously reflectance techniques are ideally suited for determining the color stabilities of adsorbed species [300]. Occasionally, it is possible to form a colored product by heating [98], or, if the substance in question fluoresces, it can be observed under ultraviolet illumination [301]. Many compounds which absorb in the ultraviolet show up as dark areas on a fluorescent background when irradiated with ultraviolet light. The fluorescent background is provided by adding a luminous pigment to the adsorbent [99], or simply by taking advantage of the natural fluorescence of adsorbents.

Another approach to the detection of substances absorbing in the ultraviolet involves the scanning of chromatograms with a spectrophotometer set at the wavelength of maximum absorption of the compound of interest. One such procedure has been devised for use with the regular reflectance attachment of the Beckman Model DK2 [99]. The scanning is carried out by holding the chromatoplate, which is taped to a protective plastic plate, against the sample exit port of the reflectance attachment unit in such a manner that the adsorbent along the path of chromatographic development is exposed to the incident beam of light. A sudden decrease in reflectance occurs when the beam falls upon the spot containing the compound being sought. The location of the spot can then be marked on the reverse side of the plate. A dark cloth should be used to cover the reflectance attachment to exclude stray light. This approach, or one very much like it, can be employed with most other commercial attachments.

With specialized recording instruments now available for the evaluation of chromatograms (see pp. 299 ff), mechanical scanning provides an easy solution to the detection problem. Some discussions of direct scanning procedures are available in the literature [105—107,111—116,302,303]. Despite these technological advances, however, visual or UV lamp detection techniques still have to be used when locating the position of substances resolved on two-dimensional chromatograms.

Measurement of reflectance spectra. Once a resolved substance has been located, its reflectance spectrum can be measured in situ after the chromatogram has been positioned against the sample port of the reflectance attachment in such a way that the light beam is centered on the area of interest. A glass plate of identical size can be taped on top of the chromatogram to protect it during the recording process. For work in the ultraviolet, the glass plate must be replaced by the windowless protective plastic plate mentioned in the previous section, or by paper masks which have openings at points of interest. It is also recommended that a backing, consisting of a sheet of nontransparent material that reflects efficiently in the spectral region of interest, be mounted behind the chromatogram. This is because the adsorbent is seldom thick enough to avoid background interference.

Reflectance spectra can also be measured with the use of the reflectance cells discussed earlier (see pp. 297 ff). When this procedure is followed, the areas of interest are excised and placed in the appropriate cell, on top of 30—50 mg of adsorbent removed from the same plate in thin-layer chromatography, or 10 layers of paper in paper chromatography [304]. In this way, the requirement for infinite layer thickness is satisfied, and so a quantitative estimate of reflectance can also be made. Reference materials usually consist of the adsorbent being employed for the resolution of the mixture.

Spectra can, of course, be recorded automatically from the chromatograms with the use of scanners equipped with a mechanical wavelength drive.

The quantitative measurement of reflectance. The process of analyzing directly substances resolved chromatographically by the in situ procedure outlined for the measurement of reflectance spectra has many difficulties and often leads to inexact results. Provided that no excessive tailing occurs, one can expect an accuracy of approximately 10% with the use of commercial instruments, such as the Spectronic 505 or the Beckman DK2 spectrophotometers [27]. This can be improved somewhat by using Eastman chromatogram sheets for the resolution process, cutting out the area of interest, and reading it with a single-beam instrument, such as the Spectronic 20 or the Beckman DU. If the diameter of the spot is less than 7.5 mm it is possible to adjust the cross-sectional area of the impinging light beam so that the measurement can be made in a single reading. By neither procedure, however, is it possible to avoid problems arising from light-scattering phenomena nor from the inhomogeneous distribution of the material incorporated in the spot [214,305]. Relative errors of 3 and 4% have been reported by Jork [112] and Pataki [306] for quantitative work in the ultraviolet region using the mechanical scanning approach in conjunction with the Zeiss chromatogram spectrophotometer. Other quantitative studies by this method have been reported by a number of investigators [111—116].

In cases where the greatest accuracy is desired, the "spot-removal" technique should be employed [27]. This involves removal of the material of interest along with sufficient adsorbent to make up an analytical sample of predetermined weight, usually 20—80 mg. The mixture is then ground in a small agate mortar for a given period of time (30—60 sec) and packed in an appropriate cell (see pp. 297 ff). The reference material is adsorbent from the same chromatogram which is treated in the same manner as the analytical sample. Preparation of the sample can be expedited by removing it with a circular aluminum planchet affixed to a cork stopper [307]. The dimensions of the planchet employed are dictated by the thickness of the adsorbent layer and the area being excised. Once the sample has been cut from the adsorbent, the most direct path between it and the edge of the chromatogram is cleared of adsorbent with a brush and the planchet is moved along this path and deposited in a mortar for processing. By this technique, the analytical sample can be prepared for measurement in less than two minutes. While the accuracy of this technique depends to a great extent on the uniformity of the adsorbent thickness, it has been shown in this same study that a com-

mercial applicator is capable of laying down adsorbent of a sufficiently uniform thickness to prevent this parameter from affecting
the precision of the reflectance measurement.

The reproducibility of adsorbent coatings is also important in
methods using direct scanning procedures, though reflectance spectroscopy is somewhat less sensitive to variations in layer thickness
than are the corresponding transmission techniques [44,45]. The use
of double-beam scanning instruments offers the advantage that fluctuations arising from variations in adsorbent layer thickness are cancelled out. Klaus [308] and Huber [309] have recently conducted
critical investigations of the effect of fluctuations in layer thickness,
layer quality, and other experimental variables on the precision of
reflectance measurements.

The advantages afforded by the "spot-removal" procedure are its
simplicity, its precision, the opportunity to exercise rather close control of the experimental variables in the sample preparation process,
and its adaptability for use with practically any commercially available spectrophotometer that can be equipped with a reflectance attachment. On the other hand, it cannot be employed in some chromatographic situations (such as with prefabricated chromatosheets)
and necessitates a somewhat longer time for analysis than the direct
measurement technique. For the handling of large numbers of samples in routine analysis, the direct scanning mode will probably pre-
dominate, particularly with the advent of more and more sophisticated commercial chromatogram scanners fitted with electronic readout
equipment. The "spot-removal" procedure will still find use, however, in nonroutine situations, such as in the design of analytical
methods.

(b) Visible reflectance spectroscopy of organic compounds

Much of the early work dealing with the in situ analysis of chromatographically resolved organic compounds, which are either colored or capable of reacting with a chromogenic spray to yield a colored
product, was carried out by transmission techniques. It should be
possible to adapt many of these procedures for reflectance spectroscopic use. Spectral reflectance has also been employed in the visible
region for the in situ identification and determination of many substances following their resolution on thin-layer plates [73,310],
paper [311], and electrophoresis strips [312,313].

Dyes and pigments. Yamaguchi [196,314] and his coworkers have reported reflectance spectroscopic studies of some food dyes adsorbed on filter and chromatography paper.

Kortüm et al. [214,304,305] investigated the system malachite green adsorbed on paper. The aim was to assess the utility of spectral reflectance in paper chromatography. One of the main difficulties encountered in developing analytical procedures was the inability to achieve the uniform distribution of the adsorbed species over the whole paper sample, as is required by theory. When attention was given to the uniform distribution of the adsorbed species within spots of approximately the same area on the chromatograms, and if some 10 sheets of the same paper were used as backing material, reproducible calibration curves were obtained. By applying appropriate correction factors for variations in spot size and for the reflectance of the chromatogram paper, it was possible to arrive at an expression representing a linear relationship between the Kubelka—Munk function of the sample and the concentration of the dye.

A much simpler way of overcoming the inhomogeneity effect, however, is to scan the spot with a narrow slit or by means of the "flying spot" technique already referred to (see p.301). A detailed account of the use of the Joyce Loebl Chromoscan for the evaluation of dyes on paper chromatograms has been provided by Butler et al. [311].

The first study [27] of the application of diffuse reflectance spectroscopy to thin-layer chromatography also made use of dyes, which, being stable and easily detectable, lent themselves readily to such an investigation. The components of mixtures of dyes (Aniline Blue, Eosine B, Basic Fuchsin, Malachite Green, Naphtol Yellow S, and Rhodamine B) resolved on alumina chromatoplates were identified by direct spectral examination of the plates with a Beckman DK2 spectrophotometer. The amounts of adsorbed dye were estimated at the same time. A precision of approximately 3% was attained when the measurements of reflectance were carried out on spots removed from the plates and packed in an appropriate cell.

Plant pigments have also been analyzed with the use of the combined chromatographic—reflectance technique. Yamaguchi et al. [315] studied the relationship between reflectance and the concentration of chlorophyll in paper—partition chromatography, recommending $2-\log R$ versus $C^{1/2}$ plots for analytical purposes. Garside and Riley [316] have recently reported a thin-layer—reflectance procedure for the measurement of chlorophylls and carotenoids in marine particulate matter. Following chromatographic resolution, the pigments were measured in situ with a Joyce Loebl Chromoscan fitted with a TLC attachment operated in the reflectance mode. Peak areas were integrated mechanically, and the pigments were identified by contrasting the displacement of a particular spot (i.e. the distance between the peak and point of origin as determined from the Chromoscan trace) with the displacement of chlorophyll *a*. The integrator readings were reproducible to better than ±2% and the percentage for the total integrated area for the individual pigments never varied more than 1% within a set of six replicate chromatograms. For the chlorophylls and the major xanthophylls measured, the coefficient of variation for the method itself did not exceed 5% and good agreement was observed between the results obtained with the use of the more expeditious reflectance technique and those obtained by the transmission spectrophotometric analysis of solutions of the eluted spots.

Amino acids. Although reflectance spectroscopy was used in conjuction with paper chromatography for the analysis of amino acids as early as 1953, more recent studies have tended to employ the thin-layer technique, which has been found to afford greater speed, resolution and sensitivity [317]. Reviews dealing with the application of reflectance spectroscopy to the analysis of amino acids resolved on thin-layer plates are available [306,318].

From the standpoint of identification of adsorbed amino acids, the colors obtained with the use of conventional ninhydrin sprays are so similar that they are of little assistance. Employing a modified spray, however, it is possible to identify different amino acids following their resolution on Silica Gel G chromatoplates [319], by combining reflectance spectra, visual observation and R_f values.

Reflectance spectroscopic studies also served to investigate the stabilities of the ninhydrin complexes of amino acids [300]. Information obtained as a result of these studies indicated that a spray recommended by Bull et al. [320] was most suited for quantitative work. Not only were highly stable ninhydrin complexes formed with the use of this spray, but also the color intensity obtained was superior to that of other sprays, as may be seen in Fig. 33. In addition, the positions of maximum absorption for the complexes of most of

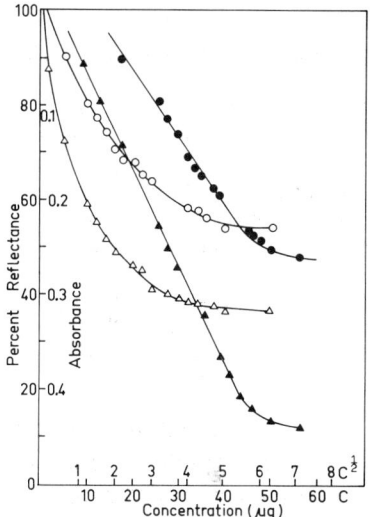

Fig. 33. Percent reflectance and absorbance at 520 nm of the ninhydrin complex of leucine adsorbed on silica gel as a function of concentration. Modified Brenner—Niederwieser spray: ○——○, % reflectance vs. C; ●——●, absorbance vs. $C^{1/2}$. Bull et al. spray: △——△, % reflectance vs. C; ▲——▲, absorbance vs. $C^{1/2}$ [321].

the acids studied were only slightly affected by changes in concentration.

A procedure employing this spray was developed whereby amino acids resolved on silica gel chromatoplates could be determined by spectral reflectance [321]. Readings made by means of this method had an average standard deviation amounting to 1.45%R. Since the factors limiting the precision of the method were found to be associated with the spraying procedure, a method in which spraying was unnecessary was subsequently devised. It involved the addition of the detecting agent to the solvent system [322]. This resulted in a significant improvement in precision, with an average standard deviation of 0.49%R being obtained for the reflectance readings for a set of 10 replicates following their one-dimensional chromatographic development. As is to be expected, the comparable figure for two-dimensional chromatography was larger, but the 0.77%R value attained is still substantially better than the 1.45%R reported for the spray method for a one-dimensional operation. All reflectance measurements were made in accordance with the "spot-removal" proce-

TABLE 2

Probable relative error in the measurement of 5 μg samples of some amino acids

	Alanine	Leucine	Serine	Valine
Range of readings obtained for 4 replicates (% R)	72.8—73.6	73.8—74.5	75.5—76.4	74.2—74.7
Mean (% R)	73.0	74.2	76.1	74.4
Standard deviation (% R)	0.39	0.30	0.41	0.27
Equivalent change in measured concentration of acid (μg)	0.18	0.15	0.25	0.14
Probable % relative error	3.6	3.8	5.0	2.8

dure. By employing such information about reproducibility in conjunction with calibration curves similar to those depicted in Fig. 33, it was possible to estimate the percent relative error that might be expected in the measurement of a given sample of the amino acids studied. The values computed for 5 μg samples of alanine, leucine, serine and valine are listed in Table 2.

Pataki [306] has reported on an amino acid analysis study which differs from the preceding one mainly in that the chromatoplates were read directly by a Zeiss Chromatogram spectrophotometer. Although plots of peak area versus the square root of the concentration were linear over an analytically useful concentration range, the precision reported was considerably poorer than that achieved by the "spot-removal" technique. This is not surprising, since the direct scanning background irregularities and tailing of spots can lead to substantial errors. Heathcote and Haworth [323] carried out a similar study using the Joyce Loebl Chromoscan with the same results.

Sugars. Bevenue and Williams [324] have employed spectral reflectance for the quantitative analysis of raffinose and mellibiose on paper chromatograms. A linear relationship between concentration and log $1/R$ for relatively low concentrations was observed. The same investigators have also devised a similar procedure for the determination of reducing sugars, such as fructose and glucose [325], and Owens et al. [326] have employed the technique in the measurement of small amounts of galactose. The overall accuracy reported for the approach by these investigators was ± 5% following one-dimensional chromatography.

Frei et al. [327] attained the same accuracy in their study of the analysis of mixtures of sugars by spectral reflectance following their resolution on cellulose thin-layer plates. With aldopentoses, the plates were developed by immersion in an ethyl acetate chromogenic solution which was 2% with respect to both aniline and trichloroacetic acid. The immersion technique, which could be employed because of the stability imparted to the thin-layers by the use of cellulose, was preferable to the spray technique from the standpoint of precision. The sugar—aniline derivatives were found, by means of in situ reflectance measurements, to be stable over a period of one week. Similar results were achieved with hexoses, and pentoses, when a naptha—resorcinol—phosphoric acid mixture was employed as the chromogenic reagent in place of the aniline—trichloroacetic acid.

(c) Ultraviolet reflectance spectroscopy of organic compounds

Diffuse reflectance spectroscopy has been used in the ultraviolet region for the measurement of the reflectance spectra of a number of 2,4-dinitrophenyl hydrazones of some ketones adsorbed on filter paper [328,329], and for the determination of 2,3,6-trimethylfluorenone on paper chromatograms [330]. The application of the technique to thin-layer chromatography was investigated by using salicylic acid—aspirin mixtures, which were resolved on silica gel plates [98]. This particular system was selected for study because it presented no difficulties during the location of its components following their separation: both appeared as yellowish-brown spots when the plates were dried. The spots were then excised from the plate and packed in a cell fitted with a quartz window for the measurement of diffuse reflectance. The standard deviation of the reflectance readings obtained for four replicates, following one-dimensional chromatography, was reported to be 0.37 reflectance units for salicylic acid and 0.47 reflectance units for aspirin. Aspirin was also employed as a model system in an error-analysis study of diffuse reflectance spectroscopy [14].

Amino acid derivatives. Another approach to amino acid analysis has been a variation on the combination reflectance—chromatographic procedure outlined in the preceding section dealing with visible reflectance spectroscopy. It involves the generation of amino acid derivatives which absorb in the ultraviolet. Two such sets of derivatives are the so-called DNP-amino acids, obtained by dinitrophenylation procedures, and the PTH-amino acids, which are 3-phenyl-2-thiohydantoins produced by the interaction of the acids with phenylisothiocyanate. Once prepared, the derivatives can be separated by one-dimensional chromatography on silica gel plates and analyzed by scanning the plates. Pataki [306] and Zürcher et al. [331] have discussed the qualitative and quantitative analysis of DNP- and PTH-amino acids, respectively, by such a method. Although the reflectance spectra thus obtained are adequate for the identification of the acids, they can be differentiated further by treating the spots with HCl or NaOH spray solutions. Scanning in both studies was accomplished with a Zeiss Chromatogram Spectrophotometer. The quantitative results obtained for the DNP-derivatives had a relative standard deviation of 6.0%; those for the PTH-derivatives, one of 5.0%.

Nucleo derivatives and related compounds. The "spot-removal" technique described earlier has also been applied to the analysis of a number of nucleotides [301]. Following their resolution on cellulose thin-layer plates, the nucleotides were located under an ultraviolet lamp (254 nm) by the fluorescence-quenching mode, excised, and then packed in a windowless cell (see pp. 297ff) for measurement. The standard deviation obtained with the use of the procedure ranged from 0.39 to 0.78 reflectance units, while the accuracy for some compounds was as low as 6% and for others as high as 2%.

A much more extensive investigation of nucleo derivatives and related compounds was carried out by Pataki [306] and Frei et al. [302], employing a chromatogram scanner in the reflectance mode. The results obtained for nucleotides were in close agreement with those reported earlier by Lieu et al. [301]. Spectra, some of which are presented in Fig. 34, of some 16 compounds were recorded directly from chromatoplates, and found to be well-suited for identification purposes. The wavelength values of maxima fluctuated by no more than 2 nm over extended periods of time (up to 6 days) and with concentrations ranging from 0.5 to 5.0 μg per spot.

Quantitative studies were restricted to hypoxanthine, uridine, inosine, uracil, and thymine. The scanning peaks obtained following two-dimensional chromatography on cellulose plates were evaluated planimetrically. Smooth curves resulted when the peak areas were plotted as a function of the square root of concentration. An improvement in the reproducibility of the procedure, amounting to

338

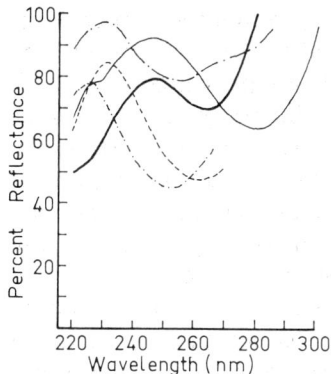

Fig. 34. Reflectance spectra of some nucleo derivatives and related compounds (∼ 5 μg per spot) adsorbed on cellulose:——, nicotinamide; ——, cytidine; ——·——, guanosine; – · – · – ·, hypoxathine; and ----uracil [302].

3—4% relative standard deviation, was achieved with the use of an internal standard technique recommended by Klaus [332] as a means of compensating for the fluctuations inherent in the chromatographic process. Although both uracil and adenine were equally good as internal standards, the latter was selected for use because of the composition of the biological samples which were later to be

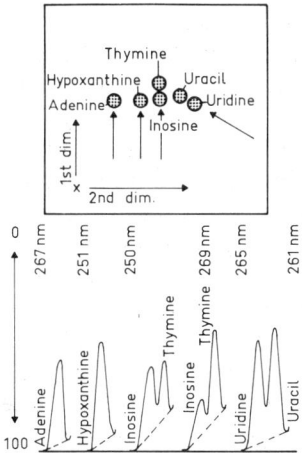

Fig. 35. Thin-layer chromatogram of some nucleo derivatives and related compounds resolved on cellulose, and chromoscan tracings obtained for the absorption peak portions of the reflectance spectra of the various spots [302]. Arrows mark the direction of scan.

analyzed. A typical chromatogram, resulting from a two-dimensional development, as well as the tracings obtained for the absorption peak portions of the spectra for the various spots are presented in Fig. 35. The peak area evaluation technique also illustrated in this figure proved to be the most suitable of a number tried in conjunction with this study, particularly for poorly resolved peaks. Calibration plots of peak area ratios of the substance being analyzed to adenine as a function of the square root of concentration were linear in the concentration range of analytical interest. Results obtained in the analysis of artificial mixtures of the 5 compounds had an average deviation of 4—5%. About 2 days were required for a complete analysis of these components in complex biological samples, such as cartilage red bone marrow extract. The first was required for the chromatographic resolution of the mixture while the second was needed for the reflectance aspect of the procedure. An attempt to reduce this time requirement by employing a set of calibration curves prepared at a time different from that of the actual analysis resulted in an increase in the average percent error to approximately 16%.

Vitamins and hormones. A procedure has also been developed by which 5 vitamins of the B-group (thiamine hydrochloride, pyridoxine hydrochloride, nicotinic acid, nicotinamide, and *p*-aminobenzoic acid) resolved on silica gel thin-layer plates can be analyzed by means of ultraviolet reflectance spectrometry [99]. Two methods of locating the resolved vitamins prior to analysis were investigated. In the one which involved observation under ultraviolet light of plates prepared with silica gel, to which fluorescent material had been added, it was easier to locate the spots. The other, which involved scanning of the plates with a spectrometer set at an appropriate wavelength, was somewhat slower but afforded a greater sensitivity. Except for these minor differences, the results obtained with the two variations were similar. Three of the vitamins could be identified unequivocally by means of reflectance spectra recorded directly from the plate. Nicotinic acid and nicotinamide, whose spectra were similar, could be distinguished with the aid of their R_f values. When excised from the plate and packed in a windowless cell, the adsorbed vitamins can be determined with a precision of 2—3%.

The reflectance spectroscopic analysis of nicotinic acid and nicotinamide by the scanning of cellulose thin-layer plates was subsequently investigated by Pataki [306] and Frei et al. [333] as outlined in the preceding section on nucleo derivatives and with the results given therein. Nicotinamide has also served as the model compound for a comparison study of the use of the fluorimetric, reflectance spectrophotometric and fluorescence-quenching techniques in conjunction with the scanning of chromatographs. In the case of the two last mentioned optical methods, a linear relationship was observed over essentially the same concentration ranges when the square of the peak areas, which approximates the Kubelka—Munk function, was plotted as a function of the concentration of nicotinamide. Ideally the sensitivity of both methods of measurement should be equivalent, but in fact the results obtained indicated that reflectance spectroscopy was preferable to fluorescence quenching, which suffered from such disadvantages as lower specificity and precision.

Struck et al. [115], using a chromatogram spectrophotometer, devised an in situ reflectance procedure for the determination of Δ^4-androstene-3,17-dione and testosterone following their two-dimensional resolution on silica gel to which a fluorescence indicator had been added. The hormones were estimated with an accuracy of ± 10% in the 0.2—0.5 μg range.

Herbicides and pesticides. Frei and Nomura [334] employed ultraviolet reflectance spectroscopy for the determination of triazine herbicides following their resolution on plates coated with silica gel to which a fluorescent indicator had been added. An accuracy of ± 3% was attained with the use of the "spot-removal" technique, and spectra recorded directly from the plots were suitable for identification purposes. A subsequent study [335] of various methods for the in situ determination of s-triazines resolved on chromatoplates indicated the spectral reflectance technique to be preferable to fluorescence quenching from the standpoint of accuracy.

s-Triazene promethryne has been used as a test substance for the investigation of the parameters affecting in situ determinations by ultraviolet reflectance spectroscopy [309], and chlorinated and thiophosphate pesticides resolved on thin-layer plates have been analyzed to demonstrate the use of a chromatogram scanner fitted with fiber optics [110]. In the latter study, variation attributable to chromatographic technique was between 11 and 16%, while instrumental variation ranged from 1 to 2%.

Miscellaneous. Stahl and his coworkers have investigated a number of systems with the use of a chromatogram spectrophotometer, including caffeine, phenazetine [336], and the alkaloids, strychnine and brucine [111]. The analysis of 5 opium alkaloids in drug samples by the ultraviolet reflectance technique has also been reported [337]; so also has been the identification of phthalides in essential oils [338]. Other applications include an investigation of aromatic, hydroxy, aldehydo derivatives such as vanillin, anisaldehyde and asarylaldehyde [336]. Kraus et al. [339] have determined scylla glycosides by spectral reflectance following their separation on silica gel plates, and Schunack and his coworkers [340] have employed a similar procedure for the determination of some xanthine derivatives in blood.

(d) Cation analysis by reflectance spectroscopy

The first application of spectral reflectance to the chromatography of inorganic substances was reported by Vaeck [341,342], who employed the approach for the determination of microgram quantities of Ni^{2+} separated on paper chromatograms. A critical comparison [343] of reflectance and transmittance techniques for the analysis of rubeanic acid complexes of copper on paper chromatograms indicated that, optically, paper appeared more uniform in reflected light and that reflectance spectroscopy is therefore preferable for in situ analysis. A precision of $\pm$ 0.43%R was reported for 4 replicate determinations carried out by reflectance measurement. Although most of the subsequent developments in this particular field have dealt with thin-layer chromatography, there has been a recent report [344] of a procedure for the determination by spectral reflectance of iron, manganese, zinc and copper in plant materials following the separation of these cations on paper. Results obtained with the use of this procedure were in good agreement with those obtained by other methods.

One of the earliest investigations of the potential of the thin-layer—spectral reflectance approach for trace metal analysis dealt with the separation and determination of the rubeanic acid complexes of cobalt, copper and nickel on silica gel, alumina and cellulose chromatoplates [345]. The influence of such experimental variables as temperature, humidity and pH upon the stability and intensity of the spots was studied by means of reflectance spectroscopy. Under optimal conditions thus deduced, instrumental detection limits for 20 mg samples, prepared according to the "spot removal" method,

342

were reported as 0.05 μg for nickel and copper and 0.1 μg for cobalt. The precision achieved depended on whether the plates were sprayed with the ethanolic rubeanic acid chromogenic solution, or, as with the cellulose plates, dipped in the solution. With the former technique reproducibilities ranged from 0.55 to 0.77%R standard deviation, while for the latter the comparable values were 0.39 to 0.44%R. The probable percentage error for the procedure ranged from 2 to 5%, depending on the metal, the adsorbent and the color-developing technique used. All precision and accuracy data are for optimal conditions for analysis (see pp.266 ff).

The use of a relatively new chelating agent, pyridine-2-aldehyde-2-quinolylhydrazone (PAQH), for the determination of copper, cobalt and nickel by the combination reflectance—chromatographic method has been developed in very much the same way as was the use of rubeanic acid [253,346,347]. Although the precision was no better than that achieved with the rubeanic acid, the detection limits were improved to approximately 0.01 μg/spot at the 50% accuracy level.

The value of the technique for the rapid identification of cations comprising a mixture was demonstrated by a method whereby 14 cations (Al^{3+}, Bi^{3+}, Cd^{2+}, Cr^{3+}, Co^{2+}, Cu^{2+}, Fe^{3+}, Pb^{2+}, Mn^{2+}, Hg^{2+}, Ni^{2+}, Ag^+, Sn^{2+} and Zn^{2+}) which had been separated on one-dimensional, cellulose thin-layer plates were subsequently identified by means of their reflectance spectra [348]. The procedure requires micro amounts of sample and only one spray reagent: a mixture of dithizone and 8-hydroxyquinoline. At the conclusion of the chromatographic process, the various spots were excised from the plate, packed into a windowless cell, and their reflectance spectra recorded. The three spectra depicted in Fig. 36(a), are representative of the spectra thus obtained. Although it was possible by means of such spectra to distinguish between many of the ions, there were several ion pairs whose spectra were so similar that it was impossible to make an unequivocal identification on the basis of spectral data alone. This situation was resolved by taking advantage of the fact that the structure of the spectra was pH-dependent. After the first set of reflectance spectra had been recorded, the reference and sample cells were exposed to ammonia fumes for some 2 min in a chamber containing 15 M ammonia and a second set of spectra were recorded. An indication of the spectral changes that were induced by the ammonia treatment is provided by Fig. 36(b), which presents the

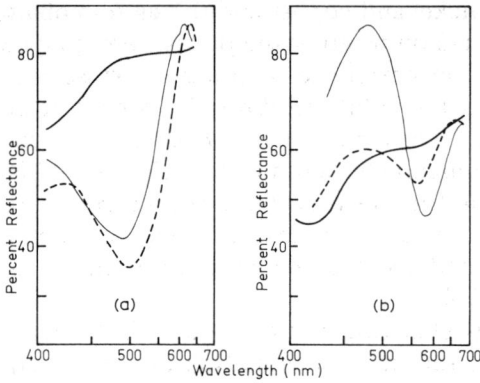

Fig. 36. (a) Reflectance spectra obtained for 10 μg of chromium (——), lead (----), and silver (——) cations adsorbed on cellulose after chromatoplates had been sprayed with dithizone-oxine reagent. (b) Reflectance spectra obtained for 10 μg of chromium (——), lead (----), and silver (——) cations adsorbed on cellulose after chromatoplates had been sprayed with dithizone-oxine reagent and subsequently exposed to ammonia fumes [348].

second set of spectra obtained for the same 3 cations whose first spectra are depicted in Fig. 36(a). By making use of information provided by both sets of reflectance spectra, it is possible to identify all fourteen of the cations without too much difficulty.

A wide variety of procedures for the detection and determination of inorganic substances by means of the combined chromatographic—reflectance technique has been developed subsequent to the above investigations. One is a method devised for the analysis of micro quantities of copper, nickel and zinc resolved on cellulose plates [349]. Copper and nickel were determined in the presence of eleven other cations without interference by employing neocuproine and dimethylglyoxime, respectively, as chromogenic reagents. With zinc, the use of 3,3'-dimethylnaphthadine was equally successful except in the presence of tin, cadmium and iron. Deviations to be expected when the procedure is employed routinely were estimated to be 2.1, 2.8 and 5.6% for nickel, copper and zinc, respectively.

Galik and Vincourova [350] have employed the combined technique as the basis for the determination of 1-(2-pyridylazo)-2-naphthol, or PAN, complexes of cobalt, copper, nickel and iron following their resolution on silica gel plates. Quantification of the reflectance from spots was achieved by the manual integration of the

344

peak areas of tracings obtained by direct recording of spectra from chromatograms. Graham et al. [351] have also used PAN as a chromogenic spray reagent for a number of cations, separating the complexes on cellulose layers impregnated with a liquid ion-exchanger, Primene-JM-T hydrochloride. Plates were evaluated by means of a Joyce Loebl Chromoscan fitted with a TLC attachment used in the reflectance mode, and the concentrations were then determined in terms of the integrated peak areas of the tracings thus obtained. This procedure was first worked out for Zn^{2+}, and then extended to the determination of Bi^{3+}, Cd^{2+}, Co^{2+}, Pb^{2+} and UO_2^{2+}. The reproducibility of the technique was found to be restricted by the inhomogeneous distribution of the spot material and factors associated with the spraying operation. Nevertheless, for the cations studied, the accuracy was found to be comparable to the approximate ± 4% value that was attained with the use of the more conventional and time-consuming approach of spot removal followed by transmittance measurement. Preliminary studies have also indicated the feasibility of using PAN, and closely related reagents, in the determination by diffuse reflectance of a large number of heavy metal cations separated on a variety of adsorbent thin-layers [352].

Diffuse reflectance spectroscopy has recently been employed in an investigation of various chromium(III) complexes resolved on prefabricated thin-layer sheets [353].

The determination of traces of heavy metals by spectral reflectance following their 10^3- to 10^4-fold concentration in ion-exchange resins has been studied by Fujimoto and Kortüm [354]. $Cu(H_2O)_4^{2+}$ was concentrated in Dowex 50W-X8(H-form), and $[Co(II)(NCS)_4]^{2-}$ was concentrated in Dowex 1-X8 anion exchange resin. The Kubelka—Munk function computed with absolute reflectance values was found to be linearly dependent on the concentration of the complexes between 0.4 and 30% of the exchange capacity of the resins. The pure resin served as the reference standard in each case.

REFERENCES

1. D.B. Judd and G. Wyszecki, Color in Business, Science and Industry, 2nd edn., Wiley, New York, 1963.
2. A.H. Taylor, J. Opt. Soc. Amer., 4 (1919) 9.

3. F. Benford, Gen. Elec. Rev., 23 (1920) 72.
4. A.C. Hardy, J. Opt. Soc. Amer., 18 (1929) 96.
5. T. Ulbricht, Elektrotech. Z., 21 (1900) 595.
6. J. Fahrenfort, Spectrochim. Acta, 17 (1961) 698.
7. G. Kortüm, Reflectance Spectroscopy, Springer-Verlag, New York, 1969.
8. W.W. Wendlandt and H.G. Hecht, Reflectance Spectroscopy, Interscience, New York, 1966.
9. W.W. Wendlandt (Ed.), Modern Aspects of Reflectance Spectroscopy, Plenum Press, New York, 1968.
10. P. Kubelka and F. Munk, Z. Tech. Phys., 12 (1931) 593.
11. P. Kubelka, J. Opt. Soc. Amer., 38 (1948) 448.
12. G. Kortüm and H. Schöttler, Z. Electrochem., 57 (1953) 353.
13. G. Kortüm, W. Braun and G. Herzog, Angew. Chem. Int. Ed. Engl., 2 (1963) 333.
14. V.T. Lieu and M.M. Frodyma, Talanta, 13 (1966) 1319.
15. A. Ringbom, Z. Anal. Chem., 115 (1939) 332.
16. G.H. Ayres, Anal. Chem., 21 (1949) 652.
17. C.N. Reilley and C.M. Crawford, Anal. Chem., 27 (1955) 716.
18. V.T. Lieu, Unpublished data.
19. E.P. Labinowich, The Determination by Spectral Reflectance of Copper Concentrated on Chromatographic Columns, M.S. Thesis, University of Hawaii, 1966.
20. V.T. Lieu, D.F. Zaye and M.M. Frodyma, Talanta, 16 (1969) 1289.
21. D.F. Zaye, Cation Analysis by Thin-Layer Chromatography and Reflectance Spectroscopy, Ph.D. Thesis, University of Hawaii, 1968.
22. R.W. Frei, D.E. Ryan and V.T. Lieu, Can. J. Chem., 44 (1966) 1945.
23. G. Kortüm, S. Vogel and W. Braun, Angew. Chem., 70 (1958) 651.
24. R.W. Frei and M.M. Frodyma, Anal. Chim. Acta, 32 (1965) 501.
25. C.A. Lermond and L.B. Rogers, Anal. Chem., 27 (1955) 340.
26. M.M. Frodyma and R.W. Frei, J. Chem. Educ., 46 (1969) 522.
27. M.M. Frodyma, R.W. Frei and D.J. Williams, J. Chromatogr., 13 (1964) 61.
28. G. Kortüm and G. Schreyer, Angew. Chem., 67 (1955) 694.
29. G. Kortüm and D. Oelkrug, Z. Phys. Chem. (Frankfurt), 34 (1962) 58.
30. G. Kortüm and J. Vogel, Chem. Ber., 93 (1960) 706.
31. G. Kortüm, Trans. Faraday Soc., 58 (1962) 1624.
32. G. Kortüm and W. Braun, Ann. Chem., 632 (1960) 104.
33. G. Kortüm, W. Braun and G. Herzog, Angew. Chem., 75 (1963) 653.
34. R.W. Frei, I.T. Fukui, V.T. Lieu and M.M. Frodyma, Chimia, 20 (1966) 23.
35. J. Goldman and R.R. Goodall, J. Chromatogr., 40 (1969) 345.
36. W.E. Sumpner, Proc. Phys. Soc. (London), 12 (1892) 10.
37. G. Kortüm and D. Oelkrug, Z. Naturforsch., 19a (1964) 28.
38. H. Schulz, Z. Phys. 31 (1925) 496.
39. H.J. Hellwig, Licht, 7 (1937) 99.
40. C. Fragstein, Optik, 12 (1955) 60.
41. W.L. Derksen, T.I. Monahan and A.J. Lawes, J. Opt. Soc. Amer., 47 (1957) 995.
42. O.E. Miller and A.J. Sant, J. Opt. Soc. Amer., 48 (1958) 828.

43. R.S. Longhurst, Geometrical and Physical Optics, Wiley, New York, 1957, pp. 388ff.
44. G. Kortüm, Kolorimetrie, Photometrie and Spektrometrie, 4th edn., Springer-Verlag, Berlin, 1962, pp. 345ff.
45. A.H. Taylor, Sci. Papers Nat. Bur. Stand. U.S., No. 391 (1920) 421.
46. E. Karrer, Sci. Papers Nat. Bur. Stand. U.S., No. 415 (1921) 203.
47. E.B. Rosa and A.H. Taylor, Sci. Papers Nat. Bur. Stand. U.S., No. 447 (1922) 281.
48. H.J. McNicholas, J. Res. Nat. Bur. Stand. U.S., 1 (1928) 29.
49. J.S. Preston, Trans. Opt. Soc. (London), 31 (1929—30) 15.
50. A.C. Hardy and O.W. Pineo, J. Opt. Soc. Amer., 21 (1931) 502.
51. J.A. Jacquez and H.F. Kuppenheim, J. Opt. Soc. Amer., 45 (1955) 460.
52. J.A. Jacquez and H.F. Kuppenheim, J. Opt. Soc. Amer., 46 (1956) 428.
53. Commission Internationale de l'Eclairage, Proceedings of the 8th Session, Cambridge, 1931, Cambridge University Press, Cambridge, 1932, p. 23.
54. W. Budde, J. Opt. Soc. Amer., 50 (1960) 217.
55. P.A. Tellex and J.R. Waldron, J. Opt. Soc. Amer., 45 (1955) 19.
56. E.A. Schatz, J. Opt. Soc. Amer., 56 (1966) 389.
57. W.E.K. Middleton and C.L. Sanders, J. Opt. Soc. Amer., 41 (1951) 419.
58. Nat. Bur. Std. U.S. Circ. LC-547, 1939.
59. J.G. Priest and J.O. Riley, J. Opt. Soc. Amer., 20 (1930) 156.
60. K. Miescher and R. Rometsch, Experientia, 6 (1950) 302.
61. W.E.K. Middleton and C.L. Sanders, Illum. Eng. New York, 48 (1953) 254.
62. G. Kortüm and G. Haug, Z. Naturforsch., 8a (1953) 372.
63. G. Kortüm and J. Vogel, Z. Phys. Chem. (Frankfurt), 18 (1958) 110.
64. J.S. Laufer, J. Opt. Soc. Amer., 49 (1959) 1135.
65. J.A. Jacquez, W. McKeehan, J. Huss, J.M. Dimitroff and H.F. Kuppenheim, J. Opt. Soc. Amer., 45 (1955) 971.
66. H.J. Keegan and K.S. Gibson, J. Opt. Soc. Amer., 34 (1944) 77.
67. J.W. Gabel and E.I. Stearns, J. Opt. Soc. Amer., 39 (1949) 481.
68. F.W. Billmeyer, Jr., J. Opt. Soc. Amer., 46 (1956) 72.
69. J.A. Jacquez, W. McKeehan, J. Huss, J.M. Dimitroff and H.F. Kuppenheim, J. Opt. Soc. Amer., 45 (1955) 781.
70. Nat. Bur. Std. U.S. Circ. LC-1017, 1948.
71. Zeiss Information Pamphlet No. 50-660.
72. A.S. Stenius, J. Opt. Soc. Amer., 45 (1955) 727.
73. R.W. Frei, in G. Pataki and A. Niederwieser (Eds.), Progress of TLC and Related Methods, Vol. II, Ann Arbor-Humphrey Science Publishers, Ann Arbor, 1970, Chap. 1.
74. Commission Internationale de l'Eclairage, Proceedings of the 14th Session, Brussels, 1959, CIE Publication No. 4, 1960, p. 36.
75. Commission Internationale de l'Eclairage, Proceedings of the 16th Session, Washington, 1967.
76. J.M. Dimitroff and D.W. Swanson, J. Opt. Soc. Amer., 46 (1956) 555.
77. A.H. Taylor, J. Opt. Soc. Amer., 25 (1935) 51.
78. A.H. Taylor, J. Opt. Soc. Amer., 21 (1931) 776.
79. F. Benford, J. Opt. Soc. Amer., 25 (1935) 332.

80. D.F. Wilcock and W. Soller, Ind. Eng. Chem., 32 (1940) 1446.
81. W.W. Wendlandt, J. Chem. Educ., 45 (1968) A861, A947.
82. H.J. Hoefert, Z. Instrumentenk., 67 (1959) 3.
83. L.R. Dearth, W.M. Shillcox and J.A. van den Akker, Tappi, 43 (1960) 230A.
84. J.A. van den Akker, L.R. Dearth, O.H. Olson and W.M. Shillcox, Tappi, 35 (1952) 141A.
85. R.S. Hunter, Nat. Paint Varn. Lacquer Assoc. Circ., No. 456, (1934) 69.
86. R.S. Hunter, J. Opt. Soc. Amer., 30 (1940) 536.
87. R.S. Hunter, J. Res. Nat. Bur. Stand. U.S., 25 (1940) 581.
88. J.A. Sanderson, J. Opt. Soc. Amer., 37 (1947) 771.
89. W.L. Derksen and T.I. Monahan, J. Opt. Soc. Amer., 42 (1952) 263.
90. A.G. Tweet, Rev. Sci. Instrum., 34 (1963) 1412.
91. J.L. Michaelson, J. Opt. Soc. Amer., 28 (1938) 365.
92. A.C. Hardy, J. Opt. Soc. Amer., 28 (1938) 360.
93. H.K. Hammond and I. Nimeroff, J. Opt. Soc. Amer., 42 (1952) 367.
94. R.E. Anacreon and R.H. Noble, Appl. Spectrosc., 14 (1960) 29.
95. A. son Stenius, J. Opt. Soc. Amer., 45 (1955) 727.
96. G. Tonnquist, J. Opt. Soc. Amer., 45 (1955) 528.
97. L. Barnes, H. Goya and H. Zeitlin, Rev. Sci. Instrum., 34 (1963) 292.
98. M.M. Frodyma, V.T. Lieu and R.W. Frei, J. Chromatogr., 18 (1965) 520.
99. M.M. Frodyma and V.T. Lieu, Anal. Chem., 39 (1967) 814.
100. R.W. Asmussen and P. Andersen, Acta Chem. Scand., 12 (1958) 939.
101. W.W. Wendlandt, Thermal Methods of Analysis, Interscience, New York, 1964, Chap. 10.
102. W.W. Wendlandt, P.H. Franke and J.P. Smith, Anal. Chem., 35 (1963) 105.
103. W.W. Wendlandt and T.D. George, Chemist-Analyst, 53 (1964) 100.
104. M.C.R. Symons and P.A. Travalion, Unicam Spectrovision, 10 (1961) 8.
105. R. Klaus, J. Chromatogr., 16 (1964) 311.
106. R. Klaus, Pharm. Zt., 112 (1967) 480.
107. L. DeGallan, J. van Leeuwen and K. Camstra, Anal. Chim. Acta, 35 (1966) 395.
108. H.T. Gordon, J. Chromatogr., 22 (1966) 60.
109. B.L. Hamman and M.M. Martin, Anal. Biochem., 15 (1966) 305.
110. M. Beroza, K.R. Hill and K.H. Norris, Anal. Chem., 40 (1968) 1608.
111. H. Jork, Cosmo Pharma, 1 (1967) 33.
112. H. Jork, Z. Anal. Chem., 236 (1968) 310.
113. E. Stahl and H. Jork, Zeiss Inform. 16 (1968) 52.
114. H. Jork, J. Chromatogr., 33 (1968) 297.
115. H. Struck, H. Karg and H. Jork, J. Chromatogr., 36 (1968) 74.
116. H. Jork, Cosmo Pharma, 4 (1968) 12.
117. A. Niederwieser, Chromatographia, 1 (1968) 23.
118. A. Niederwieser, Chromatographia, 2 (1969) 362.
119. J. Goldman and R.R. Goodall, J. Chromatogr., 32 (1968) 24.
120. M.S. Lefar and A.D. Lewis, Anal. Chem., 42 (1970) 79A.
121. G. Wyszecki, Farbsysteme, Musterschmidt Verlag, Berlin, 1960.
122. D.B. Judd and G. Wyszecki, Color in Business, Science and Industry, 2nd edn., Wiley, New York, 1963, pp. 264—361.

123. D.B. Judd, J. Opt. Soc. Amer., 23 (1933) 359.
124. D.B. Judd, Nat. Bur. Stand. U.S. Circ. 478, 1950.
125. A.C. Hardy, Handbook of Colorimetry, MIT Press, Cambridge, Mass., 1936.
126. R.M. Evans, An Introduction to Color, Wiley, New York, 1948.
127. Optical Society of American Committee on Colorimetry, The Science of Color, Thomas Y. Crowell Co., New York, N.Y., 1953.
128. W.D. Wright, The Measurement of Color, 2nd edn. MacMillan, New York, 1958.
129. F.W. Billmeyer and M. Saltzman, Principles of Color Technology, Interscience, New York, 1967.
130. H. Grassmann, Poggendorf's Ann., 89 (1853) 69.
131. R. Davis and K.S. Gibson, Nat. Bur. Std. U.S. Circ. 114, 1931.
132. M. Planck, Ann. Phys. 4 (1901) 553.
133. H.J. McNicholas, J. Res. Nat. Bur. Stand. U.S., 1 (1928) 793.
134. I.G. Priest, J. Res. Nat. Bur. Stand. U.S., 15 (1935) 529.
135. R. Donaldson, Proc. Phys. Soc., 59 (1947) 554.
136. J. Guild, J. Sci. Instrum., 11 (1939) 69.
137. R.S. Hunter, Nat. Bur. Std. U.S. Circ. C429, 1942.
138. T. Smith and J. Guild, Trans. Opt. Soc. (London), 33 (1931) 73.
139. T. Smith, Proc. Phys. Soc. (London), 46 (1934) 372.
140. D.B. Judd, in M.G. Mellon (Ed.), Analytical Absorption Spectroscopy, Wiley, New York, 1950, Chap. 9.
141. J.L. Saunderson, J. Opt. Soc. Amer., 32 (1942) 727.
142. D.R. Duncan, J. Oil Colour Chem. Ass., 32 (1949) 296.
143. D.R. Duncan, J. Oil Colour Chem. Ass., 45 (1962) 300.
144. D.R. Duncan and H. Mesner, Paint Res. Station, Tech. Paper No. 177, 1952.
145. D.R. Duncan, Photoelec. Spectrom. Group Bull. No. 16, 1965, p. 483.
146. L.S. Pratt, Chemistry and Physics of Organic Pigments, Chapman and Hall, London, 1947, pp. 301—310.
147. E.I. Stearns, Amer. Dyest. Rep., 40 (1951) 562.
148. H.R. Davidson, Amer. Paint J., 46 (1962) 9.
149. H.R. Davidson and H. Hemmendinger, J. Opt. Soc. Amer., 48 (1958) 281.
150. H.R. Davidson, H. Hemmendinger and I.L.R. Landry, J. Soc. Dyers Colour., 79 (1963) 577.
151. A.H. Taylor, J. Opt. Soc. Amer., 24 (1934) 192.
152. D.F. Wilcock and W. Soller, Ind. Eng. Chem., 32 (1940) 1446.
153. V.C. Vesce, Off. Dig. Fed. Paint Varn. Prod. Clubs, No. 227 (1943) 217.
154. F.F. Rupert, J. Opt. Soc. Amer., 20 (1930) 661.
155. D.H. Parker, Paint Ind. Mag., 72 (1957) 18.
156. D. Tough, J. Oil Colour Chem. Ass., 39 (1956) 169.
157. A.W. Bruins, Chem. Weekbl. 46 (1950) 282.
158. J.A. Meacham, Amer. Paint J., 23 (1938) 23.
159. A. DiBernardo and P. Resnick, J. Opt. Soc. Amer., 49 (1959) 480.
160. A.R. Hannam and D. Patterson, J. Soc. Dyers Colour., 79 (1963) 192.
161. W.F. Ulrich, F. Kelly and D.C. Nelson, Beckman Reprint R-6134, Beckman Instruments, Inc., 1959.

162. D.L. Tilleard, Paint Res. Station Memorandum No. 197, 1952.
163. H.H. Weber, Farbe Lack, 63 (1957) 586.
164. A.E. Jacobsen, J. Opt. Soc. Amer., 38 (1948) 442.
165. I. Brinkman and W.G. Zijlstra, Arch. Chir. Neer., 1 (1949) 177.
166. F.A. Rodrigo, Amer. Heart J., 45 (1953) 809.
167. M.L. Polanyi and R.M. Hehir, Rev. Sci. Instrum., 31 (1960) 401.
168. L. Lubnow, Zeiss Inform., 51 (1964) 10.
169. J. Dorst, Mem. Mus. Nat. Hist. Natur., Ser. A. Zool., 1 (1951) 125
170. C.H. Greenewalt, W. Brandt and D.D. Friel, J. Opt. Soc. Amer., 50 (1960) 1005.
171. W.L. Derksen and T.I. Monahan, J. Opt. Soc. Amer., 42 (1952) 263.
172. M. Luckiesh, L.L. Holladay and A.H. Taylor, J. Opt. Soc. Amer., 20 (1930) 423.
173. C.L. Wong and W.R. Blevin, Aust. J. Biol. Sci., 20 (1967) 501.
174. K. Shibata, J. Biochem (Tokyo), 45 (1958) 599.
175. P. Moon, J. Opt. Soc. Amer., 32 (1942) 238, 243.
176. P. Moon, J. Opt. Soc. Amer., 31 (1941) 317, 482, 723.
177. A.I. Andrews and C.H. Zwermann, J. Amer. Ceram. Soc., 22 (1939) 65.
178. J.A. Pask, Bull. Amer. Ceram. Soc., 20 (1941) 50.
179. F.H. Emery, Bull. Amer. Ceram. Soc., 20 (1941) 381.
180. W.L. Peskin, Bull. Amer. Ceram. Soc., 20 (1941) 402.
181. W.H. Merry, Bull. Amer. Ceram. Soc., 36 (1956) 236.
182. H.J. Bautsch, Silikat. Tech., 9 (1958) 552.
183. N.B. Guerrant, J. Agr. Food. Chem., 5 (1957) 207.
184. J.J. Naughton, M.M. Frodyma and H. Zeitlin, Science, 125 (1957) 121.
185. J.J. Naughton, H. Zeitlin and M.M. Frodyma, J. Agr. Food. Chem., 6 (1958) 933.
186. A.A. Kraft and J.C. Ayres, Food Technol., 8 (1954) 290.
187. P.C. Pirko and J.C. Ayres, Food Technol., 11 (1957) 461.
188. I.D. Ginger, V.T. Lewis and B.S. Schweigert, J. Agr. Food. Chem., 3 (1955) 156.
189. A.L. Tappel, Food Res., 22 (1957) 479.
190. W.B. Robinson, T. Wishnetsky, J.R. Ransford, W.L. Clark and D.B. Hand, Food Technol., 6 (1952) 269.
191. J.N. Yeatman, A.P. Sidwell and K.N. Norris, Food Technol., 14 (1960) 16.
192. R.S. Hunter and J.N. Yeatman, J. Opt. Soc. Amer., 51 (1961) 552.
193. T.R. Gillett and A.L. Holven, Ind. Eng. Chem., 35 (1943) 210.
194. H. Burton, J. Dairy Res., 21 (1954) 194.
195. N.J. Morris, I.W. Lohmann, R.T. O'Connor and A.F. Freeman, Food Technol., 7 (1953) 393.
196. K. Yamaguchi, S. Fujii, T. Tabata and S. Kato, J. Pharm. Soc. Jap., 74 (1954) 1322.
197. H. Piller, Mineralium Deposita, 1 (1966) 175.
198. S.H.U. Bowie and N.F.M. Henry, Trans. Inst. Mining Met., 73 (1963/64) 467.
199. H. Ehrenberg, Z. Wiss. Mikrosk., 66 (1964) 32.
200. K. v. Gehlen and H. Piller, Petrography, 10 (1964) 94.

201. J.H. Leow, Econ. Geol., 61 (1966) 598.
202. E.V. Ashburn and R.G. Weldon, J. Opt. Soc. Amer., 46 (1956) 583.
203. R.E. Anacreon and R.H. Noble, Appl. Spectrosc., 14 (1960) 29.
204. A.S. Stenius, J. Opt. Soc. Amer., 45 (1955) 727.
205. P. Luner and D. Chen, Tappi, 46 (1963) 98.
206. S.V. Vaeck, Ann. Sci. Text. Belg., 1 (1966) 95.
207. S.V. Vaeck, Textilveredlung, 1 (1966) 658.
208. H. Loof, Papier, 21 (1967) 297.
209. A.S. Stenius, Tappi, 46 (1963) 183A.
210. A.S. Stenius, Zeiss Inform., 16 (1968) 21.
211. W. Budde and S.M. Chapman, Pulp Paper Mag. Can., 67 (1968) T206.
212. K.G. Schmidt, Papier, 12 (1959) 141.
213. R.W. Hisey and H.W. Cobb, Tappi, 42 (1959) 122.
214. W. Braun and G. Kortüm, Zeiss Inform., 16 (1968) 27.
215. C.W. McKeehan and J.E. Christian, J. Amer. Pharm. Ass., 46 (1957) 631.
216. L. Lochman and J. Cooper, J. Amer. Pharm. Ass., 48 (1959) 226.
217. T. Urbanyi, C.J. Swartz, and L. Lochman, J. Amer. Pharm. Ass., 49 (1960) 163.
218. L. Lochman, S. Weinstein, C.J. Swartz, and J. Cooper, J. Amer. Pharm. Ass., 50 (1961) 141.
219. C.J. Swartz, L. Lochman, T. Urbanyi and J. Cooper, J. Amer. Pharm. Ass., 50 (1961) 145.
220. M.E. Everhard and F.W. Goodhart, J. Pharm. Sci., 52 (1963) 281.
221. M.E. Everhard, D.A. Dickcius and F.W. Goodhart, J. Pharm. Sci., 53 (1964) 173.
222. R.E. Derby, Jr., Amer. Dyest. Rep., 41 (1952) 550.
223. L. Müller-Gerbes, Spinner Weber, 78 (1960) 2.
224. O.K. Dobozy, Zh. Prikl. Khim. (Leningrad), 34 (1961) 204.
225. H.F. Launer, Text. Res. J., 33 (1963) 351.
226. L. Fourt and A.M. Sookne, Text. Res. J., 21 (1951) 469.
227. M.L. Hurwitz, Am. Dyest. Rep., 35 (1946) 83.
228. G.E. Barker and C.R. Kern, J. Amer. Oil Chem. Soc., 27 (1950) 113.
229. A. Kling, Fette, Seifen, Anstrichm., 65 (1963) 285.
230. P.J. Lucchesi, J.L. Carter and D.J.C. Yates, J. Phys. Chem., 66 (1962) 1451.
231. G. Kortüm and H. Koffer, Ber. Bunsenges. Phys. Chem., 67 (1967) 67.
232. E. Weitz, F. Schmidt and J. Singer, Z. Elektrochem., 46 (1940) 222.
233. E. Weitz, F. Schmidt and J. Singer, Z. Elektrochem., 47 (1941) 65.
234. J.H. deBoer and G.M.M. Houben, Proc. Kon. Ned. Akad. Wetensch., Ser. B, 54 (1951) 421.
235. G. Kortüm and H. Delfs, Spectrochim. Acta, 20 (1964) 405.
236. G. Kortüm and G. Bayer, Z. Phys. Chem. (Frankfurt), 33 (1962) 254.
237. G. Kortüm, M. Kortüm-Seiler and S.D. Bailey, J. Phys. Chem., 66 (1962) 2439.
238. G. Kortüm, W. Theilacker and G. Schreyer, Z. Phys. Chem. (Frankfurt), 11 (1957) 182.
239. G. Kortüm, Spectrochim. Acta Suppl., (1957) 534.
240. G.M. Schwab and E. Schneck, Z. Phys. Chem. (Frankfurt), 18 (1958) 206.

241. G.M. Schwab, B.C. Dadlhuber and E. Wall, Z. Phys. Chem. (Frankfurt), 37 (1963) 99.
242. H. Zeitlin, H. Goya and J.L.T. Waugh, Nature, 198 (1963) 178.
243. H. Zeitlin and H. Goya, Nature, 183 (1959) 1041.
244. H. Goya, J.L.T. Waugh and H. Zeitlin, J. Phys. Chem., 66 (1962) 1206.
245. H. Zeitlin, N. Kondo and W. Jordan, J. Phys. Chem. Solids, 25 (1964) 641.
246. H. Zeitlin, R.W. Frei and M. McCarter, J. Catal. 4 (1965) 77.
247. R.W. Frei, H. Zeitlin and G. Fujie, Can. J. Chem., 44 (1966) 3051.
248. P. Anthony and H. Zeitlin, Nature, 187 (1960) 936.
249. H. Zeitlin, P. Anthony and W. Jordan, Science, 141 (1963) 423.
250. R.S. Mulliken, J. Phys. Chem., 56 (1952) 801.
251. G. Kortüm and H. Vogele, Ber. Bunsenges. Phys. Chem., 72 (1968) 401.
252. G. Kortüm and M. Grathwohl, Ber. Bunsenges. Phys. Chem., 72 (1968) 500.
253. R.W. Frei, R. Liiva and D.E. Ryan, Can. J. Chem., 46 (1968) 167.
254. R.W. Frei and H. Zeitlin, Can. J. Chem., 47 (1969) 3902.
255. V.A. Barachevski and A.N. Terenin, Opt. Spectrosc. USSR, 17 (1964) 161.
256. G. Kortüm and V. Schlichenmaier, Z. Phys. Chem. (Frankfurt), 48 (1966) 267.
257. W. Braun and G. Kortüm, Z. Phys. Chem. (Frankfurt), 61 (1968) 167.
258. G. Kortüm and W. Braun, Z. Phys. Chem. (Frankfurt), 18 (1958) 242.
259. G. Kortüm and W. Braun, Z. Phys. Chem. (Frankfurt), 48 (1966) 282.
260. G. Kortüm and W. Braun, Z. Phys. Chem. (Frankfurt), 28 (1961) 362.
261. R. Baistrccchi, Ann. Chim. (Rome), 49 (1959) 1824.
262. W.W. Wendlandt and H.G. Hecht, Reflectance Spectroscopy, Interscience, New York, 1966, Chap. 7.
263. W.W. Wendlandt (Ed.), Modern Aspects of Reflectance Spectroscopy, Plenum Press, New York, 1968, Chap. 4.
264. W.W. Wendlandt, Science, 140 (1963) 1085.
265. W.W. Wendlandt, J. Inorg. Nucl. Chem., 25 (1963) 833.
266. W.W. Wendlandt, W.R. Robinson and W.Y. Yang, J. Inorg. Nucl. Chem., 25 (1963) 1495.
267. R.E. Cathers and W.W. Wendlandt, Chemist-Analyst, 53 (1964) 110.
268. E.L. Simmons and W.W. Wendlandt, J. Inorg. Nucl. Chem., 28 (1966) 2187.
269. W.W. Wendlandt, Chemist-Analyst, 53 (1964) 71.
270. W.E. Hatfield, T.S. Piper and U. Klabunde, Inorg. Chem., 2 (1963) 629.
271. P.L. Hartmann, J.R. Nelson and J.G. Siegfried, Phys. Rev., 105 (1957) 123.
272. R.J.H. Clark, J. Chem. Soc. (London), (1964) 417.
273. G. Kortüm and D. Oelkrug, Naturwissenschaften, 53 (1966) 600.
274. D. Oelkrug, Ber. Bunsenges. Phys. Chem., 71 (1967) 697.
275. R.J.H. Clark, J. Chem. Educ., 41 (1964) 488.
276. G. Kortüm, Reflectance Spectroscopy, Springer-Verlag, New York, 1969, pp. 285—288.
277. J.A. Anysas and A.L. Companion, J. Chem. Phys., 40 (1964) 1205.
278. R.W. Asmussen and O. Bostrup, Acta Chem. Scand., 11 (1957) 745, 1097.
279. W.B. White, Appl. Spectrosc., 21 (1967) 167.

280. R.C. Ropp, Appl. Spectrosc., 23 (1969) 235.
281. E. Loh, Phys. Rev., 154 (1967) 270, 158 (1971) 273.
282. R.J. Ackermann, R.J. Thorn and G.H. Winslow, J. Opt. Soc. Amer., 49 (1959) 1107.
283. A. Companion and G.H. Winslow, J. Opt. Soc. Amer., 50 (1960) 1043.
284. Z. Urbanec and D. Imrisova, Z. Anorg. Allgem. Chem., 323 (1963) 300.
285. T.R. Griffiths, K.A.K. Lott and M.C.R. Symons, Anal. Chem., 31 (1959) 1338.
286. T.R. Griffiths, Anal. Chem., 35 (1963) 1077.
287. G. Kortüm and G. Herzog, Z. Anal. Chem., 190 (1962) 239.
288. H.H. Homer, R.M. Rulon and K.H. Butler, J. Electrochem. Soc., 100 (1953) 566.
289. P.D. Johnson, J. Opt. Soc. Amer., 42 (1952) 978.
290. J.H. Schulman and C.C. Klick, J. Opt. Soc. Amer., 43 (1953) 516.
291. J.D.S. Goulden, Chem. Ind. (London), (1957) 142.
292. R.B. Fischer and F. Vratny, Anal. Chim. Acta, 13 (1955) 588.
293. W.P. Doyle and F. Forbes, Anal. Chim. Acta, 33 (1965) 108.
294. E.H. Winslow and H.A. Liebhafsky, Anal. Chem., 21 (1949) 1338.
295. C.W. Ayers, Mikrochim. Acta, (1956) 85.
296. K. Takagi, E. Nakano and K. Konemura, J. Chem. Soc. Jap., Ind. Chem. Sect., 54 (1951) 706.
297. F. Mizuniwa, T. Umino and K. Sakai, Bunseki Kagaku, 16 (1967) 1373.
298. L. Palalau, Rev. Chim. (Bucharest), 19 (1968) 54.
299. I.N. Ermolenko, M.L. Longin and M.Z. Gavrilov, Zh. Anal. Khim., 17 (1962) 1035.
300. R.W. Frei, and M.M. Frodyma, Anal. Biochem., 9 (1964) 310.
301. V.T. Lieu, M.M. Frodyma, L.S. Higashi, and L.H. Kunimoto, Anal. Biochem., 19 (1967) 454.
302. R.W. Frei, H. Zürcher and G. Pataki, J. Chromatogr., 43 (1969) 551; 45 (1969) 284.
303. H. Gänshirt, in E. Stahl (Ed.), Dünnschicht Chromatographie, Springer Verlag, Berlin, (1967) p. 142.
304. G. Kortüm and J. Vogel, Angew. Chem., 71 (1959) 451.
305. W. Braun and G. Kortüm, Zeiss Mitt. Fortschr. Tech. Opt., 4 (1968) 379.
306. G. Pataki, Chromatographia, 1 (1968) 406.
307. V.T. Lieu, R.W. Frei, M.M. Frodyma and I.T. Fukui, Anal. Chim. Acta, 33 (1965) 639.
308. R. Klaus, J. Chromatogr., 34 (1968) 539.
309. W. Huber, J. Chromatogr., 33 (1968) 378.
310. R.W. Frei, H. Zeitlin and M.M. Frodyma, Chem. Rundsch., 19 (1966) 411.
311. C.G. Butler, P.A. Linley and J.M. Rowson, Scientiae Pharmaceuticae-II, Proceedings of the 25th Congress of Pharmacautical Sciences, Prague, 1965.
312. A.L. Latner and D.C. Park, Clin. Chim. Acta, 11 (1965) 538.
313. B. Kremers, R.O. Briere and J.G. Batsakis, Amer. J. Med. Technol., 33 (1) (1967) 28.
314. K. Yamaguchi, S. Fukushima and M. Ho, J. Pharm. Soc. Jap., 75 (1955) 556.

315. K. Yamaguchi, S. Fukushima and M. Ito, J. Pharm. Soc. Jap., 76 (1965) 339.
316. C. Garside and J.P. Riley, Anal. Chim. Acta, 46 (1969) 179.
317. G. Pataki, Techniques of Thin-Layer Chromatography in Amino Acid and Peptide Chemistry, Ann Arbor Science Publishers, Ann Arbor, 1968.
318. R.W. Frei and M.M. Frodyma, Chem. Rundsch., 19 (1966) 26.
319. R.W. Frei, I.T. Fukui, V.T. Lieu and M.M. Frodyma, Chimia, 20 (1966) 23.
320. H.B. Bull, J.W. Hahn and V.R. Baptist, J. Amer. Chem. Soc., 71 (1949) 550.
321. M.M. Frodyma and R.W. Frei, J. Chromatogr., 15 (1964) 501.
322. M.M. Frodyma and R.W. Frei, J. Chromatogr., 17 (1965) 131.
323. J.G. Heathcote and C. Haworth, J. Chromatogr., 43 (1969) 84.
324. A. Bevenue and K.T. Williams, Arch. Biochem. Biophys., 73 (1958) 291.
325. A. Bevenue and K.T. Williams, J. Chromatogr., 2 (1959) 199.
326. H.S. Owens, E.A. McComb and G.W. Deming, Proc. Amer. Soc. Sugar Beet Technol., 1955.
327. R.W. Frei, F.J. Thaller and M.M. Frodyma, unpublished data.
328. H. Zeitlin and A. Niimoto, Nature, 181 (1958) 1616.
329. H. Zeitlin and A. Niimoto, Anal. Chem., 31 (1959) 1167.
330. F. Korte and H. Weitkamp, Angew. Chem., 70 (1958) 434.
331. H. Zürcher, G. Pataki, J. Borko and R.W. Frei, J. Chromatogr., 43 (1969) 457.
332. R. Klaus, J. Chromatogr., 40 (1969) 235.
333. R.W. Frei, A. Kunz, G. Pataki, T. Prims and H. Zürcher, Anal. Chim. Acta, 49 (1970) 527.
334. R.W. Frei and N.S. Nomura, Mikrochim. Acta, (1968) 565.
335. R.W. Frei, and C.D. Freeman, Mikrochim. Acta, (1968) 1214.
336. H. Jork, IVth International Symposium of Chromatography and Electrophoresis, Brussels, 1966.
337. E. Stahl, and H. Jork, Arzeim. Forsch., 18 (1968) 1231.
338. E. Stahl, and H. Bohrman, unpublished data.
339. K. Kraus, E. Mutschler and H. Rochelmeyer, J. Chromatogr., 40 (1969) 244.
340. W. Schunack, E. Eich, E. Mutschler and H. Rochelmeyer, Arzeim. Forsch., 19 (1969) 1754.
341. S.V. Vaeck, Nature, 172 (1953) 213.
342. S.V. Vaeck, Anal. Chim. Acta, 10 (1954) 48.
343. R.B. Ingle, and E. Minshall, J. Chromatogr., 8 (1962) 369.
344. R.A. Webb, D.G. Hallas and H.M. Stevens, Analyst, 94 (1969) 794.
345. R.W. Frei and D.E. Ryan, Anal. Chim. Acta, 37 (1967) 187.
346. R.W. Frei, J. Chromatogr., 34 (1968) 563.
347. R.W. Frei, D.E. Ryan and C.A. Stockton, Anal. Chim. Acta, 42 (1968) 159.
348. D.F. Zaye, R.W. Frei and M.M. Frodyma, Anal. Chim. Acta, 39 (1967) 13.
349. M.M. Frodyma, D.F. Zaye and V.T. Lieu, Anal. Chim. Acta, 40 (1968) 451.
350. A. Galik and A. Vincourova, Anal. Chim. Acta, 46 (1969) 113.
351. R.J.T. Graham, L.S. Bark and D.A. Tinsley, J. Chromatogr., 39 (1969) 211, 218.
352. R.W. Frei, unpublished data.
353. A.D. Kirk, K.C. Moss and J.G. Valentin, J. Chromatogr., 36 (1968) 332.
354. M. Fujimoto and G. Kortüm, Ber. Bunsenges. Phys. Chem., 68 (1964) 488.

The following two abbreviations have been used in entries: AAS for atomic absorption spectroscopy and AFS for atomic fluorescence spectroscopy

AAS, see: "Atomic absorption spectroscopy"
Abney mount, 45
Absolute analysis, by AAS, 221, 222, 225—226
Absolute detection limit, in AAS, 217
Absorbance, definition of, 7
—, —, in AAS, 112
—, derivation of expression for, in AAS, with sharp line source, 118, 119
Absorption, definition of, 5
Absorption coefficient, 101, 108
—, definition of, 109
—, Voigt expression for, 110
Absorption oscillator strength, 100
—, importance of, in AAS, 212
Accuracy, of AAS and AFS measurements, 228
Adenine, determination of, by diffuse reflectance spectroscopy, 339
Adsorbent — adsorbate interactions, study of, by diffuse reflectance spectroscopy, 318—322
AFS, see: "Atomic fluorescence spectroscopy"
Air, refractive index of, at various wavelengths, 2, 3
Aldopentoses, determination of, by diffuse reflectance spectroscopy, 337
Alkaline earth elements, determination of, by AAS, aluminium interference in, 238
—, interference of phosphate on, in AAS, 236
Alternating current arc, description of, 18

Alternating current spark, description of, 19
Alumina, as reflectance standard, 285
Aluminium, determination of, by AAS, flame gas composition for, 211
—, —, —, spectral interference in, 248
—, enhancement effects of, in AAS, 244
—, interference by, in AAS, 238
—, —, —, suppression of, 242
Aluminium fluoride, thermal decomposition of, in AAS, 201
Aluminium nitrate, thermal decomposition of, in AAS, 200—201
Amici prism, 37
Amino acids, determination of, by diffuse reflectance spectroscopy, 334—336
—, identification of, by diffuse reflectance spectroscopy, 279
—, relative error in measurement of, by diffuse reflectance spectroscopy, 336
Ammonium chloride, use of, as protecting agent in AAS, 242
Amplifier units, for AAS and AFS, general requirements of, 169, 170
Angle of deviation, 27
—, variation with wavelength, 28
Anisaldehydes, determination of, by diffuse reflectance spectroscopy, 342
Anomalous dispersion, 31
Anthraquinone, reflection spectrum of, 276
Antimony, interference by, in AAS, suppression of, 242

356

—, enhancement effects in, 244
—, flame emission interference in, 250
—, flame requirements for, 156, 157
—, flame sources in, 156
—, flame temperature and composition in, choice of, 240, 241
—, formation of molecular hydroxides in, 203
—, furnaces for, 157, 158
—, heated nebuliser chambers, use of, in, 233
—, height of observation in, 240
—, high-dissolved-content samples in, 234
—, high intensity lamps for, 148
—, history and development of 95, 96
—, hollow cathode atom cells for, 159, 160
—, hollow cathode lamp sources for, 22, 23, 140—150; see also: "hollow cathode lamps"
—, hot filament atom cells for, 158, 159
—, hydrogen flames in, 193—194
—, importance of absorption oscillator strength in, 212
—, incomplete atomisation in, 204
—, instrumental systems for, 134—137
—, instrument operation in, 214, 215
—, integrated absorption coefficient, 109
—, inter-element effects in, 247
—, interference in, by beryllium, 239
—, —, by boron, 239
—, —, by chromium, 239
—, —, by iron, 239
—, —, by molybdenum, 239
—, —, by silicon, 239
—, —, by titanium, 239
—, —, by uranium, 239
—, —, by vanadium, 239
—, —, from unwanted source radiation, 251
—, —, owing to incomplete vaporisation, 201
—, internal standardisation techniques in, 172

—, ionisation in, 206—209
—, ionisation equilibria in, 246
—, ionisation interference effects in, 246, 247
—, isolation of spectral lines in, by filters, 162, 163
—, monochromators for, review of types of, 163—165
—, nebulisation as source of interference in, 231—233
—, nitrous oxide-acetylene flames in, 191—192
—, nitrous oxide-hydrogen flames in, 192
—, non-pneumatic nebulisers for, 186, 187
—, non-specific background effects in, 235
—, oxy-acetylene flames in, 192—193
—, percentage absorption, definition of, 112
—, physical interferences in, sources of, 229—236
—, plasmas as atom cells in, 160
—, pneumatic nebulisers for, 183—186
—, precision of, 227—228
—, premixed flames for, burners for, 179—182
—, —, burning velocities of, 175
—, —, introduction of samples into, 182—187
—, —, stoichiometric reactions in, 175
—, —, structures of, 173—179
—, —, temperatures of, 175
—, —, utility of interconal zones of, 178
—, principles of standard preparation for, 216
—, radiation scatter interference in, 248—250
—, rate of desolvation in, 233
—, Rayleigh scattering in, 234
—, relationship between sample concn. and concn. of atoms for absorption in, 132
—, releasing and protective agents in, use of, 241—243

359

Brewster angle, 14

Brucine, determination of, by diffuse reflectance spectroscopy, 342

"Buffers", use of, in AAS, 241—243

Building materials, study of, by reflectance techniques, 312

Burners, for premixed flames, 179—182

Burning velocity, 174

Burning velocities, table of, 175

Cadmium, determination of, by AAS, 158

—, —, by high frequency excitation, 23

—, vapour discharge source of, for AAS and AFS, 22, 154

Cadmium red line, historical involvement of, in definition of the metre, 2

Caesium, vapour discharge source of, for AAS and AFS, 154

Caesium iodide, transparency of, 32

Caffeine, determination of, by diffuse reflectance spectroscopy, 342

Calcium, determination of, by AAS, aluminium interference in, 238

—, —, —, effect of phosphate on, 237

—, —, —, interference in, 201

—, —, —, use of protecting agents in, 242

—, use of, as releasing agent in AAS, 242

Calcium atoms, spatial distribution of, in flames for AAS, 210, 211

Calcium fluoride, short-wave cut-off wavelength of, 32

—, transparency of, 32

Calcium hollow cathode lamps, importance of filler-gas in, 144, 145

Calcium silicate phosphors, examination of, by diffuse reflectance spectroscopy, 328

Calcium sulphate, reflectance spectrum of, 328

Carbon arc, spectral energy distribution from, at 3900°K, 9

Carbon arc sources, 25

Carbon filament cell, 159

Carbon rods, as atom cells for AAS and AFS, 158, 159

Carbon rod sources, 25

Carotenoids, determination of, by diffuse reflectance spectroscopy, 334

Carpets, study of, by reflectance techniques, 312

Cathodes, photo-sensitive, coatings for, 68

—, —, response curves of, 68

Cations, determination of, by diffuse reflectance spectroscopy, 342—345

Charge-transfer bonds, study of, by diffuse reflectance spectroscopy, 322

Chelates, determination of, by diffuse reflectance spectroscopy, 342—345

Chemisorption, studies of, by diffuse reflectance spectroscopy, 318—322

Chlorophylls, determination of, by diffuse reflectance spectroscopy, 334

Choppers, application of, in AAS and AFS, 170

Chopping, application of, in spectrophotometry, 84

Chromates, reflectance spectra of, 327

Chromaticity coordinates, 306

Chromaticity diagram, 307

Chromatograms, measurement of reflectance spectra on, 330

—, quantitative measurement of reflectance of, 331—332

Chromium, determination of, by AAS, spectral interference in, 249

—, —, —, use of protecting agents in, 242

—, —, in steel, by laser microprobe, 15

—, interference by, in AAS, 239

Chromium atoms, spatial distribution of, in flames for AAS, 210, 211

CIE colour measurement system, 308

CIE reflectance standards, 283

CIE standard illuminant, 307—308

CIE standard observer, 308

Cobalt, determination of PAN complex of, by diffuse reflectance spectroscopy, 344

360

362

Emission oscillator strength, 100

Emulsion, photographic, *see*: "photographic emulsion"

Eosine B, reflectance spectra of, 278

Equilibrium constants, determination of, by diffuse reflectance studies, 323

Erbium borate, reflectance spectrum of, 327

Erbium oxide, reflectance spectrum of, 327

Ethylene, study of adsorbed, by diffuse reflectance spectroscopy, 317, 321

Ethylene glycol, use of, as protecting agent, in AAS, 242

Europium, spectral interference by, in AAS, 248

Eye, response curve of, 83

Fabric, dyed, study of, by reflectance techniques, 316

Far IR spectral region, description of, 5

Far IR spectrophotometry, filters for, 32

Fastie-Ebert spectrograph, 49

Feathers, study of, by reflectance, 311

Fermat's principle, 30

Féry prism, 36

Field broadening, 104

Filament atom reservoir, 159

Filaments, hot, as atom cells for AAS and AFS, 158, 159

Filters, for far IR spectrophotometry, 32

—, interference, 85

—, —, application of, in AAS and AFS, 162, 163

—, photometric, 85

—, Reststrahlen, 32

—, transmission curves of, 162

Fishtail diaphragm, 57

Flames, air-hydrogen, 193

—, as atom cells in AAS and AFS, 156, 157

—, dissociation equilibria in, 201—206

—, formation of atoms in, 196—211

—, fuel-rich, 17

—, hydrocarbon, 17

—, hydrogen, 193—194

—, ionisation in, 206—209

—, nitrous oxide-acetylene, 191—192

—, —, degrees of ionisation in, 209

—, —, suppression of interference by, in AAS, 241

—, nitrous oxide-hydrogen, 193

—, origin of electrons in, 206

—, oxy-acetylene, 192—193

—, oxy-hydrogen, 17

—, premixed, *see*: "premixed flames"

—, principal ions in, 206

—, separated, 195

—, solute vaporisation in, 233, 234

—, solvent vaporisation in, 233

—, spatial distribution of elements in, in AAS, 210, 211

—, unpremixed, *see*: "unpremixed flames"

Flame length, effect of, on absorbance, in AAS, 131

—, importance of, in AAS, 180

Flame sources, background spectrum of, 17, 18

—, choice of fuel for, 17

—, spectroscopic, 16—17

Fluorescence, atomic, intensity of, 123

—, description of, 6

—, direct line, 120

—, quenching, reduction of, 124

—, resonance, 120

—, sensitised, 122

—, stepwise excitation, 122

—, stepwise line, 121

—, from thin-layer chromatograms, measurement of, 300

Fluorescence cells, 123

Fluorescence intensity, actual, 128—130

—, ideal, 124—128

Fluorescence yield, 124

"Flying spot", scanning of TLC plates by, 301—302

"Flying spot" technique, 333

—, interference by, in AAS, 239

Kayser, the, 3

Krypton-86, involvement in definition of the metre, 2

Kubelka-Munk equation 264—266

—, application of, in colour mixing and matching, 310

—, validity of, in paper reflectance studies, 315

Lambert-Beer law, applicability to AAS, 130

Lamps, hollow-cathode, 22, 23

—, quartz-iodine, 24

—, tungsten filament, 23, 24

Lanthanum, use of, as releasing agent, in AAS, 242

Laser microprobe, 15

Laser radiation, 11—15

Laser Raman spectroscopy, 14, 15

Lasers, free atom formation by, in AAS, 160

—, giant-pulse, 14

—, helium-neon gas, 13, 15

—, ruby, 12

Laundering processes, study of, by reflectance techniques, 316

Lead, determination of, by AAS, 158

Lead selenide, as photoconductive detector, 74

Lead sulphide, as IR detector, 65, 74, 75

Lead telluride, as photoconductive detector, 74

Leaves, reflectance studies of, 311

Line width, definition of, 101

Lithium, determination of, by AAS, spectral interference in, 249

Lithium carbonate, as reflectance standard, 285

Lithium fluoride, short-wave cut-off wavelength of, 32

—, transparency of, 32

Littrow spectrometer, 33—35

Lorentz broadening, 102, 103

Lyman series, for hydrogen, 10

Magnesium, determination of, by AAS, aluminium interference in, 238

—, —, —, use of protecting agents in, 242

Magnesium carbonate, as reflectance standard, 284

Magnesium chloride, thermal decomposition of, in AAS, 200

Magnesium nitrate, thermal decomposition of, in AAS, 200

Magnesium oxide, as reflectance standard, 283

Magnesium sulphate, thermal decomposition of, in AAS, 200

Malachite Green, study of, adsorbed on paper, by diffuse reflectance spectroscopy, 333

Manganese, determination of, by AAS, spectral interference in, 248

—, —, on cation-exchange paper, by diffuse reflectance spectroscopy, 328

Mannitol, use of, as protecting agent, in AAS, 242

Mellibiose, determination of, by diffuse reflectance spectroscopy, 336

Mercury, determination of, by AAS, 158

—, —, —, in air, 95

—, —, —, spectral interference in, 248

—, —, by diffuse reflectance spectroscopy, in air, 328

—, —, by high frequency excitation, 23

—, stepwise excitation fluorescence by, 122

—, vapour discharge source of, for AAS and AFS, 21, 154

Mercury-198 vapour lamp, as wavelength standard, 2

Metmyoglobin, determination of, in beef, by reflectance techniques, 312

Metre, definition of, 2

Microdensitometer, 83

Microscopic reflectance techniques, 314

Microwave plasma, as atom cell, in AAS, 160

Modulation, selective, in AAS and AFS, 166, 167

367

Peltier effect, in thermocouples, 76

Percentage absorption, definition of, in AAS, 112

Periodates, reflectance spectra of, 327

Pesticides, determination of, by diffuse reflectance spectroscopy, 341

Pfund-Hardy mounting, 50

Pfund series, for hydrogen, 10

Pfund spectrometer, 34

Phenazetine, determination of, by diffuse reflectance spectroscopy, 342

Phosphate, interference by, in AAS, suppression by releasing agents, 242

—, interference of, on alkaline earth elements in AAS, 236, 237

Phosphorescence, description of, 6

Phosphorus, determination of, by AAS, in copper 160

Photo-emissive detectors, 67—74

Photographic emulsion, ASA rating of, 81

—, as a radiation detector, 79—83

—, contrast of, 80, 81

—, DIN rating of, 81

—, extension of short wavelength limit of, 82

—, grain of, 82, 83

—, "gross fog level" of, 80

—, Hurter-Driffield curve of, 80

—, reciprocity failure of, 81

—, sensitivity of, 81

—, speed of, 81

Photometer, definition of, 6

Photometers, principle of, 84, 85

Photometry, filters for, 85

Photomultiplier tubes, 69—73

—, adaptation to near IR, 72

—, current output of, 169

—, effect of cooling, 71

—, for detection of vacuum UV radiation, 72

—, gain of, 168

—, linearity of, 70

—, matching of, in scanning spectrophotometers, 87

—, over-exposure of, 72

—, sensitivity of, 168

—, signal-to-noise ratio of, 71

—, solar-blind, 65, 71

—, speed of response of, 70

—, variation of dark current of, 70

—, variation of sensitivity of, 70

Photopic vision, 83

Photo-sensitive cathodes, coatings for, 68

Phthalides, determination of, in essential oils, by diffuse reflectance spectroscopy, 342

Physical interference, sources of, in AAS and AFS, 229—236

Pigments, analysis of, by diffuse reflectance techniques, 310

—, white, investigations of, by diffuse reflectance, 311

Planck's constant, 9

Plant pigments, study of chromatographically separated, by diffuse reflectance spectroscopy, 333

Plasmas, radiofrequency or microwave, as atom cells in AAS, 160

Platinum, determination of, by AAS 158

—, spectral interference by, in AAS, 248

Polyethylene, black, use of, as far-IR filter, 21

Porcelain enamel, as reflectance standard, 284

Potassium, vapour discharge source of, for AAS and AFS, 154

Potassium bromide, transparency of, 32

Potassium bromide phosphors, examinations of, by diffuse reflectance spectroscopy, 328

Potassium chloride, transparency of, 32

Potassium chloride phosphors, examination of, by diffuse reflectance spectroscopy, 328

Potassium iodide, transparency of, 32

Powders, determination of surface areas of, by diffuse reflectance spectroscopy, 323

Precision, of AAS and AFS, measurements, 227—228

369

Rumsey prism, 37
Runge and Paschen mounting, 47
Rutile, determination of, in presence of anatase, 327
Rydberg constant, 10
Saha equation, 206, 207
Scattering, problems of light, in AAS and AFS, 234—236
Scotopic vision, 83
Scylla glycosides, determination of, by diffuse reflectance spectroscopy, 342
Secondary emission coefficient, 69
Secondary reaction zone, 178—179
Sectors, spectrographic, 57, 58
Selenium, determination of, by high frequency excitation, 23
—, electrodeless discharge tube source of, 23
Self-reversal, 106
—, loss of, 107
Self-reversal broadening, 105
Sensitised fluorescence, 122
Sensitivity, definition of, AAS, 217
—, of photographic emulsions, 81
—, of photomultiplier tubes, 70
—, spectral, of photographic emulsions, 82
Sensitivities, for AAS, table of, 218—219
Sharp line source, absorbance in AAS with, 118, 119
Shot-noise, photomultiplier tube, 71
Signal noise, in AAS and AFS, 172
Signal-to-noise ratio, of a photomultiplier tube, 71
Silica (fused), transparency of, 32
Silicate, interference by, in AAS, suppression by releasing agents, 242
Silicon, determination of, by AAS, in aluminium, 160
—, —, —, in steel, 160
—, —, —, spectral interference in, 248
—, interference by, in AAS, 239
Silver, determination of, by AAS, 158
—, —, —, in copper, 159
Silver atoms, spatial distribution of, in flames for AAS, 210, 211

Silver chloride, transparency of, 32
Skin, human, reflectance studies of, 311
Slits, bilateral, 55
—, spectrometer, 52
—, —, care of, 57
Slit-width, choice of, 56
—, effect of, on width of spectral lines, 53
Smithells separator, 195
Snell's law, 27
Sodium, determination of, by AAS, spectral interference in, 249
—, —, —, use of protecting agents in, 242
—, stepwise line fluorescence by, 122
—, vapour discharge source of, for AAS and AFS, 154
Sodium atoms, spatial distribution of, in flames for AAS, 210, 211
Sodium chloride, as reflectance standard, 285
—, transparency of, 32
Sodium chloride phosphors, examination of, by diffuse reflectance spectroscopy, 328
Sodium fluoride, as reflectance standard, 285
—, short-wave cut-off wavelength of, 32
Sodium salicylate, as phosphor, 72
Solar-blind photomultipliers, 65, 71
—, application of, in AAS and AFS, 169
—, spectral response curve of, 72
Spectra, observation of, 5, 6
Spectral band-width, 52
—, maintenance of constant, 55
Spectral half-band pass, 161
Spectral interference, in AAS, 247—253
Spectral line, total width of, 107
Spectral lines, curvature of, 54, 57
—, isolation of, in AAS and AFS, 161—167
Spectral line shape, 101
Spectral line width, 100—108
—, contribution to, by Doppler broadening, 102